Universitext

Springer

Berlin
Heidelberg
New York
Hong Kong
London
Milan
Paris
Tokyo

Sebastià Xambó-Descamps

Block Error-Correcting Codes

A Computational Primer

 Springer

Sebastià Xambó-Descamps
Universitat Politecnica de Catalunya
C. Pau Gargallo 5
08028 Barcelona
Spain
e-mail: sebastia.xambo@upc.es

Cataloging-in-Publication Data applied for

A catalog record for this book is available from the Library of Congress.

Bibliographic information published by Die Deutsche Bibliothek
Die Deutsche Bibliothek lists this publication in the Deutsche Nationalbibliografie;
detailed bibliographic data is available in the Internet at <http://dnb.ddb.de>.

ISBN 3-540-00395-9 Springer-Verlag Berlin Heidelberg New York

---

Mathematics Subject Classification (2000): 94Bxx, 68P30, 11T71

Springer-Verlag Berlin Heidelberg New York
a member of BertelsmannSpringer Science+Business Media GmbH

http://www.springer.de

© Springer-Verlag Berlin Heidelberg 2003
Printed in Germany

Cover design: *design & production,* Heidelberg
Typesetting by the author using a TeX macro package
Printed on acid-free paper   40/3142ck-5 4 3 2 1 0

Preface

> There is a unique way to teach: to lead the other person through
> the same experience with which you learned.
>
> Óscar Villarroya, cognitive scientist.[1]

In this book, the mathematical aspects in our presentation of the basic theory
of block error-correcting codes go together, in mutual reinforcement, with
computational discussions, implementations and examples of all the relevant
concepts, functions and algorithms. We hope that this approach will facilitate
the reading and be serviceable to mathematicians, computer scientists and
engineers interested in block error-correcting codes.[2]

In its digital form, which is a **pdf** document with hyperlinks, the examples included in the book can be run with just a mouse click. Moreover, the
examples can be modified by users and saved to their local facilities for later
work. For the convenience of readers, the site

   http://www.wiris.com/cc/

has been set up not only to allow a free downloading of the digital version,
but also, and primarily, in order to provide direct and free access to the examples, and to the services provided to run them.

The program that handles the computations in the examples, together
with the interface that handles the editing (mathematics and text), here will
be called WIRIS/*cc*. More specifically, WIRIS stands for the interface, the
(remote) computational engine and the associated language, and *cc* stands
for the extension of WIRIS that takes care of the computational aspects that
are more specific of error-correcting codes.

WIRIS/*cc* is an important ingredient in our presentation, but we do not
presuppose any knowledge of it. As it is very easy to learn and use, it is

---

[1]Cognitive Research Center at the Universitat Autònoma de Barcelona, interviewed by
Lluís Amiguet in "la contra", LA VANGUARDIA, 22/08/2002.

[2]A forerunner of this approach was outlined in [31].

introduced gradually, according to the needs of our presentation. A succinct description of the functions used can be found in the Index-Glossary at the end. The Appendix is a summary of the main features of the system.

This book represents the author's response to the problem of teaching a one-semester course in coding theory under the circumstances that will be explained in a moment. Hopefully it will be useful as well for other teachers faced with similar challenges.

The course has been taught at the Facultat de Matemàtiques i Estadística (FME) of the Universitat Politècnica de Catalunya (UPC) in the last few years. Of the sixty sessions of fifty minutes each, nearly half are devoted to problem discussions and problem solving. The students are junior or senior mathematics majors (third and fourth year) and some of them are pursuing a double degree in mathematics and telecommunications engineering.

The nature of the subject, the curriculum at the FME and the context at the UPC advise that, in addition to sound and substantial mathematical concepts and results, a reasonable weight should be given to algorithms and the effective programming of them. In other words, learning should span a wide spectrum ranging from the theoretical framework to aspects that may be labeled as 'experimental' and 'practical'.

All these various boundary conditions, especially the stringent time constraints and the huge size of the subject, are to be pondered very carefully in the design of the course. Given the prerequisites that can be assumed (say linear and polynomial algebra, some knowledge of finite fields, and basic facts about probability theory), it is reasonable to aim at a good understanding of some of the basic algebraic techniques for constructing block error-correcting codes, the corresponding coding and decoding algorithms and a good experimental and practical knowledge of their working. To give a more concrete idea, this amounts to much of the material on block error correcting codes included in, say, chapters 3-6 in [25] (or chapters 4 to 8 in [20]), supplemented with a few topics that are not covered in these books, plus the corresponding algorithmics and programming in the sense explained above.

Of course, this would be impossible unless a few efficiency principles are obeyed. The basic one, some sort of Ockam's razor, is that in the theoretical presentation mathematics is introduced only when needed. A similar principle is applied to all the computational aspects.

The choice of topics is aimed at a meaningful and substantial climax in each section and chapter, and in the book as a whole, rather than at an obviously futile attempt at completeness (in any form), which would be senseless anyway due to the immense size of the fields involved.

WIRIS/cc makes it possible to drop many aspects of the current presentations as definitely unnecessary, like the compilations of different sorts of tables (as for example those needed for the hand computations in finite fields).

As a consequence, more time is left to deal with conceptual matters.

It is perhaps also interesting to point out that the examples provided at

http://www.wiris.com/cc/

can be a key tool for the organization of laboratory sessions, both for individual work assignments and for group work during class hours. The system, together with the basic communications facilities that today can be assumed in most universities, makes it easier for students to submit their homework in digital form, and for teachers to test whether the computational implementations run properly.

The basic pedagogical assumptions underlying the whole approach are that the study of the algorithms leads to a better understanding of the mathematics involved, and that the discipline of programming them in an effective way promotes a better understanding of the algorithms. It could be argued that the algorithms and programs are not necessary to solve problems, at least not in principle, but it can also be argued that taking them into account reinforces learning, because it introduces an experimental component in the mathematical practice, and because it better prepares learners for future applications.

Of course, we do not actually know whether these new tools and methods will meet the high expectations of their capacity to strengthen the students' understanding and proficiency. But part of the beauty of it all, at least for the author, stems precisely from the excitement of trying to find out, by experimenting together with our students, how far those tools and methods can advance the teaching and learning of mathematics, algorithms and programs.

> The growing need for mathematicians and computer scientists in industry will lead to an increase in courses in the area of discrete mathematics. One of the most suitable and fascinating is, indeed, coding theory.
>
> J. H. van Lint, [25], p. ix.

Acknowledgements

The character of this book, and especially the tools provided at

http://www.wiris.com/cc/,

would hardly be conceivable without the WIRIS system, which would not have been produced without the sustained, industrious and unflappable collaboration of Daniel Marquès and Ramon Eixarch, first in the Project OMEGA at the FME and afterwards, since July 1999, as partners in the firm **Maths for More**.

On the Springer-Verlag side, the production of this book would never have been completed without the help, patience and know-how of Clemens Heine. Several other people at Springer have also been supportive in different phases of the project. My gratitude to all, including the anonymous people that provided English corrections in the final stage.

It is also a pleasant duty to thank the FME and the Departament de Matemàtica Aplicada II (MA2) of the UPC for all the support and encouragement while in charge of the Coding Theory course, and especially the students and colleagues for the everyday give-and-take that certainly has influenced a lot the final form of this text. Joan Bruna, for example, suggested an improvement in the implementation of the Meggitt decoder that has been included in Section 3.4.

Part of the material in this text, and especially the software aspects, were presented and discussed in the EAGER school organized by Mina Teicher and Boris Kunyavski at Eilat (Israel, January 12-16, 2003). I am grateful to the organizers for this opportunity, and to all the participants for their eager inquiring about all aspects of the course. In particular, I have to thank Shmulik Kaplan for suggesting an improvement of the alternant decoder implementation presented in Section 4.3.

Grateful thanks are due to Thomas Hintermann for sharing his enlightening views on several aspects of writing, and especially on English writing. They have surely contributed to improve this work, but of course only the author is to be found responsible for the mistakes and blunders that have gone through undetected.

And thanks to my wife, Elionor Sedó. Without her enduring support and love it would not have been possible to dedicate this work to her on the occasion of our thirtieth wedding anniversary, for it would hardly have been finished.

The author
L'Escala
20/1/03

Contents

Introduction

Error-correcting codes have been incorporated in numerous
working communication and memory systems.

W. Wesley Peterson and E. J. Weldon, Jr., [16], p. v.

This chapter is meant to be an informal introduction to the main ideas of error-correcting block codes.

We will see that if redundancy is added in a suitable form to the information to be sent through a communications channel (this process is called coding), then *it is possible to correct some of the errors caused by the channel noise* if the symbols received are processed appropriately (this process is called decoding).

The theoretical limit of how good coding can be is captured by Shannon's concept of (channel) *capacity* and Shannon's epoch-making channel coding theorem.[3]

At the end we will also explain how are we going to present the computational discussions and materials.

How do we correct text mistakes?

'A trap df ond milrian cileq begens wifh a sinqle soep',
Chinese saying

M. Bossert, [5], p. xiii.

Assume that you are asked to make sense of a 'sentence' like this: "Coding theary zs bodh wntelesting and challenning" (to take a simpler example than the Chinese saying in the quotation).

---

[3]"While his incredible inventive mind enriched many fields, Claude Shannon's enduring fame will surely rest on his 1948 paper *A methematical theory of Communication* and the ongoing revolution in information technology it engendered" (Solomon W. Golomb, in "Claude Elwood Shannon (1916-2001)", Notices of the AMS, volume 49, number 1, p. 8).

You realize that this proposed text has mistakes and after a little inspection you conclude that most likely the writer meant "Coding theory is both interesting and challenging".

Why is this so? Look at 'challenning', for example. We recognize it is a mistake because it is not an English word. But it differs from the English word 'challenging' in a single letter. In fact, we do not readily find other English words that differ from 'challenning' in a single letter, and thus any other English word that could be meant by the writer seems much more unlikely than 'challenging'. So we 'correct' 'challenning' to 'challenging'. A similar situation occurs with the replacement of 'theary' by 'theory'. Now in 'bodh', if we were to change a single letter to get an English word, we would find 'bode, 'body" or 'both', and from the context we would choose 'both', because the other two do not seem to make sense.

There have been several implicit principles at work in our correction process, including contextual information. For the purposes of this text, however, the main idea is that most English words contain *redundancy* in the sense that altering one letter (or sometimes even more) will usually still allow us to identify the original word. We take this word to be the correct one, because, under reasonable assumptions on the source of errors, we think that it is the *most likely*.

The repetition code Rep(3)

Similar principles are the cornerstones of the theory and practice of error-correcting codes. This can be illustrated with a toy example, which, despite its simplicity, has most of the key features of channel coding.

Suppose we have to transmit a stream of *bits* $u_1u_2u_3\cdots$ (thus $u_i \in \{0,1\}$ for all i) through a 'noisy channel' and that the probability that one bit is altered is p. We assume that p is independent of the position of the bit in the stream, and of whether it is 0 or 1. This kind of channel is called a *binary symmetric channel* with *bit-error rate* p.

In order to try to improve the quality of the information at the receiving end, we may decide to repeat each bit three times (this is the redundancy we add in this case). Since for each information bit there are three transmission bits, we say that this coding scheme has *rate* $1/3$ (it takes three times longer to transmit a coded message than the uncoded one). Thus every information bit u is coded into the *code word* uuu (or $[u\,u\,u]$ if we want to represent it as a vector). In particular, we see that there are two code words, 000 and 111. Since they differ in all three positions, we say that the *minimum distance* of the code is 3. The fact that the code words have 3 bits, that each corresponds to a single information bit, and that the two code-words differ in

3 positions is summarized by saying that this repetition code, which we will denote $Rep(3)$, has type $[3, 1, 3]$.

If we could assume that at most one error is produced per each block of three bits, then the redundancy added by coding suggests to decode each block of three bits by a 'majority vote' decision: $000, 100, 010, 001$ are decoded as 0 and $111, 011, 101, 110$ are decoded as 1. Of course, if two or three errors occurred —a possibility that usually is much more unlikely than having at most one error—, then the decoder would give the wrong answer.

All seems very good, but nevertheless we still have to ask whether this scheme really improves the quality of the received information. Since there are three times more bits that can be in error, are we sure that this coding and decoding is helping?

To decide this, notice that p represents the proportion of error bits if no coding is used, while if we use $Rep(3)$ the proportion of bit errors will equal the proportion p' of received 3-bit blocks with 2 or 3 errors (then we may call the decoded bit a *code-error*). Since

$$p' = 3p^2(1 - p) + p^3,$$

we see that p'/p, which can be called the *error-reduction factor* of the code, is equal to

$$3p(1 - p) + p^2 = 3p - 2p^2 = 3p(1 - \tfrac{2}{3}p)$$

If p is small, as it usually is, this expression is also small, and we can use the (excess) approximation $3p$ as its value. For example, $3p = 0.003$ for $p = 0.001$, which means that, on average, for every 1000 errors produced without coding we will have not more than 3 with coding.

This is definitely a substantial improvement, although we are paying a price for it: the number of transmitted bits has been tripled, and we also have added computing costs due to the coding and decoding processes.

Can we do better than in the $Rep(3)$ example? Yes, of course, and this is what the theory of error-correcting codes is about.

Remarks. There is no loss of generality in assuming, for a binary symmetric channel, that $p \leqslant 1/2$ (if it were $p > 1/2$, we could modify the channel by replacing any 0 by 1 and any 1 by 0 at the receiving end). If this condition is satisfied, then $p' \leqslant p$, with equality only for $p = 0$ and $p = 1/2$ (see Figure 1). It is clear, therefore, that we can always assume that $Rep(3)$ reduces the proportion of errors, except for $p = 1/2$, in which case it actually does not matter because the channel is totally useless. In Figure 1 we have also drawn the graph of p'/p (the error reduction factor). Note that it is < 1 for $p < 1/2$ and $< 1/2$ for $p < 0.19$.

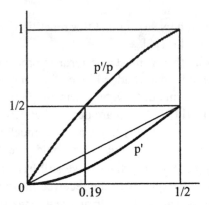

Figure 1: *The graph of* $p' = 3p^2 - 2p^3$ *as a function of* p *in the interval* $[0, \frac{1}{2}]$ *is the concave arc. The convex arc is the graph in the same interval of* p'/p *(the error-reduction factor). It is* < 1 *for* $p < 1/2$ *and* $< 1/2$ *for* $p < 0.19$. *Note that in the interval* $[0, 0.19]$ *it decreases almost linearly to* 0.

Channel capacity and Shannon's channel coding theorem

Shannon, in his celebrated paper [21], introduced the notion of *capacity* of a channel — a number C in the interval $[0, 1]$ that measures the maximum fraction of the information sent through the channel that is available at the receiving end. In the case of a binary symmetric channel it turns out that

$$C = 1 + p \log_2(p) + (1 - p) \log_2(1 - p),$$

where $\log_2$ is the base 2 logarithm function. Notice that $C = C(p)$ is a strictly decreasing function in the interval $[0, 1/2]$, with $C(0) = 1$ and $C(1/2) = 0$, and that $C(p) = C(1 - p)$ (see Figure 2).

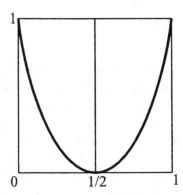

Figure 2: *Graph of the capacity* $C(p)$, $p \in [0, 1]$, *for a binary symmetric channel*

Shannon's *channel coding theorem* states that if R is a positive real number less than the *capacity* C of a binary symmetric channel (so $0 < R < C$), and ε is any positive real number, then there are 'codes' with *rate* at least R and with a probability of code-error less than ε (we refer to [25], [12] or [20] for details).

Shannon's theorem shows that in theory it is possible to transmit information with sufficient confidence and with a transmission time increase by a factor that can be as close to $1/C$ as desired. Unfortunately, his methods only show the existence of such codes, but do not produce them, nor their coding and decoding, in an *effective* way. It can be said that the main motivation of the theory of error-correcting codes in the last half century has been, to a great extent, to find explicit codes with good rates, small code-error probabilities and with fast coding and decoding procedures.

Comments on our computational approach and Web services

> Algorithms ... are a central part of both digital signal processing and decoders for error-control codes...
>
> R. E. Blahut, [4], p. vii.

Let us return to the *Rep*(3) example in order to explain how we are going to deal with the algorithmic and computing issues. In this case we would like to define two functions, say f and g, that express, in a computational sense, the coding and the decoding we have been discussing. The result could be the listing **The code *Rep*(3)**. In this case it is a Library, with a Library Area (the area inside the rectangle below the title) and a Computation Area (the area below the Library Area).

```
▲ Library | The code Rep(3)                                        ▲
  Coding function
  f(u) := [u, u, u]
  Majority decoder
  g([a, b, c]) :=
     if a=b | a=c  then
        a
     else if b=c  then
        b
     else
        "Decoder error"
     end

  Write the expressions you want to evaluate
  in this area, starting at the block below
  ∐
```

The syntax and semantics of the expressions in the listing is explained in more detail in the Appendix, but it should be clear enough at this point. The relevant comment to be made here is that in the digital version we do not get a listing, but a label like The code Rep(3) . This label is linked to a suitable file and when we click on the label, that file is displayed on an **WIRIS** screen. The result is equivalent to the listing **The code *Rep*(3)**, but now with all the power of **WIRIS**/*cc* is available to us.

There are two main kinds of tasks that we can do.

One is modifying the code in the Library area. This is illustrated in the listing **An alternative decoder for *Rep*(3)**, where we have added the function h, which is another implementation of the majority decoder for length 3 binary vectors. In this case we are using the function weight that returns the number of nonzero entries of a vector (it is just one of the many functions known to **WIRIS**/*cc*).[4]

```
▲ Library | An alternative decoder for Rep(3)                              ▲
  Coding function
  f(u) := [u, u, u]
  Majority decoder
  g([a, b, c]) :=
      if a=b | a=c  then
         a
      else if b=c  then
         b
      end
  h(x) :=
      if weight(x)>1  then
         1
      else
         0
      end
```

The other kind of tasks we can do is writing expressions in the Computation Area (in this case the empty block underneath the Library Area) and ask for their evaluation. This is illustrated in the listings **Some coding and decoding requests** and **Results of the requests**.

A **WIRIS**/*cc* document can be saved at any time as an html document in the facilities available to the user. These html files can be loaded and read with a Web browser with Java, or can be 'pasted', without loosing any of their functionalities, to other html documents. In fact, this is how we have generated the listings we have been considering so far, and all the remaining similar listings in this book.

---

[4]Here for simplicity we assume that the vector $[a\,b\,c]$ is binary. Thus we have also omitted the error handling message "Decoder error" when a, b, c are different.

```
▲ Library | Some coding and decoding requests                          ▲
  Coding function
  f(u) := [u, u, u]
  Majority decoder
  g([a, b, c]) :=
     if a=b | a=c  then
        a
     else if b=c  then
        b
     end
  h(x) := if weight(x)>1 then
             1
          else
             0
          end
```

```
  Regardless of whether the Library folder is open or closed,
  the functions defined in it, and all other WIRIS/cc functions,
  can be used in this area
  f(0)
  g([1,0,1])
  h([1,0,1])
```

```
  Results of evaluating the requests
  f(0)  →  [0, 0, 0]
  g( [1,0,1] )  →  1
  h( [1,0,1] )  →  1
```

WIRIS/cc

As indicated before, we use **WIRIS/cc** for all the algorithmic and programming aspects. It may be thought of as the conjunction of the general purpose language of **WIRIS** and the special library *cc* for the treatment of block error-correcting codes. This language has been designed to be close to accepted mathematical practices and at the same time encompass various programming paradigms. As with the mathematics, its features will be described when they are needed to understand the examples without assuming programming skills on the part of the reader.

As an illustration, let us look at the decoder function g for the $Rep(3)$ code which, for the reader's convenience, we reproduce in the listing **Decoder for Rep(3)**. The formal parameter of g is a length 3 vector $[a\,b\,c]$. Since the components of this vector are named symbolically, and the names can be used in the body of the function (that is, the code written after :=), we see that **WIRIS** supports 'pattern-matching'. In this case, the body of the function is an if ... then ... else ... end expression. The value of this expression is a when a and b are equal or when a and c are equal, and it is b when b and c are equal.

```
┌─────────────────────────────────────────────────────────────────────┐
│ ▲ Library │ Decoder for Rep(3)                                     ▲ │
├─────────────────────────────────────────────────────────────────────┤
║ g([a, b, c])                                                         │
║ := if a=b | a=c  then                                                │
║      a                                                               │
║    else if b=c  then                                                 │
║      b                                                               │
║    end                                                               │
└─────────────────────────────────────────────────────────────────────┘
║ \caret
```

Of course, **WIRIS** can be used for many purposes other than codes. Figure 2 on page 4, for example, has been generated as indicated in the listing **The capacity function**.[5]

```
┌─────────────────────────────────────────────────────────────────────┐
│ ▲ Library │ The capacity function C(p)                             ▲ │
├─────────────────────────────────────────────────────────────────────┤
║ capacity(p:ℝ)                                                        │
║ check 0⩽p & p⩽1                                                      │
║ := if p=0 | p=1  then                                                │
║      1                                                               │
║    else                                                              │
║      1+p·log₂(p)+(1−p)·log₂(1−p)                                      │
║    end                                                               │
└─────────────────────────────────────────────────────────────────────┘
```

$$1+p \cdot \log_2(p) + (1-p) \cdot \log_2(1-p)$$

```
┌
│ Drawing the graph of the capacity function
║ C=curve(capacity,0..1);
║ plotter({width=1.2, height=1.2, center=point(0.5,0.5),
║          window_width=150, window_height=150});
║ plot(C);
```

Goals of this book

The goal of this text is to present the basic mathematical theory of block error-correcting codes together with its algorithmic and computational aspects. Thus we will deal not only with the mathematics involved, but also with the related algorithms and with effective ways of programming these algorithms. This treatment has been illustrated already in the case of the repetition code $Rep(3)$ above: we have listed there, after explaining the concepts of coder and decoder, a translation of them as algorithms (the functions

---

[5]In the listing we have included only the parts that are essential for the graph. For example, we have omitted how labels, or the auxiliary segments, have been included in the final picture. In the digital book, the graph is generated by evaluating the expressions that appear on the **WIRIS** screen when we click on the link.

f and g) and programs (in this case, the fact that f and g can be loaded and run by a computer).

The book has four chapters. The first is a broad introduction to the subject of block error-correcting codes. After a first section devoted to the presentation of the basic concepts, we study linear codes and the basic computational issues of the syndrome-leader decoder. The third section is devoted to Hadamard codes, which in general are non-linear, and in the fourth section we study the more outstanding bounds on the parameters of codes.

The second chapter is an independent introduction to finite fields and their computational treatment to the extent that they are needed later. Usually it is prefarable to cover its contents as the need arises while working on the material of the last two chapters.

Chapter 3 is devoted to cyclic codes, and to one of their practical decoding schemes, namely the Meggitt decoder. This includes a presentation of the two Golay codes, including its Meggitt decoding. The important Bose–Chaudhuri–Hocquenghem codes, a subfamily of the cyclic codes, are also considered in detail.

Chapter 4 is the culmination of the text in many respects. It is devoted to alternant codes, which are generally not cyclic, and to their main decoders (basically the Berlekamp–Massey–Sugiyama decoder, based on the Euclidean division algorithm, and the Peterson–Gorenstein–Zierler algorithm, based on linear algebra, as well as some interesting variations of the two). It is to be noted that these decoders are applicable to the classical Goppa codes, the Reed–Solomon codes (not necessarily primitive), and the Bose–Chaudhuri–Hocquenghem codes (that include the primitive Reed–Solomon codes).

Chapter summary

- We have introduced the *repetition code* Rep(3) for the *binary symmetric channel*, and Shannon's celebrated formula for its *capacity*, $C = 1 + p\log_2(p) + (1 - p)\log_2(1 - p)$, where p is the *bit-error rate* of the channel.

- Assuming p is small, the quotient of the code-error rate p' of Rep(3) over the bit-error rate p is aproximately $3p$. Hence, for example, if there is one bit error per thousand bits transmited (on average), then using the code will amount to about three bit errors per million bits transmitted (on average).

- The repetition code Rep(3) shows that error correction is possible at the expense of the information transmission rate and of computational costs. Shannon's channel coding theorem, which says that if R and ε denote positive real numbers such that $0 < R < C$, where C is the capacity, then there are 'codes' with a rate not less than R and with a code-error probability not higher than ε, expresses the theoretical limits of coding.

- Shannon only showed the existence of such codes, but no efficient ways for constructing, coding and decoding them.

- The language in which the functions and algorithms are programmed, **WIRIS/cc**, is very easy to learn and use, is explained gradually along the way and does not require any programming skills.

- There is a **pdf** hypertext version of this book freely accessible at http://www.wiris.com/cc/.

  This version allows the reader to run all the examples in the book, as well as her own examples and programs. The examples are also accessible directly at the same site.

- The goal of this text is to present in the simplest possible terms, and covering systematically the most relevant computational aspects, some of the main breakthroughs that have occurred in the last fifty years in the quest of explicit codes, and efficient procedures for coding and decoding them, that approach Shannon's theoretical limit ever more closely.

Conventions and notations

Exercises and Problems. They are labeled $\mathbf{E}.m.n$ and $\mathbf{P}.m.n$, respectively, where n is the n-th exercise or problem within chapter m. Problems are grouped in a subsection at the end of each section, whereas exercises are inserted at any place that seems appropriate.

Mathematics and *WIRIS/cc.* In the text, the expression of mathematical objects and the corresponding **WIRIS/cc** expression are written in different types. This translation may not be done if the mathematical and **WIRIS** syntaxes are alike, especially in the case of symbolic expressions. For example, we do not bother in typesetting a..b or x|x' for the **WIRIS/cc** expressions of the mathematical range $a..b$ or the concatenation $x|x'$ of the vectors x and x'.

Note that there may be **WIRIS/cc** objets (like Range, a type) that are not assigned an explicit formal name in the mathematical context.

The ending symbol □ .   In general this symbol is used to signal the end of a block (for example a Remark) that is followed by material that is not clearly the beginning of another block. An exception is made in the case of proofs: we always use the symbol to denote the end of a proof, even if after it there is a definite beginning of a new block. On the other hand, since the end of theorems without proof is always clear, the symbol is never used in that case.

Quality improvement.   We would be very appreciative if errors and suggestions for improvements were reported to the author at the following email address: sebastia.xambo@upc.es.

1 Block Error-correcting Codes

Channel coding is a very young field. However, it has gained
importance in communication systems, which are almost incon-
ceivable without channel coding.

M. Bossert, [5], p. xiv.

The goal of this chapter is the study of some basic concepts and constructions pertaining to block error-correcting codes, and of their more salient
properties, with a strong emphasis on all the relevant algorithmic and computational aspects.

The first section is meant to be a general introduction to the theory of
block error-correcting codes. Most of the notions are illustrated in detail
with the example of the Hamming [7,4,3] code. Two fundamental results on
the parameters of a block code are proved: the Hamming upper-bound and
the Gilbert lower-bound.

In the second section, we look at linear codes, also from a general point
of view. We include a presentation of syndrome decoding. The main examples are the Hamming codes and their duals and the (general) Reed–Solomon
codes and their duals. We also study the Gilbert–Varshamov condition on the
parameters for the existence of linear codes and the MacWilliams identities
between the weight enumerators of a linear code and its dual.

Special classes of non-linear codes, like the Hadamard and Paley codes,
are considered in the third section. Some mathematical preliminairies on
Hadamard matrices are developed at the beginning. The first order Reed–
Muller codes, which are linear, but related to the Hadamard codes, are also
introduced in this section.

The last section is devoted to the presentation of several bounds on the
parameters of block codes and the corresponding asymtotic bounds. Several procedures for the construction of codes out of other codes are also
discussed.

1.1 Basic concepts

> Although coding theory has its origin in an engineering prob-
> lem, the subject has developed by using more and more so-
> phisticated mathematical techniques.
>
> F. J. MacWilliams and N. J. A. Sloane, [11], p. vi.

Essential points

- The definition of block code and some basic related notions (code-
  words, dimension, transmission rate, minimum distance, equivalence
  criteria for block codes).

- The definition of decoder and of the correcting capacity of a decoder.

- The minimum distance decoder and its error-reduction factor.

- The archtypal Hamming code [7,4,3] and its computational treatment.

- Basic dimension upper bounds (Singleton and Hamming) and the notions of MDS codes and perfect codes.

- The dimension lower bound of Gilbert.

Introductory remarks

> The fundamental problem of communication is that of repro-
> ducing at one point either exactly or approximately a message
> selected at another point.
>
> C. E. Shannon 1948, [21].

Messages generated by an *information source* often can be modelled as a
stream $s_1, s_2, \dots$ of *symbols* chosen from a finite set S called the *source alphabet*. Moreover, usually we may assume that the time t_s taken by the
source to generate a symbol is the same for all symbols.

To send messages we need a *communications channel*. A communications channel can often be modelled as a device that takes a stream of symbols chosen from a finite set T, that we will call the *channel* (or *transmission*)
alphabet, and pipes them to its destination (called the *receiving end* of the

channel) in some physical form that need not concern us here.[1] As a result, a stream of symbols chosen from T arrives at the receiving end of the channel.

The channel is said to be *noiseless* if the sent and received symbols always agree. Otherwise it is said to be *noisy*, as real channels almost always are due to a variety of physical phenomena that tend to distort the physical representation of the symbols along the channel.

The transfer from the source to the *sending end* of the channel requires an *encoder*, that is, a function $f : S \to T^*$ from the set of source symbols into the set T^* of finite sequences of transmission symbols. Since we do not want to loose information at this stage, we will always assume that encoders are injective. The elements of the image of f are called the *code-words* of the encoder.

In this text we will consider only *block encoders*, that is, encoders with the property that there exists a positive integer n such that $C \subseteq T^n$, where $C = f(S)$ is the set of code words of the encoder. The integer n is called the *length* of the block encoder.

For a block encoding scheme to make sense it is necessary that the source generates the symbols at a rate that leaves enough time for the opperation of the encoder and the channel transmission of the code-words. We always will assume that this condition is satisfied, for this requirement is taken into account in the design of communications systems.

Since for a block encoder the map $f : S \to C$ is bijective, we may consider as equivalent the knowledge of a source symbol and the corresponding code word. Generally speaking this equivalence holds also at the algorithmic level, since the computation of f or of its inverse usually can be efficiently managed. As a consequence, it will be possible to phrase the main issues concerning block encoders in terms of the set C and the properties of the channel. In the beginning in next section, we adopt this point of view as a general starting point of our study of block codes and, in particular, of the decoding notions and problems.

1.1 Remarks. The set $\{0, 1\}$ is called the *binary alphabet* and it is very widely used in communications systems. Its two symbols are called *bits*. With the addition and multiplication modulo 2, it coincides with the field $\mathbb{Z}_2$ of *binary digits*. The transposition of the two bits $(0 \mapsto 1, 1 \mapsto 0)$ is called *negation*. Note that the negation of a bit b coincides with $1 + b$.

1.2 *Example*. Consider the $Rep(3)$ encoder f considered in the Introduction (page 2). In this case S and T are the binary alphabet and $C = \{000, 111\}$, respectively. Note that the inverse of the bijection $f : S \to C$ is the map $000 \mapsto 0, 111 \mapsto 1$, which can be defined as $x \mapsto x_1$.

---

[1]The interested reader may consult [19, 5].

Block codes

> Only *block* codes for correcting *random* errors are discussed
>
> F. J. MacWilliams and N. J. A. Sloane, [11], p. vii.

Let $T = \{t_1, \ldots, t_q\}$ ($q \geqslant 2$) be the channel alphabet. By a (block) *code* of length n we understand any non-empty subset $C \subseteq T^n$. If we want to refer to q explicitly, we will say that C is *q-ary* (*binary* if $q = 2$, *ternary* if $q = 3$). The elements of C will be called *vectors*, *code-words* or simply *words*.

Usually the elements of T^n are written in the form $x = (x_1, \ldots, x_n)$, where $x_i \in T$. Since the elements of T^n can be seen as length n sequences of elements of T, the element x will also be written as $x_1 x_2 \ldots x_n$ (concatenated symbols), especially when $T = \{0, 1\}$.

If $x \in T^n$ and $x' \in T^{n'}$, the expression $x|x'$ will be used as an alternative notation for the element $(x, x') \in T^{n+n'}$. Similar notations such as $x|x'|x''$ will be used without further explanation.

Dimension, transmission rate and channel coding

If C is a code of length n, we will set $k_C = \log_q(|C|)$ and we will say that k_C is the *dimension* of C. The quotient $R_C = k_C/n$ is called *transmission rate*, or simply *rate*, of C.

1.3 Remarks. These notions can be clarified in terms of the basic notions explained in the first subsection (Introductory remarks). Indeed, $\log_q(|T^n|) = \log_q(q^n) = n$ is the number of transmission symbols of any element of T^n. So $k_C = \log_q(|C|) = \log_q(|S|)$ can be seen as the number of transmission symbols that are needed to capture the information of a source symbol. Consequently, to send the information content of k_C transmission symbols (which, as noted, amounts to a source symbol) we need n transmission symbols, and so R_C represents the proportion of source information contained in a code word before transmission. An interesting and useful special case occurs when $S = T^k$, for some positive integer k. In this case the source symbols are already resolved explicitly into k transmission symbols and we have $k_C = k$ and $R_C = k/n$.   □

If C has length n and dimension k (respectively $|C| = M$), we say that C has type $[n, k]$ (respectively type (n, M)). If we want to have q explicitly in the notations, we will write $[n, k]_q$ or $(n, M)_q$. We will often write $C \sim [n, k]$ to denote that the type of C is $[n, k]$, and similar notations will be used for the other cases.

Minimum distance

> Virtually all research on error-correcting codes has been based
> on the Hamming metric.
>
> W.W. Peterson and E.J. Weldon, Jr., [16], p. 307.

Given $x, y \in T^n$, the *Hamming distance* between x and y, which we will
denote $hd(x, y)$, is defined by the formula

$$hd(x, y) = |\{i \in 1..n \mid x_i \neq y_i\}|.$$

In other words, it is the number of positions i in which x and y differ.

E.1.1. Check that hd is a distance on T^n. Recall that $d : X \times X \to \mathbb{R}$ is said
to be a distance on the set X if, for all $x, y, z \in X$,

1) $d(x, y) \geqslant 0$, with equality if and only if $x = y$;

2) $d(y, x) = d(x, y)$; and

3) $d(x, y) \leqslant d(x, z) + d(z, y)$.

The last relation is called *triangle inequality*.

E.1.2. Let $x, y \in \mathbb{Z}_2^n$ be two binary vectors. Show that

$$hd(x, y) = |x| + |y| - 2|x \cdot y|$$

where $|x|$ is the number of non-zero entries of x (it is called the *weight* of x)
and $x \cdot y = (x_1 y_1, \cdots, x_n y_n)$. □

We will set $d = d_C$ to denote the minimum of the distances $hd(c, c')$, where
$c, c' \in C$ and $c \neq c'$, and we will say that d is the *minimum distance* of C.
Note that for the minimum distance to be defined it is necessary that $|C| \geqslant 2$.
For the codes C that have a single word, we will see that it is convenient to
put $d_C = n + 1$ (see the Remark 1.12).

We will say that a code C is of type $[n, k, d]$ (or of type (n, M, d)), if
C has length n, minimum distance d, and its dimension is k (respectively
$|C| = M$). If we want to have q explicitly in the notations, we will write
$[n, k, d]_q$ or $(n, M, d)_q$. In the case $q = 2$ it is usually omitted. Sometimes
we will write $C \sim [n, k, d]_q$ to denote that the type of C is $[n, k, d]_q$, and
similar notations will be used for the other cases.

The rational number $\delta_C = d_C/n$ is called *relative distance* of C.

E.1.3. Let C be a code of type (n, M, d). Check that if $k = n$, then $d = 1$.
Show also that if $M > 1$ (in order that d is defined) and $d = n$ then $M \leqslant q$,
hence $k_C \leqslant 1$.

1.4 *Example* (A binary code (8,20,3)). Let C be the binary code $(8, 20)$ consisting of 00000000, 11111111 and all cyclic permutations of 10101010, 11010000 and 11100100. From the listing **Cyclic shifts of a vector** it is easy to infer that $d_C = 3$. Thus C is a code of type $(8, 20, 3)$.

```
▲│Library │ Cyclic shifts of a vector                                         ▲
 │ cyclic_shift( x:Vector ) := [ x_length(x) ] | take(x,length(x)−1)
 │ cyclic_shifts( x:Vector ) :=
 │ begin
 │    local X={x}, y=cyclic_shift(x)
 │    while y≠x do
 │       X=X|{y}
 │       y=cyclic_shift(y)
 │    end
 │    X
 │ end
```

```
│ \careta=[1,1,0,1,0,0,0,0]; X=cyclic_shifts(a);
│ b=[1,1,1,0,0,1,0,0]; Y=cyclic_shifts(b);
│ c=[1,0,1,0,1,0,1,0]; Z=cyclic_shifts(c);
│ {hd(a, x) with x in tail(X)}  ⟶  {4, 4, 4, 6, 4, 4, 4}
│ {hd(a,y) with y in Y}  ⟶  {3, 3, 5, 3, 5, 7, 3, 3}
│ {hd(a,z) with z in Z}  ⟶  {5, 3}
│ {hd(b,y) with y in tail(Y)}  ⟶  {4, 6, 4, 4, 4, 6, 4}
│ {hd(b,z) with z in Z}  ⟶  {4, 4}
│ {hd(c,z) with z in tail(Z)}  ⟶  {8}
```

1.5 Remark. A code with minimum distance d *detects* up to $d-1$ errors, in the sense that the introduction of a number of errors between 1 and $d-1$ in the transmission gives rise to a word that is not in C. Note that $d-1$ is the highest integer with this property (by definition of d).

> However, [the Hamming distance] is not the only possible and indeed may not always be the most appropriate. For example, in $(F_{10})^3$ we have $d(428, 438) = d(428, 468)$, whereas in practice, e.g. in dialling a telephone number, it might be more sensible to use a metric in which 428 is closer to 438 than it is to 468.
>
> R. Hill, [9], p. 5.

Equivalence criteria

We will say that two codes C and C' of length n are *strictly equivalent* if C' can be obtained by permuting the entries of all vectors in C with some fixed permutation. This relation is an equivalence relation on the set of all codes of length n and by definition we see that it is the equivalence relation that corresponds to the natural action of S_n on T^n, and hence also on subsets of T^n.

In the discussion of equivalence it is convenient to include certain permutations of the alphabet T in some specified positions. This idea can be formalized as follows. Let $\Gamma = (\Gamma_1, \ldots, \Gamma_n)$, where Γ_i is a subgroup of permutations of T, that is, a subgroup of S_q ($1 \leqslant i \leqslant n$). Then we say that two codes C and C' are Γ-*equivalent* if C' can be obtained from C by a permutation $\sigma \in S_n$ applied, as before, to the entries of all vectors of C, followed by permutations $\tau_i \in \Gamma_i$ of the symbols of each entry i, $i = 1, \ldots, n$. If $\Gamma_i = S_q$ for all i, instead of Γ-equivalent we will say S_q-*equivalent*, or simply *equivalent*. In the case in which T is a finite field $\mathbb{F}$ and $\Gamma_i = \mathbb{F} - \{0\}$, acting on $\mathbb{F}$ by multiplication, instead of Γ-equivalent we will also say $\mathbb{F}$-*equivalent*, or $\mathbb{F}^*$-*equivalent*, or *scalarly equivalent*.

Note that the identity and the transposition of $\mathbb{Z}_2 = \{0, 1\}$ can be represented as the operations $x \mapsto x+0$ and $x \mapsto x+1$, respectively, and therefore the action of a sequence $\tau_1, \ldots, \tau_n$ of permutations of $\mathbb{Z}_2$ is equivalent to the addition of the vector $\tau \in \mathbb{Z}_2^n$ that corresponds to those permutations.

In general it is a relatively easy task, given Γ, to obtain from a code C other codes that are Γ-equivalent to C, but it is much more complicated to decide whether two given codes are Γ-equivalent.

E.1.4. Check that two (strongly) equivalent codes have the same parameters n, k and d.

E.1.5. Show that given a code $C \subseteq T^n$ and a symbol $t \in T$, there exists a code equivalent to C that contains the constant word t_n.

E.1.6. Can there be codes of type $(n, q, n)_q$ which are not equivalent to the q-ary repetition code? How many non-equivalent codes of type $(n, 2, d)_2$ there are?

Decoders

The essential ingredient in order to use a code $C \subseteq T^n$ at the receiving end of a channel to reduce the errors produced by the channel noise is a *decoding function*. In the most general terms, it is a map

$$g \colon D \to C, \text{ where } C \subseteq D \subseteq T^n$$

such that $g(x) = x$ for all $x \in C$. The elements of D, the domain of g, are said to be g-*decodable*. By hypothesis, all elements of C are decodable. In case $D = T^n$, we will say that g is a *full decoder* (or a *complete decoder*).

We envisage g working, again in quite abstract terms, as follows. Given $x \in C$, we imagine that it is sent through a communications channel. Let

$y \in T^n$ be the vector received at the other end of the channel. Since the channel may be noisy, y may be different from x, and in principle can be any vector of T^n. Thus there are two possibilities:

- if $y \in D$, we will take the vector $x' = g(y) \in C$ as the decoding of y;
- otherwise we will say that y is *non-decodable*, or that a *decoder error* has occurred.

Note that the condition $g(x) = x$ for all $x \in C$ says that when a code word is received, the decoder returns it unchanged. The meaning of this is that the decoder is assuming, when the received word is a code-word, that it was the transmitted code-word and that no error occurred.

If we transmit x, and y is decodable, it can happen that $x' \neq x$. In this case we say that an (undetectable) *code error* has occurred.

Correction capacity

We will say that the decoder g has *correcting capacity* t, where t is a positive integer, if for any $x \in C$, and any $y \in T^n$ such that $hd(x, y) \leqslant t$, we have $y \in D$ and $g(y) = x$.

1.6 Example. Consider the code $Rep(3)$ and its decoder g considered in the Introduction (page 2). In this case $T = \{0, 1\}$, $C = \{000, 111\}$ and $D = T^3$, so that it is a full decoder. It corrects one error and undetectable code errors are produced when 2 or 3 bit-errors occur in a single code-word.

1.7 Remark. In general, the problem of decoding a code C is to construct D and g by means of efficient algorithms and in such a way that the correcting capacity is as high as possible.

Minimum distance decoder

Let us introduce some notation first. Given $w \in T^n$ and a non-negative integer r, we set

$$B(w, r) = \{z \in T^n \mid hd(w, z) \leqslant r\}.$$

The set $B(w, r)$ is called the *ball of center* w and *radius* r.

If $C = \{x^1, \ldots, x^M\}$, let $D_i = B(x^i, t)$, where $t = \lfloor (d-1)/2 \rfloor$, with d the minimum distance of C. It is clear that $C \cap D_i = \{x^i\}$ and that $D_i \cap D_j = \emptyset$ if $i \neq j$ (by definition of t and the triangular inequality of the Hamming distance). Therefore, if we set $D_C = \bigsqcup_i D_i$, there is a unique map $g \colon D_C \to C$ such that $g(y) = x^i$ for all $y \in D_i$. By construction, g is a decoder of C and it corrects t errors. It is called the *minimum distance decoder* of C.

E.1.7. Check the following statements:

1. $g(y)$ is the word $x' \in C$ such that $hd(y, x') \leqslant t$, if such an x' exists, and otherwise y is non-decodable for g.

2. If y is decodable and $g(y) = x'$, then

$$hd(x, y) > t \quad \text{for all} \quad x \in C - \{x'\}.$$

1.8 Remarks. The usefulness of the minimum distance decoder arises from the fact that in most ordinary situations the transmissions $x \mapsto y$ that lead to a decoder error ($y \notin D$), or to undetectable errors ($y \in D$, but $hd(y, x) > t$) will in general be less likely than the transmissions $x \mapsto y$ for which y is decodable and $g(y) = x$.

To be more precise, the minimum distance decoder maximizes the likelihood of correcting errors if all the transmission symbols have the same probability of being altered by the channel noise and if the $q - 1$ possible errors for a given symbol are equally likely. If these conditions are satisfied, the channel is said to be a (q-ary) *symmetric channel*. Unless otherwise declared, henceforth we will understand that 'channel' means 'symmetric channel'.

From the computational point of view, the minimum distance decoder, as defined above, is inefficient in general, even if d_C is known, for it has to calculate $hd(y, x)$, for $x \in C$, until $hd(y, x) \leqslant t$, so that the average number of distances that have to be calculated is of the order of $|C| = q^k$. Note also that this requires having generated the q^k elements of C.

But we also have to say that the progress in block coding theory in the last fifty years can, to a considerable extent, be seen as a series of milestones that signal conceptual and algorithmic improvements enabling to deal with the minimum distance decoder, for large classes of codes, in ever more efficient ways. In fact, the wish to collect a representative sample of these brilliant achievements has guided the selection of the decoders presented in subsequent sections of this book, starting with next example.

1.9 *Example* (The *Hamming code* [7,4,3]). Let $K = \mathbb{Z}_2$ (the field of binary digits). Consider the K-matrix

$$R = \begin{pmatrix} 1 & 1 & 1 & 0 \\ 1 & 1 & 0 & 1 \\ 1 & 0 & 1 & 1 \end{pmatrix}$$

Note that the columns of R are the binary vectors of length 3 whose weight is at least 2. Writing I_r to denote the identity matrix of order r, let $G = I_4 | R^T$ and $H = R | I_3$ (concatenate I_4 and R^T, and also R and I_3, by rows). Note that the columns of H are precisely the seven non-zero binary vectors of length 3.

Let $S = K^4$ (the source alphabet) and $T = K$ (the channel alphabet). Define the block encoding $f : K^4 \to K^7$ by $u \mapsto uG = u|uR^T$. The image of this function is $C = \langle G \rangle$, the K-linear subspace spanned by the rows of G, so that C is a $[7, 4]$ code. Since

$$GH^T = (I_4|R^T)\Big(\frac{R^T}{I_3}\Big) = R^T + R^T = 0,$$

because the arithmetic is mod 2, we see that the rows of G, and hence the elements of C, are in the kernel of H^T. In fact,

$$C = \{y \in K^7 \,|\, yH^T = 0\},$$

as the right-hand side contains C and both expressions are K-linear sub-spaces of dimension 4. From the fact that all columns of H are distinct, it is easy to conclude that $d_C = 3$. Thus C has type $[7, 4, 3]$. Since $|C| \cdot vol(7, 1) = 2^4(1 + 7) = 2^8 = |K^8|$, we see that $D_C = K^7$.

As a decoding function we take the map $g : K^7 \to C$ defined by the following recipe:

1. Let $s = yH^T$ (this length 3 binary vector is said to be the *syndrome* of the vector y).

2. If $s = 0$, return y (as we said above, $s = 0$ is equivalent to say that $y \in C$).

3. If $s \neq 0$, let j be the index of s as a row of H^T.

4. Negate the j-th bit of y.

5. Return the first four components of y.

Let us show that this decoder, which by construction is a full decoder $(D = K^7)$, *coincides with the minimum distance decoder*. Indeed, assume that $x \in C$ is the vector that has been sent. If there are no errors, then $y = x$, $s = 0$, and the decoder returns x. Now assume that the j-th bit of $x \in C$ is changed during the transmission, and that the received vector is y. We can write $y = x + e_j$, where e_j is the vector with 1 on the j-th entry and 0 on the remaining ones. Then $s = yH^T = xH^T + e_jH^T = e_jH^T$, which clearly is the j-th row of H^T. Hence $g(y) = y + e_j = x$ (note that the operation $y \mapsto y + e_j$ is equivalent to negating the j-th bit of y). $\qquad\square$

The expression of this example in **WIRIS/cc** is explained in the listing **Hamming code [7,4,3]**. The **WIRIS** elements that are used may be consulted in the Appendix, or in the Index-Glossary.

┌───┐
│ ▲ Library │ Hamming code [7,4,3]   ▲ │
├───┤
Let R be the binary matrix whose columns are the binary vectors of length three
and weight at least two

$$R = \begin{pmatrix} 1\,1\,1\,0 \\ 1\,1\,0\,1 \\ 1\,0\,1\,1 \end{pmatrix} : \text{Matrix}(\mathbb{Z}_2)$$

The matrices G and H

$G = I_4 \mid R^T; \quad H = R \mid I_3;$

Encoding function

hamming_encoder(u) := u·G

Decoding function

```
hamming_decoder(y) :=
  begin
    local r, n, s, j
    (r,n)=dimensions(G)
    s=y·H^T
    if not zero?(s) then
      j=index(s, H^T)
      y_j=y_j-1
    end
    take(y, r)
  end
```

└───┘

\caretExample
u=[1, 1, 0, 1] ⟶ [1, 1, 0, 1]
x=hamming_encoder(u) ⟶ [1, 1, 0, 1, 0, 1, 0]
hamming_decoder(x) ⟶ [1, 1, 0, 1]
Let us simulate an error in position 4
e=epsilon_vector(7,4); y=x+e ⟶ [1, 1, 0, 0, 0, 1, 0]
hamming_decoder(y) ⟶ [1, 1, 0, 1]

The notion of an correcting code was introduced by Hamming
[in 1950]

V.D. Goppa, [8], p. 46.

Error-reduction factor

Assume that p is the probability that a symbol of T is altered in the transmission. The probability that j errors occur in a block of length n is

$$\binom{n}{j} p^j (1 - p)^{n-j}.$$

Therefore

$$P_e(n,t,p) = \sum_{j=t+1}^{n} \binom{n}{j} p^j (1 - p)^{n-j} = 1 - \sum_{j=0}^{t} \binom{n}{j} p^j (1 - p)^{n-j} \quad [1.1]$$

gives the probability that $t + 1$ or more errors occur in a code vector, and this is the probability that the received vector is either undecodable or that an undetectable error occurs.

If we assume that N blocks of k symbols are transmitted, the number of expected errors is pkN if no coding is used, while the expected number of errors if coding is used is at most $(N \cdot P_e(n, t, p)) \cdot k$ (in this product we count as errors all the symbols corresponding to a code error, but of course some of those symbols may in fact be correct). Hence the quotient

$$\rho(n, t, p) = P_e(n, t, p)/p,$$

which we will call the *error reduction factor* of the code, is an upper bound for the average number of errors that will occur in the case of using coding per error produced without coding (cf. E.1.8). For p small enough, $\rho(n, t, p) < 1$ and the closer to 0, the better error correction resulting from the code. The value of $\rho(n, t, p)$ can be computed with the function erf(n,t,p).

1.10 *Example* (Error-reduction factor of the Hamming code). According to the formula [1.1], the error-reduction factor of the Hamming code $C \sim [7, 4, 3]$ for a bit error probability p has the form

$$\binom{7}{2} p(1 - p)^5 + \cdots$$

which is of the order $21p$ for p small. Note that this is 7 times the error-reduction factor of the code $\mathsf{Rep}(3)$, so that the latter is more effective in reducing errors than C, even if we take into account that the true error reduction factor of C is smaller than $21p$ (because in the error-reduction factor we count all four bits corresponding to a code error as errors, even though some of them may be correct). On the other hand the rates of C and $\mathsf{Rep}(3)$ are $4/7$ and $1/3$, respectively, a fact that may lead one to prefer C to $\mathsf{Rep}(3)$ under some circumstances.

E.1.8. Let $p' = p'(n, t, p)$ be the probability of a symbol error using a code of length n and correcting capacity t. Let $\bar{p} = P_e(n, t, p)$. Prove that $\bar{p} \geqslant p' \geqslant \bar{p}/k$.

Elementary parameter bounds

In order to obtain efficient coding and decoding schemes with small error probability, it is necessary that k and d are high, inasmuch as the maximum number of errors that the code can correct is $t = \lfloor (d - 1)/2 \rfloor$ and that the transmission rate is, given n, proportional to k.

```
┌─┬──────────────────────────────────────────────────────────────────────┬─┐
│▲│ Library │ Probabilaty of code error (an upper bound) and error reduction factor │▲│
├─┴──────────────────────────────────────────────────────────────────────┴─┤
```

$$P\_e(n,t,p) := \sum_{j=t+1}^{n} \binom{n}{j} \cdot p^j \cdot (1-p)^{n-j}$$

```
erf(n,t,p) := if p=0 then
                  0
              else
                  P_e(n,t,p)/p
              end
```

ρ=erf

n=3; t=1; p=0.01;
P_e(n,t, p), ρ(n,t,p) → 0.000298, 0.0298
n=7; t=1; p=0.01;
P_e(n,t, p), ρ(n,t,p) → 0.002031, 0.2031
n=23; t=3; p=0.01;
P_e(n,t, p), ρ(n,t,p) → $7.6053 \cdot 10^{-5}$, 0.0076

It turns out, however, that k and d cannot be increased independently and arbitrarily in their ranges (for the extreme cases $k = n$ or $d = n$, see E.1.3). In fact, the goal of this section, and of later parts of this chapter, is to establish several non trivial restrictions of those parameters. In practice these restrictions imply, for a given n, that if we want to improve the rate then we will get a lower correcting capability, and conversely, if we want to improve the correcting capability, then the transmission rate will decrease.

Singleton bound and MDS codes

Let us begin with the simplest of the restrictions we will establish.

1.11 Proposition (Singleton bound). *For any code of type* $[n, k, d]$,

$$k + d \leqslant n + 1.$$

Proof: Indeed, if C is any code of type (n, M, d), let us write $C' \subseteq T^{n-d+1}$ to denote the subset obtained by discarding the last $d - 1$ symbols of each vector of C. Then C' has the same cardinal as C, by definition of d, and so $q^k = M = |C| = |C'| \leqslant q^{n-d+1}$. Hence $k \leqslant n - d + 1$, which is equivalent to the stated inequality. □

MDS codes. Codes that satisfy the equality in the Singleton inequality are called *maximum distance separable* codes, or *MDS codes* for short. The repetition code $Rep(3)$ and the Hamming code $[7,4,3]$ are MDS codes. The *repetition code* of any length n on the alphabet T, which by definition is $Rep_q(n) = \{t_n \mid t \in T\}$, is also an MDS code (since it has q elements, its dimension is 1, and it is clear that the distance between any two distinct code-words is n).

Values of $A_2(n, d)$			
n	$d = 3$	$d = 5$	$d = 7$
5	4	2	—
6	8	2	—
7	16	2	2
8	20	4	2
9	40	6	2
10	72–79	12	2
11	144–158	24	4
12	256	32	4
13	512	64	8
14	1024	128	16
15	2048	256	32
16	2720–3276	256–340	36–37

Table 1.1: *Some known values or bounds for $A_2(n, d)$*

1.12 Remark. If C is a code with only one word, then $k_C = 0$, but d_C is undefined. If we want to assign a conventional value to d_C that satisfies the Singleton bound, it has to satisfy $d_C \leqslant n + 1$. On the other hand, the Singleton bound tells us that $d_C \leqslant n$ for all codes C such that $|C| \geqslant 2$. This suggests to put $d_C = n + 1$ for one-word codes C, as we will do henceforth. With this all one-word codes are MDS codes.

The function $A_q(n, d)$

Given positive integers n and d, let $R_q(n, d)$ denote the maximum of the rates $R_C = k_C/n$ for all codes C of length n and minimum distance d.

We will also set $k_q(n, d)$ and $A_q(n, d)$ to denote the corresponding number of information symbols and the cardinal of the code, respectively, so that $R_q(n, d) = k_q(n, d)/n$ and $A_q(n, d) = q^{k(n,d)}$.

A code C (of length n and mimimum distance d) is said to be *optimal* if $R_C = R_q(n, d)$ or, equivalently, if either $k_C = k_q(n, d)$ or $M_C = A_q(n, d)$.

E.1.9. If $q \geqslant 2$ is an integer, show that $A_q(3, 2) = q^2$. In particular we have that $A_2(3, 2) = 4$ and $A_3(3, 2) = 9$. *Hint:* if $T = \mathbb{Z}_q$, consider the code $C = \{(x, y, x + y) \in T^3 \mid x, y \in T\}$.

E.1.10. Show that $A_2(3k, 2k) = 4$ for all integers $k \geqslant 1$.

E.1.11. Show that $A_2(5, 3) = 4$.

The exact value of $A_q(n, d)$ is unknown in general (this has often been called the *main problem* of coding theory). There are some cases which are very easy, like $A_q(n, 1) = q^n$, $A_2(4, 3) = 2$, or the cases considered in E.1.9 and E.1.11. The values $A_2(6, 3) = 8$ and $A_2(7, 3) = 16$ are also fairly easy to determine (see E.1.19), but most of them require, even for small n, much work and insight. The table 1.1 gives a few values of $A_2(n, d)$, when they are known, and the best known bounds (lower and upper) otherwise. Some of the facts included in the table will be established in this text; for a more comprehensive table, the reader is referred to the table 9.1 on page 248 of the book [6]. It is also interesting to visit the web page

http://www.csl.sony.co.jp/person/morelos/ecc/codes.html

in which there are links to pages that support a dialog for finding the best bounds known for a given pair (n, d), with indications of how they are ascertained.

E.1.12. In the Table 1.1 only values of $A_2(n, d)$ with d odd are included. Show that if d is odd, then $A_2(n, d) = A_2(n + 1, d + 1)$. Thus the table also gives values for even minimum distance. *Hint:* given a binary code C of length n, consider the code $\overline{C}$ obtained by adding to each vector of C the binary sum of its bits (this is called the *parity completion* of C).

The Hamming upper bound

Before stating and proving the main result of this section we need an auxiliary result.

1.13 Lemma. *Let r be a non-negative integer and $x \in T^n$. Then*

$$|B(x, r)| = \sum_{i=0}^{r} \binom{n}{i} (q - 1)^i.$$

Proof: The number of elements of T^n that are at a distance i from an element $x \in T^n$ is

$$\binom{n}{i} (q - 1)^i$$

and so

$$|B(x, r)| = \sum_{i=0}^{r} \binom{n}{i} (q - 1)^i,$$

as claimed.   □

The lemma shows that the cardinal of $B(x, r)$ only depends on n and r, and not on x, and we shall write $vol_q(n, r)$ to denote it. By the preceeding lemma we have

$$vol_q(n, r) = |B(x, r)| = \sum_{i=0}^{r} \binom{n}{i} (q - 1)^i. \qquad [1.2]$$

| ▲ Library \| Computation of $vol_q(n,r)$ | ▲ |

$$\text{volume}(n,r,q) := \sum_{i=0}^{r} \binom{n}{i} \cdot (q-1)^i$$

$$\text{volume}(n,r) := \sum_{i=0}^{r} \binom{n}{i}$$

\caretvolume(9,3,3) ⟶ 835
volume(9,3) ⟶ 130

1.14 Theorem (Hamming upper bound). *If* $t = \lfloor (d-1)/2 \rfloor$, *then the following inequality holds:*

$$A_q(n,d) \leqslant \frac{q^n}{vol_q(n,t)}.$$

Proof: Let C be a code of type $(n, M, d)_q$. Taking into account that the balls of radius $t = \lfloor (d-1)/2 \rfloor$ and center elements of C are pair-wise disjoint (this follows from the definition t and the triangular inequality of the Hamming distance), it turns out that $\sum_{x \in C} |B(x,t)| \leqslant |T^n| = q^n$.

On the other hand we know that

$$|B(x,t)| = vol_q(n,t)$$

and hence

$$q^n \geqslant \sum_{x \in C} |B(x,t)| = M \cdot vol_q(n,t).$$

Now if we take C optimal, we get

$$A_q(n,d)vol_q(n,t) \leqslant q^n,$$

which is equivalent to the inequality in the statement.   □

1.15 Remark. The Hamming upper bound is also called *sphere-packing upper bound*, or simply *sphere upper bound*.

E.1.13. Let m and s be integers such that $1 \leqslant s \leqslant m$ and let $c_1, \ldots, c_m \in \mathbb{F}^n$, where $\mathbb{F}$ is a finite field with q elements. Show that the number of vectors that are linear combinations of at most s vectors from among $c_1, \ldots, c_m$ is bounded above by $vol_q(m, s)$.

[Hamming] established the upper bound for codes
V.D. Goppa, [8], p. 46.

▲ Library	Hamming upper bound (also called sphere upper bound)	▲

$$\text{ub\_sphere}(n,d,q) := \left\lfloor \frac{q^n}{\text{volume}(n, \lfloor (d-1)/2 \rfloor, q)} \right\rfloor$$

$$\text{ub\_sphere}(n,d) := \left\lfloor \frac{2^n}{\text{volume}(n, \lfloor (d-1)/2 \rfloor)} \right\rfloor$$

\caretub_sphere(8,3) ⟶ 28
ub_sphere(9,3) ⟶ 51
ub_sphere(10,3) ⟶ 93
ub_sphere(11,3) ⟶ 170
ub_sphere(11,3,3) ⟶ 7702
ub_sphere(21,5) ⟶ 9039

Perfect codes

In general D_C is a proper subset of T^n, which means that there are elements $y \in T^n$ for which there is no $x \in C$ with $hd(y, x) \leqslant t$. If $D_C = T^n$, then C is said to be *perfect*. In this case, for every $y \in T^n$ there is a (necessarily unique) $x \in C$ such that $hd(y, x) \leqslant t$.

Taking into account the reasoning involved in proving the sphere-bound, we see that the necessary and sufficient condition for a code C to be perfect is that

$$\sum_{i=0}^{t} \binom{n}{i} (q-1)^i = q^n/M \quad (= q^{n-k}),$$

where $M = |C| = q^k$ (this will be called the *sphere* or *perfect* condition).

The total code T^n and the binary repetion code of **odd** length are examples of perfect codes, with parameters $(n, q^n, 1)$ and $(2m + 1, 2, 2m + 1)$, respectively. Such codes are said to be *trivial perfect codes*. We have also seen that the Hamming code [7,4,3] is perfect (actually this has been checked in Example 1.9).

E.1.14. In next section we will see that if q is a prime-power and r a positive integer, then there are codes with parameters

$$[(q^r - 1)/(q - 1), (q^r - 1)/(q - 1) - r, 3]).$$

Check that these parameters satisfy the condition for a perfect code. Note that for $q = 2$ and $r = 3$ we have the parameters [7,4,3].

E.1.15. Show that the parameters $[23, 12, 7]$, $[90, 78, 5]$ and $[11, 6, 5]_3$ satisfy the perfect condition.

E.1.16. Can a perfect code have even minimum distance?

E.1.17. If there is a perfect code of length n and minimum distance d, what is the value of $A_q(n, d)$? What is the value of $A_2(7, 3)$?

E.1.18. Consider the binary code consisting of the following 16 words:

0000000	1111111	1000101	1100010
0110001	1011000	0101100	0010110
0001011	0111010	0011101	1001110
0100111	1010011	1101001	1110100

Is it perfect? *Hint:* show that it is equivalent to the Hamming [7,4,3] code.

1.16 Remarks. The parameters of any non-trivial q-ary perfect code, with q a prime power, must be those of a Hamming code, or $[23, 12, 7]$, or $[11, 6, 5]_3$ (van Lint and Tietäväinen (1975); idependently established by Zinovi'ev and Leont'ev (1973)). There are non-linear codes with the same parameters as the Hamming codes (Vasili'ev (1962) for binary codes; Schönheim (1968) and Lindström (1969) in general). Codes with the parameters $(23, 2^{12}, 7)$ and $(11, 3^6, 5)_3$ exist (binary and ternary Golay codes; see Chapter 3), and they are unique up to equivalence (Pless (1968), Delsarte and Goethals (1975); see [25], Theorem 4.3.2, for a rather accessible proof in the binary case, and [11], Chapter 20, for a much more involved proof in the ternary case).

One of the steps along the way of characterizing the parameters of q-ary perfect codes, q a prime power, was to show (van Lint, H.W. Lenstra, A.M. Odlyzko were the main contributors) that the only non-trivial parameters that satisfy the perfect condition are those of the Hamming codes, the Golay codes (see E.1.15, and also P.1.7 for the binary case), and $(90, 2^{78}, 5)_2$. The latter, however, cannot exist (see P.1.8).

Finally let us note that it is conjectured that there are no non-trivial q-ary perfect codes for q not a prime power. There are some partial results that lend support to this conjecture (mainly due to Pless (1982)), but the remaining cases are judged to be very difficult.

The Gilbert inequality

We can also easily obtain a lower bound for $A_q(n, d)$. If C is an optimal code, any element of T^n lies at a distance $\leqslant d - 1$ of an element of C, for otherwise there would be a word $y \in T^n$ lying at a distance $\geqslant d$ from all elements of C and $C \cup \{y\}$ would be a code of length n, minimum distance d and with a greater cardinal than $|C|$, contradicting the optimality of C. This means that the union of the balls of radius $d - 1$ and with center the elements of C is the whole T^n. From this it follows that $A_q(n, d) \cdot vol_q(n, d-1) \geqslant q^n$. Thus we have proved the following:

1.17 Theorem (Gilbert lower bound). *The function $A_q(n, d)$ satisfies the fol-*

lowing inequality:

$$A_q(n, d) \geqslant \frac{q^n}{vol_q(n, d-1)}.$$

What is remarkable about the Gilbert lower bound, with the improvement we will find in the next chapter by means of linear codes, is that it is the only known general lower bound. This is in sharp contrast with the variety of upper bounds that have been discovered and of which the Singleton and sphere upper bounds are just two cases that we have already established.

▲ Library | Gilbert lower bound | ▲

$$\text{lb\_gilbert}(n,d,q) := \left\lceil \frac{q^n}{\text{volume}(n,d-1,q)} \right\rceil$$

$$\text{lb\_gilbert}(n,d) := \left\lceil \frac{2^n}{\text{volume}(n,d-1)} \right\rceil$$

\caretlb_gilbert(8,3) → 7
lb_gilbert(9,3) → 12
lb_gilbert(10,3) → 19
lb_gilbert(11,3) → 31
lb_gilbert(11,3,3) → 729
lb_gilbert(21,5) → 278

1.18 Remark. The Hamming and Gilbert bounds are not very close. For example, we have seen that $7 \leqslant A_2(8,3) \leqslant 28$, $12 \leqslant A_2(9,3) \leqslant 51$, $19 \leqslant A_2(10,3) \leqslant 93$ and $31 \leqslant A_2(11,3) \leqslant 170$. But in fact $A_2(8,3) = 20$ (we will get this later, but note that we already have $A_2(8,3) \geqslant 20$ by E.1.4), $A_2(9,3) = 40$ and the best known intervals for the other two are $72 \leqslant A_2(10,3) \leqslant 79$ and $144 \leqslant A_2(11,3) \leqslant 158$ (see Table 1.1).

E.1.19. The sphere upper bound for codes of type $(6, M, 3)$ turns out to be $M \leqslant 9$ (check this), but according to the table 1.1 we have $A_2(6,3) = 8$, so that there is no code of type $(6, 9, 3)$. Prove this. *Hint:* assuming such a code exists, show that it contains three words that have the same symbols in the last two positions.

Summary

- Key general ideas about information generation, coding and transmission.

- The definition of block code and some basic related notions (codewords, dimension, transmission rate, minimum distance and equivalence criteria between codes).

- Abstract decoders and error-correcting capacity.
- The minimum distance decoder and its correcting capacity $t_C = \lfloor (d-1)/2 \rfloor$.
- The Hamming code $[7, 4, 3]$ (our computational approach includes the construction of the code and the coding and decoding functions).
- Error-reduction factor of a code.
- Singleton bound: $k \leqslant n + 1 - d$.
- The function $A_q(n, d)$, whose determination is sometimes called the 'main problem' of block error-correcting codes. Table 1.1 provides some information on $A_2(n, d)$.
- Hamming upper bound:
  $$A_q(n, d) \leqslant q^n / vol_q(n, t), \text{ where } t = \lfloor (d-1)/2 \rfloor.$$
- Perfect codes: a code $(n, M, d)_q$ is perfect if and only if
  $$\sum_{i=0}^{t} \binom{n}{i} (q-1)^i = q^n / M \quad (= q^{n-k}).$$
- Gilbert lower bound: $A_q(n, d) \geqslant q^n / vol_q(n, d-1)$.

Problems

P.1.1 (Error-detection and correction). We have seen that a code C of type (n, M, d) can be used to detect up to $d - 1$ errors or to correct up to $t = \lfloor (d-1)/2 \rfloor$ errors. Show that C can be used to simultaneously detect up to $s \geqslant t$ errors and correct up to t errors if $t + s < d$.

P.1.2 Prove that $A_2(8, 5) = 4$ and that all codes of type $(8, 4, 5)_2$ are equivalent. *Hint:* by replacing a binary optimal code of length 8 and minimum distance 5 by an equivalent one, we may assume that 00000000 is a code word and then there can be at most one word of weight $\geqslant 6$.

P.1.3. Show that for binary codes of odd minimum distance the Hamming upper bound is not worse than the Singleton upper bound. Is the same true for even minimum distance? And for q-ary codes with $q > 2$?

P.1.4 Prove that $A_2(n, d) \leqslant 2A_2(n - 1, d)$. *Hint:* if C is an optimal binary code of length n and minimum distance d, we may assume, changing C into an equivalent code if necessary, that both 0 and 1 appear in the last position of elements of C, and then it is useful to consider, for $i = 0, 1$, the codes $C_i = \{x \in C \mid x_n = i\}$.

P.1.5 (Plotkin construction, 1960). Let C_1 and C_2 be binary codes of types (n, M_1, d_1) and (n, M_2, d_2), respectively. Let C be the length $2n$ code whose words have the form $x|(x + y)$, for all $x \in C_1$ and $y \in C_2$. Prove that $C \sim (2n, M_1 M_2, d)$, where $d = \min(2d_1, d_2)$.

P.1.6 Show that $A_2(16, 3) \geqslant 2560$. *Hint:* use Example E.1.4 and the Plotkin construction.

P.1.7 If $C \sim (n, M, 7)$ is a perfect binary code, prove that $n = 7$ or $n = 23$. *Hint:* use the sphere upper bound.

P.1.8. Show that there is no code with parameters $(90, 2^{78}, 5)$. *Hint:* Extracted from [9], proof of Theorem 9.7: if C were a code with those parameters, let X be the set of vectors in C that have weight 5 and begin with two 1s, Y the set of vectors in $\mathbb{Z}_2^{90}$ that have weight 3 and begin with two 1s, and $D = \{(x, y) \in X \times Y \mid S(y) \subset S(x)\}$, and count $|D|$ in two ways ($S(x)$ is the support of x, that is, the set of indices i in $1..90$ such that $x_i = 1$).

1.2 Linear codes

> In an attempt to find codes which are simultaneously good in
> the sense of the coding theorem, and reasonably easy to im-
> plement, it is natural to impose some kind of structure on the
> codes.
>
> R. J. McEliece, [12], p. 133.

Essential points

- Basic notions: linear code, minimum weight, generating matrix, dual code, parity-check matrix.
- Reed–Solomon codes and their duals.
- Syndrome-leader decoding algorithm.
- Gilbert–Varshamov condition on q, n, k, d for the existence of a linear code $[n, k, d]_q$.
- Hamming codes and their duals.
- Weight enumerators and the MacWilliams theorem ($\sim$ identities).

Introduction

The construction of codes and the processes of coding and decoding are computationally intensive. For the implemetation of such opperations any additional structure present in the codes can be of much help.

In order to define useful structures on the codes, a first step is to enrich the alphabet T with more structure than being merely a set. If we want, for example, to be able to add symbols with the usual properties, we could use the group $\mathbb{Z}_q$ as T. Since this is also a ring, the alphabet symbols could also be multiplied.[2] If we also want to be able to divide symbols, then we are faced with an important restriction, because division in $\mathbb{Z}_q$ by nonzero elements is possible without exception if and only if q is prime.

We will have more room in that direction if we allow ourselves to use not only the fields $\mathbb{Z}_p$, p prime, but also all other finite fields (also called *Galois fields*). As we will see in the next chapter in detail, there is a field of q elements if and only if q is a power of a prime number, and in that case the

---

[2]For an interesting theory for the case $T = \mathbb{Z}_4$, see Chapter 8 of [25].

field is unique up to isomorphism. Henceforth this field will be denoted $\mathbb{F}_q$ (another popular notation is $GF(q)$).

```
Prime powers less than 50
N=50;
{n in 2..N where is_prime_power?(n)}
   → {2, 3, 4, 5, 7, 8, 9, 11, 13, 16, 17, 19, 23, 25, 27, 29, 31, 32, 37, 41, 43, 47, 49}
```

So let us assume that the alphabet T is a finite field $\mathbb{F}_q$. Then we can consider the codes $C \subseteq \mathbb{F}_q^n$ that are an $\mathbb{F}_q$-vector subspace of $\mathbb{F}_q^n$. Such codes are called *linear codes* and we devote this section to the study some of the basic notions related to them. Note that if C is a linear code and its dimension as an $\mathbb{F}_q$-vector space is k, then $|C| = q^k$ and therefore k coincides with the dimension of C as a code: $dim_{\mathbb{F}_q}(C) = dim(C)$.

Linear codes are important for diverse reasons. The more visible are that they grant powerful computational facilities for coding and decoding. If we revisit the Hamming code [7,4,3] (example 1.9), we will realize that linear algebra over $\mathbb{Z}_2$ plays a major role there, and similar advantages will be demonstrated for several families of codes from now on. A more subtle motive, which is however beyond the scope of this introductory text, is the fact that Shannon's channel coding theorem can be achieved by means of linear codes.

1.19 Example (The International Standard Book Number). The ISBN code of a book is 10-digit codeword assigned by the publisher, like the 3-540-00395-9 of this book. For the last digit the symbol X (whose value is 10) is also allowed. For example, the ISBN of [6] is 0-387-96617-X. The hyphens divide the digits into groups that have some additional meaning, like the first 0, which indicates the language (English), and the groups 387-96617, which identify the publisher and the book number assigned by the publisher. The last digit is a *check digit* and is computed by means of the following *weighted check sum*:

$$\sum_{i=1}^{10} i \cdot x_i \equiv 0 \pmod{11},$$

or equivalently

$$x_{10} = \sum_{i=1}^{9} i \cdot x_i \pmod{11},$$

with the convention that the remainder 10 is replaced by X. For some interesing properties of this coding scheme, which depend crucially on the fact that $\mathbb{Z}_{11}$ is a field, see P.1.9. See also the listing **ISBN check digit** for the computation of the ISBN check digit and the examples of the two books quoted in this Example.

```
┌─────────────────────────────────────────────────────────────────────┐
│ ▲ Library │ ISBN check digit                                       ▲ │
├─────────────────────────────────────────────────────────────────────┤
│  isbn(x:Vector)                                                       │
│  check length(x) = 9                                                  │
│  := begin                                                             │
│                    9                                                  │
│         local r=  ∑   i·x_i mod 11                                    │
│                   i=1                                                 │
│         if r<10  then                                                 │
│             r                                                         │
│         else                                                          │
│             X                                                         │
│         end                                                          │
│     end                                                              │
└─────────────────────────────────────────────────────────────────────┘
```

```
 isbn ( [ 3, 5, 4, 0, 0, 0, 3, 9, 5 ] )  ⟶  9
 isbn ( [ 0, 3, 8, 7, 9, 6, 6, 1, 7 ] )  ⟶  X
```

Notations and conventions

To ease the exposition, in this section *code* will mean *linear code* unless explicitly stated otherwise. Often we will also write $\mathbb{F}$ instead of $\mathbb{F}_q$. We will simply write $M_n^k(\mathbb{F})$ to denote the matrices of type $k \times n$ with coefficients in $\mathbb{F}$. Instead of $M_k^k(\mathbb{F})$ we will write $M_k(\mathbb{F})$.

The usual scalar product of $x, y \in \mathbb{F}^n$ will be denoted $\langle x|y \rangle$:

$$\langle x|y \rangle = xy^T = x_1 y_1 + \ldots + x_n y_n.$$

Weights

We will let $|\ |\colon \mathbb{F}^n \to \mathbb{N}$ denote the map $x \mapsto d(x, 0)$. By definition, $|x|$ is equal to the number of non-zero symbols of $x \in \mathbb{F}^n$. We will say that $|x|$, which sometimes is also denoted $wt(x)$, is the *weight* of x.

E.1.20. The weight is a *norm* for the space $\mathbb{F}^n$: for all $x, y, z \in \mathbb{F}^n$, $|0| = 0$, $|x| > 0$ if $x \neq 0$, and $|x + y| \leqslant |x| + |y|$ (triangle inequality). $\square$

If C is a linear code, then the *minimum weight* of C, w_C, is defined as the minimum of the weights $|x|$ of the non-zero $x \in C$.

1.20 Proposition. *The minimum weight of C coincides with the minimum distance of C: $d_C = w_C$.*

Proof: Let $x, y \in C$. If C is linear, then $x - y \in C$ and we have $hd(x, y) = |x - y|$. Since $x - y \neq 0$ if and only if $x \neq y$, we see that any distance between distinct elements is the weight of a nonzero vector. Conversely, since $|x| = hd(x, 0)$, and $0 \in C$ because C is linear, the weight of any nonzero vector is the distance between two distinct elements of C. Now the claimed equality follows from the definitions of d_C and w_C. □

1.21 Remark. For a general code C of cardinal M, the determination of d_C involves the computation of the $M(M-1)/2$ Hamming distances between its pairs of distinct elements. The proposition tells us that if C is linear then the determination of d_C involves only the computation of $M-1$ weights. This is just an example of the advantage of having the additional linear structure.

It is important to take into account that the norm $|\;|$ and the scalar product $\langle\;|\;\rangle$ are **not** related as in Euclidean geometry. In the geometrical context we have the formula $|x|^2 = \langle x|x \rangle$ (by definition of $|x|$), while in the present context $|x|$ and $\langle x|x \rangle$ have been defined independently. In fact they are quite unrelated, if only because the values of the norm are non-negative integers and the values of the scalar product are elements of $\mathbb{F}$.

E.1.21. Let C be a linear code of type $[n, k]$ over $\mathbb{F} = \mathbb{F}_q$. Fix any integer j such that $1 \leqslant j \leqslant n$. Prove that either all vectors of C have 0 at the j-th position or else that every element of $\mathbb{F}$ appears there in precisely q^{k-1} vectors of C.

E.1.22. Given a binary linear code C, show that either all its words have even weight or else there are the same number of words with even and odd weight. In particular, in the latter case $|C|$ is even.

Generating matrices

Given a code C of type $[n, k]$, we will say that a matrix $G \in M_n^k(\mathbb{F})$ is a *generating matrix* of C if the rows of G form a linear basis of C.

Conversely, given a matrix $G \in M_n^k(\mathbb{F})$ the subspace $\langle G \rangle \subseteq \mathbb{F}^n$ generated by the rows of G is a code of type $[n, k]$, where k is the rank of G. We will say that $\langle G \rangle$ is *the code generated by* G.

1.22 Remark. For the knowledge of a general code we need to have an explicit list of its M vectors. In the case of linear codes of dimension k, $M = q^k$, but the M code vectors can be generated out of the k rows of a generating matrix. This is another example of the advantage provided by the additional linear structure.

1.23 *Examples*. a) The repetition code of length n is generated by $\mathbf{1}_n$.

b) A generating matrix for the Hamming code $[7, 4, 3]$ is $G = I_4 | R^T$ (notations as in the example 1.9, p. 21).

c) If $C \subseteq \mathbb{F}^n$ is a code of dimension k, let $\overline{C} \subseteq \mathbb{F}^{n+1}$ be the image of C by the linear map $\mathbb{F}^n \to \mathbb{F}^n \times \mathbb{F} = \mathbb{F}^{n+1}$ such that $x \mapsto x | - s(x)$, where $s(x) = \sum_i x_i$. Then $\overline{C}$ is a code of type $[n+1, k]$ which is called the *parity extension* of C (the symbol $-s(x)$ is called the *parity check symbol* of the vector x). If G is a generating matrix of C, then the matrix $\overline{G}$ obtained by appending to G the column consisting of the parity check symbols of its rows is a generating matrix of $\overline{C}$. The matrix $\overline{G}$ will be called the *parity completion* (or parity extension) of G. The listing **Parity completion of a matrix** shows a way of computing it.

▲ Library │   Parity completion of a matrix   ▲

The expression G | [ ] below yields the matrix G augmented with the column vector [ ]^T at the end. An expression like [ ] | G works similarly, but this time G is augmented with the column vector [ ]^T at the beginning.

```
parity_completion (G :Matrix) := G | [−sum(g) with g in G]
left_parity_completion (G :Matrix) := [−sum(g) with g in G] | G
```

$$G = \begin{pmatrix} 1100001 \\ 0110010 \\ 1010100 \\ 1111000 \end{pmatrix} : \text{Matrix}(\mathbb{Z}_2) \to \begin{pmatrix} 1100001 \\ 0110010 \\ 1010100 \\ 1111000 \end{pmatrix}$$

$$\text{parity\_completion}(G) \to \begin{pmatrix} 11000011 \\ 01100101 \\ 10101001 \\ 11110000 \end{pmatrix}$$

$$\text{left\_parity\_completion}(\text{diagonal\_matrix}([1,2,3,4]):\text{Matrix}(\mathbb{Z}_5)) \to \begin{pmatrix} 41000 \\ 30200 \\ 20030 \\ 10004 \end{pmatrix}$$

E.1.23. The elements $x \in \mathbb{F}^{n+1}$ such that $s(x) = 0$ form a code C of type $[n+1, n]$ (it is called the *zero-parity code* of length $n+1$, or of dimension n). Check that the matrix $(I_n | \mathbf{1}_n^T)$ is a generating matrix of C.

Coding

It is clear that if G is a generating matrix of C, then the map

$$f: \mathbb{F}^k \to \mathbb{F}^n, \quad u \mapsto uG$$

induces an isomorphism of $\mathbb{F}^k$ onto C and hence we can use f as a coding map for C.

If $A \in M_k(\mathbb{F})$ is an invertible matrix (in other words, $A \in GL_k(\mathbb{F})$), then AG is also a matrix of C. From this it follows that for each code C there exists an equivalent code which is generated by a matrix that has the form $G = (I_k|P)$, where I_k is the identity matrix of order k and $P \in M_{n-k}^k(\mathbb{F})$. Since in this case $f(u) = uG = (u|uP)$, for all $u \in \mathbb{F}^k$, we see that the coding of u amounts to appending the vector $uP \in \mathbb{F}^{n-k}$ to the vector u (we may think of uP as a 'redundancy' vector appended to the 'information vector' u).

The codes C (this time not necessarily linear) of dimension k for which there are k positions in which appear, when we let x run in C, all sequences of k symbols, are said to be *systematic* (with respect to those k positions). According to the preceding paragraph, each linear code of dimension k is equivalent to a systematic code with respect to the first k positions.

1.24 *Example* (Reed–Solomon codes). Let n be an integer such that $1 \leqslant n \leqslant q$ and $\alpha = \alpha_1, \ldots, \alpha_n \in \mathbb{F}$ a sequence of n distinct elements of $\mathbb{F}$. For every integer $k > 0$, let $\mathbb{F}[X]_k$ be the $\mathbb{F}$-vector space whose elements are polynomials of degree $< k$ with coefficients in $\mathbb{F}$. We have $\mathbb{F}[X]_k = \langle 1, X, \ldots, X^{k-1} \rangle_\mathbb{F}$ and $dim\,(\mathbb{F}[X]_k) = k$. If $k \leqslant n$, the map

$$\epsilon \colon \mathbb{F}[X]_k \to \mathbb{F}^n, \qquad f \mapsto (f(\alpha_1), \ldots, f(\alpha_n))$$

is injective, since the existence of a non-zero polynomial of degree $< k$ vanishing on all the α_i implies $n < k$ (a non-zero polynomial of degree r with coefficients in a field can have at most r roots). The image of ϵ is therefore a linear code C of type $[n, k]$.

We claim that the minimum distance of C is $n-k+1$. Indeed, a non-zero polynomial f of degree $< k$ can vanish on at most $k - 1$ of the elements α_i and hence the weight of $(f(\alpha_1), \ldots, f(\alpha_n))$ is not less than $n - (k - 1) = n - k + 1$. On the other hand, this weight can be exactly $n - k + 1$ for some choices of f, like $f = (X - \alpha_1) \cdots (X - \alpha_{k-1})$, and this proves the claim.

We will say that C is a Reed–Solomon (*RS*) code of length n and dimension k, and will be denoted $RS_\alpha(k)$. When the α_i can be understood by the context, we will simply write $RS(k)$, or $RS(n, k)$ if we want to display the length n. It is clear that the *RS* codes satisfy the equality in the Singleton bound, and so they are examples of MDS codes. On the other hand we have $n \leqslant q$, and so we will have to take high values of q to obtain interesting codes.

Since $1, X, \ldots, X^{k-1}$ is a basis of $\mathbb{F}[X]_k$, the *Vandermonde matrix*

$$V_k(\alpha_1, \ldots, \alpha_n) = (\alpha_i^j) \quad (1 \leqslant i \leqslant n, \, 0 \leqslant j < k)$$

is a generating matrix for $RS_\alpha(k)$.

The dual code

The linear subspace of $\mathbb{F}^n$ ortogonal to a subset $Z \subseteq \mathbb{F}^n$ will be denoted $Z^\perp$. Let us recall that the vectors of this subspace are the vectors $x \in \mathbb{F}^n$ such that $\langle x | z \rangle = 0$ for all $z \in Z$.

If C is a code, the code $C^\perp$ is called the *dual code* of C.

Since the scalar product $\langle \, | \, \rangle$ is non-degenerate, by linear algebra we know that $C^\perp$ has dimension $n - k$ if C has dimension k. In other words, $C^\perp$ is of type $[n, n - k]$ if C is of type $[n, k]$.

As $C \subseteq C^{\perp\perp}$, tautologically, and both sides of this inclusion have dimension k, we infer that $C^{\perp\perp} = C$.

Usually it happens that C and $C^\perp$ have a non-zero intersection. Even more, it can happen that $C \subseteq C^\perp$, or that $C = C^\perp$. In the latter case we say that C is *self-dual*. Note that in order to be self-dual it is necessary that $n = 2k$.

E.1.24. If C is the length n repetition code over C, check that $C^\perp$ is the zero-parity code of length n.

E.1.25. If G is a generating matrix of C, prove that $C \subseteq C^\perp$ is equivalent to the relation $GG^T = 0$. If in addition we have $n = 2k$, prove that this relation is equivalent to $C = C^\perp$. As an application, check that the parity extension of the Hamming code $[7, 4, 3]$ is a self-dual code.

E.1.26. Let $C_1, C_2 \subseteq \mathbb{F}_q^n$ be linear codes. Show that $(C_1 + C_2)^\perp = C_1^\perp \cap C_2^\perp$ and $(C_1 \cap C_2)^\perp = C_1^\perp + C_2^\perp$.

Parity-check matrices

If H is a generating matrix of $C^\perp$ (in which case H is a matrix of type $(n - k) \times n$) we have that

$$x \in C \quad \text{if and only if} \quad xH^T = 0,$$

because if h is a row of H we have $xh^T = \langle x | h \rangle$. Said in other words, the elements $x \in C$ are exactly those that satisfy the $n - k$ linear equations $\langle x | h \rangle = 0$, where h runs through the rows of H.

Given $h \in C^\perp$, the linear equation $\langle x | h \rangle = 0$, which is satisfied for all $x \in C$, is called the *check equation* of C corresponding to h. By the previous paragraph, C is determined by the $n - k$ check equations corresponding to the rows of H, and this is why the matrix H is said to be a *check matrix* of C (or also, as most authors say, a *parity-check matrix* of C, in a terminology that was certainly appropriate for binary codes, but which is arguably not that good for general codes).

The relation

$$C = \{x \in \mathbb{F}^n | x H^T = 0\}$$

can also be interpreted by saying that C is *the set of linear relations satisfied by the rows of H^T, that is, by the columns of H*. In particular we have:

1.25 Proposition. *If any $r - 1$ columns of H are linearly independent, then the minimum distance of C is at least r, and conversely.*

1.26 *Example* (The dual of an RS code). Let $C = RS_{\alpha_1, \ldots, \alpha_n}(k)$, where $\alpha = \alpha_1, \ldots, \alpha_n$ are distinct nonzero elements of a finite field K. Then we know that $G = V_k(\alpha_1, \ldots, \alpha_n)$ is a generating matrix of C (Example 1.24). Note that the rows of G have the form $(\alpha_1^i, \ldots, \alpha_n^i)$, with $i = 0, \ldots, k - 1$.

Now we are going to describe a check matrix H of C, that is, a generating matrix of $C^\perp$. Recall that if we define $D(\alpha_1, \ldots, \alpha_n)$ as the determinant of the Vandermonde matrix $V_n(\alpha_1, \ldots, \alpha_n)$ (*Vandermonde determinant*), then

$$D(\alpha_1, \ldots, \alpha_n) = \prod_{i < j} (\alpha_j - \alpha_i).$$

Define the vector $h = (h_1, \ldots, h_n)$ by the formula

$$h_i = (-1)^{i-1} D(\alpha_1, \ldots, \alpha_{i-1}, \alpha_{i+1}, \ldots, \alpha_n)/D(\alpha_1, \ldots, \alpha_n)$$
$$= 1/(\prod_{j \neq i} (\alpha_j - \alpha_i))$$

(remark that in the last product there are precisely $i - 1$ factors with the indices reversed, namely the $\alpha_j - \alpha_i$ with $j < i$). Then the matrix

$$H = V_{n-k}(\alpha_1, \ldots, \alpha_n) \text{diag} (h_1, \ldots, h_n)$$

is a check matrix of C.

To see this, it is enough to prove that any row of G is orthogonal to any row of H because H has clearly rank $n - k$. Since the rows of H have the form $(h_1 \alpha_1^j, \ldots, h_n \alpha_n^j)$, with $j = 0, \ldots, n - k - 1$, we have to prove that $\sum_{l=1}^{n} \alpha_l^{i+j} h_l = 0$ if $0 \leqslant i \leqslant k-1$ and $0 \leqslant j \leqslant n-k-1$. Thus it will be enough to prove that $\sum_{l=1}^{n} \alpha_l^s h_l = 0$ $(0 \leqslant s \leqslant n - 2)$. Multiplying throughout by the nonzero determinant $D(\alpha_1, \ldots, \alpha_n)$, and taking into account the definition of h_l, we wish to prove that

$$\sum_{l=1}^{n} \alpha_l^s (-1)^{l-1} D(\alpha_1, \ldots, \alpha_{l-1}, \alpha_{l+1}, \ldots, \alpha_n) = 0.$$

But finally this is obvious, because the left hand side coincides with the
determinant

$$
\begin{vmatrix}
\alpha_1^s & \cdots & \alpha_n^s \\
1 & \cdots & 1 \\
\alpha_1 & \cdots & \alpha_n \\
\vdots & & \vdots \\
\alpha_1^{n-2} & \cdots & \alpha_n^{n-2}
\end{vmatrix}
$$

(developed along the first row), and this determinant has a repeated row.

As we will see in Chapter 4, the form of the matrix H indicates that
$RS_\alpha(k)$ is an alternant code, and as a consequence it will be decodable with
any of the fast decoders for alternant codes studied in that chapter (sections
4.3 and 4.4).

E.1.27. Check that the dual of an $RS(n, k)$ code is scalarly equivalent to an
$RS(n, n - k)$. In particular we see that the dual of a RS code is an MDS
code. $\square$

Let us consider now the question of how can we obtain a check matrix from
a generating matrix G. Passing to an equivalent code if necessary, we may
assume that $G = (I_k|P)$ and in this case we have:

1.27 Proposition. If $G = (I_k|P)$ is a generating matrix of C, then $H =
(-P^T|I_{n-k})$ is a generating matrix of $C^\perp$ (note that $-P^T = P^T$ in the
binary case).

Proof: Indeed, H is a matrix of type $(n - k) \times n$ and its rows are clearly
linearly independent. From the expressions of G and H it is easy to check
that $GH^T = 0$ and hence $\langle H \rangle \subseteq C^T$. Since both terms in this inclusion
have the same dimension, they must coincide. $\square$

1.28 Example. In the binary case, the matrix $H = (1_k|1)$ is a check matrix
of the *parity-check code* $C \sim [k + 1, k]$, which is defined as the parity extension of the total code of dimension k. The only check equation for C is
$\sum_{i=1}^{k+1} x_i = 0$, which is equivalent to say that the weight $|x|$ is even for all x.

For the repetition code of length n relative to an arbitrary finite field $\mathbb{F}$,
the vector $1_n = (1|1_{n-1})$ is a generating matrix and $H = (1_{n-1}^T|I_{n-1})$ is a
check matrix. Note that the check equations corresponding to the rows of H
are $x_i = x_1$ for $i = 2, \ldots, n$.

1.29 Example. If H is a check matrix for a code C of type $[n, k]$, then

$$
\overline{H} = \begin{pmatrix} 1_n & 1 \\ H & 0_{n-k}^T \end{pmatrix}
$$

is a check matrix of the parity extension $\overline{C}$ of C.

1.30 *Example*. Let $\mathbb{F}$ be a field of q elements and let $\mathbb{F}^* = \{\alpha_1, \ldots, \alpha_n\}$, $n = q - 1$. Let $C = RS_{\alpha_1, \ldots, \alpha_n}(k)$. Then we know that the Vandermonde matrix $G = V_k(\alpha_1, \ldots, \alpha_n)$ is a generating matrix of C. Now we will see that the Vandermonde matrix

$$H = V_{1,n-k}(\alpha_1, \ldots, \alpha_n) = (\alpha_j^i), \quad 1 \leqslant i \leqslant n - k, 1 \leqslant j \leqslant n,$$

is a check matrix of C.

Indeed, as seen in Example 1.26, there is a check matrix of C that has the form $V_{n-k}(\alpha_1, \ldots, \alpha_n) diag(h_1, \ldots, h_n)$, with $h_i = 1/\prod_{j \neq i}(\alpha_j - \alpha_i)$, and in the present case we have $h_i = \alpha_i$ (as argued in a moment), so that $V_{1,n-k}(\alpha_1, \ldots, \alpha_n) = V_{n-k}(\alpha_1, \ldots, \alpha_n) diag(\alpha_1, \ldots, \alpha_n)$ is a check matrix of C.

To see that $h_i = \alpha_i$, first note that $\prod_{j \neq i}(\alpha_j - \alpha_i)$ is the product of all the elements of $\mathbb{F}^*$, except $-\alpha_i$. Hence $h_i = -\alpha_i/P$, where P is the product of all the elements of $\mathbb{F}^*$. And the claim follows because $P = -1$, as each factor in P cancels with its inverse, except 1 and -1, that coincide with their own inverses.

Remark that the dual of $RS_{\alpha_1, \ldots, \alpha_n}(k)$ is not an $RS_{\alpha_1, \ldots, \alpha_n}(n - k)$, because $H = V_{1,n-k}(\alpha_1, \ldots, \alpha_n)$ is not $V_{n-k}(\alpha_1, \ldots, \alpha_n)$. But on dividing the j-th column of H by α_j we see that the dual of $RS_{\alpha_1, \ldots, \alpha_n}(k)$ is scalarly equivalent to $RS_{\alpha_1, \ldots, \alpha_n}(n - k)$.

E.1.28. Let G be the generating matrix of a code $C \sim [n, k]$ and assume it has the form $G = A|B$, where A is a $k \times k$ non-singular matrix $(\det(A) \neq 0)$. Show that if we put $P = A^{-1}B$, then $H = (-P^T)|I_{n-k}$ is a check matrix of C.

E.1.29. In Proposition 1.27 the matrix I_k occupies the first k columns, and P the last $n - k$. This restriction can be easily overcome as follows. Assume that I_k is the submatrix of a matrix $G \in M_n^k(\mathbb{F})$ formed with the columns $j_1 < \cdots < j_k$, and let $P \in M_{n-k}^k(\mathbb{F})$ be the matrix left from G after removing these columns. Show that we can form a check matrix $H \in M_n^{n-k}(\mathbb{F})$ by placing the columns of $-P^T$ successively in the columns $j_1, \ldots, j_k$ and the columns of I_{n-k} successively in the remaining columns. [For an illustration of this recipe, see the Example 1.36]

E.1.30. We have seen (E.1.27; see also Example 1.30) that the dual of an RS code is scalarly equivalent to a RS code. Since RS codes are MDS, this leads us to ask whether the dual of a linear MDS code is an MDS code. Prove that the answer is affirmative and then use it to prove that a linear code $[n, k]$ is MDS if and only if any k columns of a generating matrix are linearly independent.

1.31 Remark. The last exercise includes a characterization of linear MDS codes. Several others are known. An excellent reference is [20], § 5.3. Note that Theorem 5.3.13 in [20] gives an explicit formula for the weight enumerator of a linear $[n, k]_q$, hence in particular for RS codes:

$$a_i = \binom{n}{i}(q-1)\sum_{j=0}^{i-d}(-1)^j\binom{i-1}{j}q^{i-d-j}.$$

| ▲ Library | Weight enumerator of an MDS linear code | ▲ |

```
weight_enumerator_mds(n,k,q)
check 2≤min(k,n−k) & max(k,n−k)≤q−1
:=
begin
    local d=n−k+1
          a
```

$$a = \left[\binom{n}{i}\cdot(q-1)\sum_{j=0}^{i-d}(-1)^j\cdot\binom{i-1}{j}\cdot q^{i-d-j} \text{ with i in d..n}\right]$$

```
    [1] | constant_vector(d−1,0) | a
end
```

Weight enumerator of the RS code of dimension 4
on the non−zero elements of F_9

weight_enumerator_mds(8,4,9) → [1, 0, 0, 0, 0, 448, 896, 2688, 2528]

Syndrome decoding

Let C be a code of type $[n, k]$ and H a check matrix of C. Given $y \in \mathbb{F}^n$, the element

$$yH^T \in \mathbb{F}^{n-k}$$

is called the *syndrome* of y (with respect to H). From what we saw in the last subsection, the elements of C are precisely those that have null syndrome: $C = \{x \in \mathbb{F}^n \mid xH^T = 0\}$.

More generally, given $s \in \mathbb{F}^{n-k}$, let $C_s = \{z \in \mathbb{F}^n \mid zH^T = s\}$ (hence $C_0 = C$). Since the map $\sigma : \mathbb{F}^n \to \mathbb{F}^{n-k}$ such that $y \mapsto yH^T$ is linear, surjective (because the rank of H is $n - k$) and its kernel is C, it follows that C_s is non-empty for any $s \in \mathbb{F}^{n-k}$ and that $C_s = z_s + C$ for any $z_s \in C_s$. In other words, C_s is a class modulo C (we will say that it is the class of the syndrome s).

The notion of syndrome is useful in general for the purpose of minimum distance decoding of C (as it was in particular for the Hamming [7,4,3] code in example 1.9). The key observations we need are the following. Let g be the minimum distance decoder. Let $x \in C$ be the transmitted vector and y the received vector. We know that y is g-decodable if and only if there exists $x' \in C$ such that $hd(y, x') \leqslant t$, and in this case x' is unique and $g(y) = x'$. With these notations we have:

1.32 Lemma. *The vector y is g-decodable if and only if there exists $e \in \mathbb{F}^n$ such that $yH^T = eH^T$ and $|e| \leqslant t$, and in this case e is unique and $g(y) = y - e$.*

Proof: If y is g-decodable, let $x' = g(y)$ and $e = y - x'$. Then $eH^T = yH^T - x'H^T = yH^T$, because the syndrome of $x' \in C$ is 0, and $|e| = d(y, x') \leqslant t$.

Conversely, let e satisfy the conditions in the statement and consider the vector $x' = y - e$. Then $x' \in C$, because $x'H^T = yH^T - eH^T = 0$ (by the first hypothesis on e) and $d(y, x') = |e| \leqslant t$. Then we know that x' is unique (hence e is also unique) and that $g(y) = x'$ (so $g(y) = y - e$). $\qquad\square$

E.1.31. If $e, e' \in C_s$ and $|e|, |e'| \leqslant t$, prove that $e = e'$. In other words, the class C_s can contain at most one element e with $|e| \leqslant t$.

Leaders' table

The preceeding considerations suggest the following decoding scheme. First precompute a table $E = \{s \to e_s\}_{s \in \mathbb{F}^{n-k}}$ with $e_s \in C_s$ and in such a way that e_s has minimum weight among the vectors in C_s (the vector e_s is said to be a *leader* of the class C_s and the table E will be called a *leaders' table* for C). Note that e_s is unique if $|e_s| \leqslant t$ (exercise E.1.31); otherwise we will have to select one among the vectors of C_s that have minimum weight. Now the *syndrome decoder* can be described as follows:

Syndrome-leader decoding algorithm

1) find the syndrome of the received vector y, say $s = yH^T$;

2) look at the leaders' table E to find e_s, the leader of the class corresponding to s;

3) return $y - e_s$.

1.33 *Example.* Before examining the formal properties of this decoder, it is instructive to see how it works for the code $C = \text{Rep}(3)$. The matrix $H = \begin{pmatrix} 1 & 1 & 0 \\ 1 & 0 & 1 \end{pmatrix}$ is a check matrix for the binary repetition code $C = \{000, 111\}$. For each $s \in \mathbb{Z}_2^2$, it is clear that $e'_s = 0|s$ has syndrome s. Thus we have $C_{00} = C = \{000, 111\}$, $C_{10} = e'_{10} + C = \{010, 101\}$,

$C_{01} = e'_{01} + C = \{001, 110\}$, $C_{11} = e'_{11} + C = \{011, 100\}$. The leaders of these classes are $e_{00} = e'_{00} = 000$, $e_{10} = e'_{10} = 010$, $e_{01} = e'_{01} = 001$, $e_{11} = 100$. Hence the vectors y in the class C_s are decoded as $y + e_s$ (we are in binary). For example, 011 is decoded as $011 + 100 = 111$ and 100 as $100 + 100 = 000$, like the majority-vote decoding explained in the Introduction (page 3). In fact, it can be quickly checked that this coincidence is valid for all $y \in \mathbb{Z}_2^3$.   □

The facts discovered in the preceeding example are a special case of the following general result:

1.34 Proposition. *If* $D = \bigsqcup_{x \in C} B(x, t)$ *is the set of decodable vectors by the minimum distance decoder, then the syndrome-leader decoder coincides with the minimum distance decoder for all* $y \in D$.

Proof: Let $x \in C$ and $y \in B(x, t)$. If $e = y - x$, then $|e| \leqslant t$ by hypothesis. Moreover, if $s = yH^T$ is the syndrome of y, then the syndrome of e is also s, because

$$eH^T = yH^T - xH^T = s.$$

Since e_s has minimum weight among the vectors of syndrome s, $|e_s| \leqslant |e| \leqslant t$. So $e = e_s$ (E.1.31) and $g(y) = y - e_s = y - e = x$, as claimed.   □

1.35 Remarks. With the syndrome decoder there are no docoder errors, for the table $E = \{s \to e_s\}$ is extended to all $s \in \mathbb{F}^{n-k}$, but we get an undetectable error if and only if $e = y - x$ is not on the table E.

For the selection of the table values e_s the following two facts may be helpful:

a) Let e'_s be any element with syndrome s. If the class $e'_s + C$ contains an element e such that $|e| \leqslant t$, then e is the only element in C_s with this property (E.1.31) and so $e_s = e$.

b) If the submatrix of H formed with its last $n - k$ columns is the identity matrix I_{n-k} (note that this will happen if H has been obtained from a generating matrix G of C of the form $(I_k|P)$), then the element $e'_s = (0_k|s)$ has syndrome s and hence the search of e_s can be carried out in $C_s = e'_s + C$. Moreover, it is clear that for all s such that $|s| \leqslant t$ we have $e_s = e'_s$.

1.36 *Example.* See the listing **A binary example of syndrome decoding** for how the computations have been arranged. Let $C = \langle G \rangle$, where

$$G = \begin{pmatrix} 1 & 1 & 0 & 0 & 1 \\ 0 & 0 & 1 & 0 & 1 \\ 1 & 0 & 0 & 1 & 1 \end{pmatrix} \in M_5^3(\mathbb{Z}_2).$$

In the listing, C is returned as the list of its eight vectors.

Then we compute the table $E = \{s \rightarrow C_s\}_{s \in \mathbb{Z}_2^2}$ of classes modulo C and get

$$X = \{ \; [0,0] \rightarrow \{ \; [0,0,0,0,0], [1,0,0,1,1], [0,0,1,0,1], [1,0,1,1,0],$$
$$[1,1,0,0,1], [0,1,0,1,0], [1,1,1,0,0], [0,1,1,1,1] \; \},$$
$$[0,1] \rightarrow \{ \; [0,0,0,0,1], [1,0,0,1,0], [0,0,1,0,0], [1,0,1,1,1],$$
$$[1,1,0,0,0], [0,1,0,1,1], [1,1,1,0,1], [0,1,1,1,0] \; \},$$
$$[1,0] \rightarrow \{ \; [1,0,0,0,0], [0,0,0,1,1], [1,0,1,0,1], [0,0,1,1,0],$$
$$[0,1,0,0,1], [1,1,0,1,0], [0,1,1,0,0], [1,1,1,1,1] \; \},$$
$$[1,1] \rightarrow \{ \; [1,0,0,0,1], [0,0,0,1,0], [1,0,1,0,0], [0,0,1,1,1],$$
$$[0,1,0,0,0], [1,1,0,1,1], [0,1,1,0,1], [1,1,1,1,0] \; \} \; \}$$

Note that to each s there corresponds the list of vectors of C_s. Of course, the list corresponding to $[0,0]$ is C.

```
▲ Library │    A binary example of syndrome decoding                    ▲

  min_weight(X) := min({weight(x) with x in X})
  min_weights(X)
  := begin
       local m=min_weight(X)
       {x in X such_that weight(x)=m}
     end
```

```
B=[0,1] : Vector(ℤ₂) ;

    ⎛ 1 1 0 0 1 ⎞
G = ⎜ 0 0 1 0 1 ⎟ : Matrix(ℤ₂) ;
    ⎝ 1 0 0 1 1 ⎠

H=Gᵀ_{1, 5}; H=[1, 0]| H |[0, 1];

C={ a·G₁+b·G₂+c·G₃ with (a, b, c) in (B,B,B) };

X={ [a, b]→{[a, 0, 0, 0, b]+x  with x in C} with (a,b) in (B,B) };

M={ x₁→min_weights(x₂) with x in X };

E={ x₁→x₂,₁ with x in M };

y=[0, 1, 0, 0, 1];

s=y·Hᵀ ➡ [1, 0]

y+E(s) ➡ [1, 1, 0, 0, 1]
```

Let M be the table $\{s \rightarrow M_s\}_{s \in \mathbb{Z}_2^2}$, where M_s is the set of vectors of minimum weight in C_s. This is computed by means of the function min_weights, which yields, given a list of vectors, the list of those that have minimum weight in that list. We get:

$$M = \{[0,0] \rightarrow \{[0,0,0,0,0]\}, [0,1] \rightarrow \{[0,0,0,0,1], [0,0,1,0,0]\},$$
$$[1,0] \rightarrow \{[1,0,0,0,0]\}, [1,1] \rightarrow \{[0,0,0,1,0], [0,1,0,0,0]\}\}.$$

A leaders' table $E = \{s \rightarrow e_s\}_{s \in \mathbb{Z}_2^2}$ is found by retaining the first vector of M_s. We get:

$$E = \{[0,0] \rightarrow [0,0,0,0,0], [0,1] \rightarrow [0,0,0,0,1],$$
$$[1,0] \rightarrow [1,0,0,0,0], [1,1] \rightarrow [0,0,0,1,0]\}.$$

We also need a check matrix. Since the identity I_3 is a submatrix of G, although this time at the second, third and forth columns, we can construct a check matrix of C using E.1.29, as we do in the listing:

$$H = \begin{pmatrix} 1 & 1 & 0 & 1 & 0 \\ 0 & 1 & 1 & 1 & 1 \end{pmatrix}.$$

Now we can decode any vector y as $y + E(yH^T)$. For example, if $y = [0,1,0,0,1]$, then $yH^T = [1,0]$ and the decoded vector is $[0,1,0,0,1] + [1,0,0,0,0] = [1,1,0,0,1]$.

1.37 Example. The listing **Ternary example of syndrome decoding** is similar to the previous example. Here the list C contains 3^3 elements; the table X has 9 entries, and the value assigned to each syndrome is a list of 3^3 elements; the classes of the syndromes

$$[0,1], [0,2], [1,0], [1,1], [2,0], [2,2]$$

contain a unique vector of minimum weight, namely

$$[0,0,0,0,1], [0,0,0,0,2], [0,0,0,1,0], [1,0,0,0,0], [0,0,0,2,0], [2,0,0,0,0],$$

while

$$C_{[1,2]} = \{[0,0,0,1,2], [0,0,2,0,2], [0,1,0,1,0], [0,1,2,0,0],$$
$$[1,0,0,0,1], [1,2,0,0,0], [2,0,0,2,0], [2,0,1,0,0]\}$$

($C_{[2,1]}$ is obtained by multiplying all the vectors in $C_{[1,2]}$ by 2). With all this, it is immediate to set up the syndrome-leader table E. The listing contains the example of decoding the vector $y = [1,1,1,1,1]$.

Complexity

Assuming that we have the table $E = \{e_s\}$, the decoder is as fast as the evaluation of the product $s = yH^T$ and the looking up on the table E. But for the construction of E we have to do a number of operations of the order q^n: there are q^{n-k} entries and the selection of e_s in C_s needs a number of weight evaluations of the order q^k on the average. In case we know that $H = (-P^T | I_{n-k})$, the time needed to select the e_s can be decreased a little bit for all $s \in \mathbb{F}^{n-k}$ such that $|s| \leqslant t$, for we may take $e_s = 0_k | s$ (note that this holds for $vol_q(n - k, t)$ syndromes). In any case, the space needed to keep the table E is of the order of q^{n-k} vectors of length n.

```
Ternary example of syndrome decoding
K=[0,1,2] : Vector(Z₃);
     ⎛1 0 0 2 2⎞
G=   ⎜0 1 0 0 1⎟ : Matrix(Z₃);
     ⎝0 0 1 1 0⎠
H=-Gᵀ₍₄,₅₎|I₂ ;
C=[a·G₁+b·G₂+c·G₃ with (a,b,c) in (K,K,K)] ;
X={ [a,b]→{ [0,0,0,a,b]+x with x in C} with (a,b) in (K,K) } ;
M= { x₁→min_weights(x₂) with x in X } ;
E={ x₁→x₂,₁ with x in M } ;
y=[1,1,1,1,1]; s=y·Hᵀ ➝ [1, 1]
E(s) ➝ [1, 0, 0, 0, 0]
y−E(s) ➝ [0, 1, 1, 1, 1]
```

Bounding the error for binary codes

As we noticed before (first of Remarks 1.35), an error occurs if and only if the error pattern $e = y - x$ does not belong to the table $E = \{e_s\}$. If we let α_i denote the number of e_s whose weight is i, and we are in the binary case, then

$$P_c = \sum_{i=0}^{n} \alpha_i (1-p)^{n-i} p^i$$

is the probability that $x' = x$, assuming that p is the bit error probability. Note that $(1-p)^{n-i} p^i$ is the probability of getting a given error-pattern. Hence $P_e = 1 - P_c$ is the code error probability. Moreover, since $\alpha_i = \binom{n}{i}$ for $0 \leqslant i \leqslant t$, and $1 = \sum_{i=0}^{n} \binom{n}{i} p^i (1-p)^{n-i}$,

$$P_e = 1 - P_c = \sum_{i=t+1}^{n} \left(\binom{n}{i} - \alpha_i \right) p^i (1-p)^{n-i}.$$

If p is small, the dominant term in this sum can be approximated by

$$\left(\binom{n}{t+1} - \alpha_{t+1} \right) p^{t+1}$$

and the error reduction factor by

$$\left(\binom{n}{t+1} - \alpha_{t+1} \right) p^t.$$

In general the integers α_i are difficult to accertain for $i > t$.

Incomplete syndrome-leader decoding

If asking for retransmission is allowed, sometimes it may be useful to modify the syndrome-leader decoder as follows: If s is the syndrome of the received vector and the leader e_s has weight at most t, then we return $y - e_s$, otherwise we ask for retransmission. This scheme is called *incomplete decoding*. From the preceding discussions it is clear that it will correct any error-pattern of weigt at most t and detect any error-pattern of weight $t + 1$ or higher that is not a code word.

E.1.32. Given a linear code of length n, show that the probability of an undetected code error is given by

$$P_u(n,p) = \sum_{i=1}^{n} a_i p^i (1 - p)^{n-i},$$

where a_i is the number of code vectors that have weight i $(1 \leqslant i \leqslant n)$. What is the probability of retransmission when this code is used in incomplete decoding?

Gilbert–Varshamov existence condition

1.38 Theorem. *Fix positive integers n, k, d such that $k \leqslant n$ and $2 \leqslant d \leqslant n + 1$. If the relation*

$$vol_q(n - 1, d - 2) < q^{n-k}$$

is satisfied (this is called the Gilbert–Varshamov condition), then there exists a linear code of type $[n, k, d']$ with $d' \geqslant d$.

Proof: It is sufficient to see that the condition allows us to construct a matrix $H \in M_n^{n-k}(\mathbb{F})$ of rank $n - k$ with the property that any $d - 1$ of its columns are linearly independent. Indeed, if this is the case, then the code C defined as the space orthogonal to the rows of H has length n, dimension k (from $n - (n - k) = k$), and minimum weight $d' \geqslant d$ (since there are no linear relations of length less than d among the columns of H, which is a check matrix of C).

Before constructing H, we first establish that the Gilbert–Varshamov condition implies that $d - 1 \leqslant n - k$ (note that this is the Singleton bound for the parameters d and k). Indeed,

$$vol_q(n-1, d-2) = \sum_{j=0}^{d-2} \binom{n-1}{j}(q-1)^j$$

$$\geqslant \sum_{j=0}^{d-2} \binom{d-2}{j}(q-1)^j$$

$$= (1 + (q-1))^{d-2} = q^{d-2},$$

and hence $q^{d-2} < q^{n-k}$ if the Gilbert–Varshamov condition is satisfied. Therefore $d - 2 < n - k$, which for integers is equivalent to the claimed inequality $d - 1 \leqslant n - k$.

In order to construct H, we first select a basis $c_1, \ldots, c_{n-k} \in \mathbb{F}^{n-k}$. Since $d - 1 \leqslant n - k$, any $d - 1$ vectors extracted from this basis are linearly independent. Moreover, a matrix H of type $(n - k) \times m$ with entries in $\mathbb{F}$ and which contains the columns c_j^T, $j = 1, \ldots, n - k$, has rank $n - k$.

Now assume that we have constructed, for some $i \in [n - k, n]$, vectors $c_1, \ldots, c_i \in \mathbb{F}^{n-k}$ with the property that any $d - 1$ from among them are linearly independent. If $i = n$, it is sufficient to take $H = (c_1^T, \ldots, c_n^T)$ and our problem is solved. Otherwise we will have $i < n$. In this case, the number of linear combinations that can be formed with at most $d - 2$ vectors from among $c_1, \ldots, c_i$ is not greater than $vol_q(i, d - 2)$ (see E.1.13). Since $i \leqslant n - 1$, $vol_q(i, d - 2) \leqslant vol_q(n - 1, d - 2)$. If the Gilbert–Varshamov condtion is satisfied, then there is a vector $c_{i+1} \in \mathbb{F}^{n-k}$ which is not a linear combination of any subset of $d - 2$ vectors extracted from the list $c_1, \ldots, c_i$, and our claim follows by induction. $\square$

Since $A_q(n, d) \geqslant A_q(n, d')$ if $d' \geqslant d$, the Gilbert–Varshamov condition shows that $A_q(n, d) \geqslant q^k$. This lower bound, called the *Gilbert–Varshamov bound*, often can be used to improve the Gilbert bound (theorem 1.17). In the listing **The Gilbert–Varshamov lower bound** we define a function that computes this bound (lb_gilbert_varshamov).

Hamming codes

> Hamming codes were discovered by Hamming (1950) and Golay (1949).
>
> R. Hill, [9], p. 90.

We will say that a q-ary code C of type $[n, n - r]$ is a *Hamming code of codimension* r if the columns of a check matrix $H \in M_n^r(\mathbb{F})$ of C form

---

```
lb_gilbert_varshamov(n, d, q)
:= begin
      local k=0, v=volume(n−1, d−2, q)
      while qⁿ⁻ᵏ>v do
        k=k+1
      end
      qᵏ⁻¹
   end
lb_gilbert_varshamov(n, d):= lb_gilbert_varshamov(n, d, 2)
```

```
lb_gilbert_varshamov(10,3)  →  64
```
Hence A(10,3) ⩾ 64
```
lb_gilbert(10,3)  →  19
```

a maximal set among the subsets of $\mathbb{F}^r$ with the property that any two of its elements are linearly independent (of such a matrix we will say that it is a q-ary *Hamming matrix of codimension* r).

E.1.33. For $i = 1, \ldots, r$, let H_i be the matrix whose columns are the vectors of $\mathbb{F}^r$ that have the form $(0, \ldots, 0, 1, \alpha_{i+1}, \ldots, \alpha_r)$, with $\alpha_{i+1}, \ldots, \alpha_n \in \mathbb{F}$ arbitrary. Let $H = H_1|H_2|\cdots|H_r$. Prove that H is a q-ary Hamming matrix of codimension r (we will say that it is a *normalized* q-ary Hamming matrix of codimension r). Note also that H_i has q^{r-i} columns and hence that H has $(q^r - 1)/(q - 1)$ columns. The listing **Normalized Hamming matrix** defines a function based on this procedure, and contains some examples.

1.39 Remark. Since two vectors of $\mathbb{F}^r$ are linearly independent if and only if they are non-zero and define distinct points of the projective space $\mathbb{P}(\mathbb{F}^r)$, we see that the number n of columns of a q-ary Hamming matrix of codimension r coincides with the cardinal of $\mathbb{P}(\mathbb{F}^r)$, which is $n = \dfrac{q^r - 1}{q - 1}$. This expression agrees with the observation made in E.1.33. It also follows by induction from the fact that $\mathbb{P}^{r-1} = \mathbb{P}(\mathbb{F}^r)$ is the disjoint union of an affine space $\mathbb{A}^{r-1}$, whose cardinal is q^{r-1}, and a hyperplane, which is a $\mathbb{P}^{r-2}$.   □

It is clear that two Hamming codes of the same codimension are scalarly equivalent. We will write $Ham_q(r)$ to denote any one of them and $Ham_q^\vee(r)$ to denote the corresponding dual code, that is to say, the code generated by the check matrix H used to define $Ham_q(r)$. By E.1.33, it is clear that $Ham_q(r)$ has dimension $k = n - r = (q^r - 1)/(q - 1) - r$. Its codimension, which is the dimension of $Ham_q^\vee(r)$, is equal to r.

1.40 *Example.* The binary Hamming code of codimension 3, $Ham_2(3)$, is

```
┌─┬─────────┬──────────────────────────────────────────────────────────┬─┐
│▲│ Library │ Normalized Hamming matrix of codimension r over a finite field K │▲│
├─┴─────────┴──────────────────────────────────────────────────────────┴─┤
│ normalized_hamming_matrix(K:GF, r:ℤ)                                     │
│ check r>0                                                                │
│ :=begin                                                                  │
│     local B=[constant_vector(r−1, 0)| [1]], H=B, S, s                    │
│     K={element(j, K) with j in 0 .. cardinal(K)−1}                      │
│     for j in 1 .. r−1 do                                                 │
│        S=null                                                            │
│        for b in B do                                                     │
│           s=null; b=tail(b)                                              │
│           for t in K do                                                  │
│              s=(append(b,t), s)                                          │
│           end                                                            │
│           S=(S, s)                                                       │
│        end                                                               │
│        B=[S]                                                             │
│        H=H&B                                                             │
│     end                                                                  │
│     reverse(H)^T                                                         │
│ end                                                                      │
└──────────────────────────────────────────────────────────────────────┘
```

$$\text{normalized\_hamming\_matrix}(\mathbb{Z}_2,3) \;\longrightarrow\; \begin{pmatrix} 1\,1\,1\,1\,0\,0\,0 \\ 0\,0\,1\,1\,1\,1\,0 \\ 0\,1\,0\,1\,0\,1\,1 \end{pmatrix}$$

$$\text{normalized\_hamming\_matrix}(\mathbb{Z}_2,4) \;\longrightarrow\; \begin{pmatrix} 1\,1\,1\,1\,1\,1\,1\,1\,0\,0\,0\,0\,0\,0\,0 \\ 0\,0\,0\,0\,1\,1\,1\,1\,1\,1\,1\,1\,0\,0\,0 \\ 0\,0\,1\,1\,0\,0\,1\,1\,0\,0\,1\,1\,1\,1\,0 \\ 0\,1\,0\,1\,0\,1\,0\,1\,0\,1\,0\,1\,0\,1\,1 \end{pmatrix}$$

$$\text{normalized\_hamming\_matrix}(\mathbb{Z}_3,3) \;\longrightarrow\; \begin{pmatrix} 1\,1\,1\,1\,1\,1\,1\,1\,1\,0\,0\,0\,0 \\ 0\,0\,0\,1\,1\,1\,2\,2\,2\,1\,1\,1\,0 \\ 0\,1\,2\,0\,1\,2\,0\,1\,2\,0\,1\,2\,1 \end{pmatrix}$$

the code $[7, 4]$ that has

$$H = \begin{pmatrix} 1 & 0 & 0 & 1 & 1 & 0 & 1 \\ 0 & 1 & 0 & 1 & 0 & 1 & 1 \\ 0 & 0 & 1 & 0 & 1 & 1 & 1 \end{pmatrix}$$

as check matrix. Indeed, the columns of this matrix are all non-zero binary vectors of length 3 and *in the binary case two non-zero vectors are linearly independent if and only if they are distinct.*

E.1.34. Compare the binary Hamming matrix of the previous example with a binary normalized Hamming matrix of codimension 3, and with the matrix H of the Hamming $[7, 4, 3]$ code studied in the Example 1.9.

1.41 Proposition. *If C is any Hamming code, then $d_C = 3$. In particular, the error-correcting capacity of a Hamming code is 1.*

Proof: If H is a check matrix of C, then we know that the elements of C are the linear relations satisfied by the columns of H. Since any two columns of H are linearly independent, the minimum distance of C is at least 3. On the other hand, C has elements of weight 3, because the sum of two columns of H is linearly independent of them and hence it must be proportional to another column of H (the columns of H contain all non-zero vectors of $\mathbb{F}^r$ up to a scalar factor). □

1.42 Proposition. *The Hamming codes are perfect.*

Proof: Since $Ham_q(r)$ has type $[(q^r - 1)/(q - 1), (q^r - 1)/(q - 1) - r, 3]$, we know from E.1.14 that it is perfect. It can be checked directly as well: the ball of radius 1 with center an element of $\mathbb{F}^n$ contains $1 + n(q - 1) = q^r$ elements, so the union of the balls of radius 1 with center an element of C is a set with $q^{n-r}q^r = q^n$ elements. □

1.43 Proposition. *If $C' = Ham_q^\vee(r)$ is the dual of a Hamming code $C = Ham_q(r)$, the weight of any non-zero element of C' is q^{r-1}. In particular, the distance between any pair of distinct elements of C' is q^{r-1}.*

Proof: Let $H = (h_{ij})$ be a check matrix of C. Then the non-zero vectors of C' have the form $z = aH$, $a \in \mathbb{F}^r$, $a \neq 0$. So the i-th component of z has the form $z_i = a_1 h_{1i} + \cdots + a_r h_{ri}$. Therefore the condition $z_i = 0$ is equivalent to say that the point P_i of $\mathbb{P}^{r-1} = \mathbb{P}(\mathbb{F}^r)$ defined by the i-th column of H belongs to the hyperplane of $\mathbb{P}^{r-1}$ defined by the equation

$$a_1 x_1 + \ldots + a_r x_r = 0.$$

Since $\{P_1, \ldots, P_n\} = \mathbb{P}^{r-1}$ (cf. Remark 1.39), it follows that the number of non-zero components of z is the cardinal of the complement of a hyperplane of $\mathbb{P}^{r-1}$. Since this complement is an affine space $\mathbb{A}^{r-1}$, its cardinal is q^{r-1} and so any non-zero element of C' has weight q^{r-1}. □

Codes such that the distance between pairs of distinct elements is a fixed integer d are called *equidistant* of distance d. Thus $Ham_q^\vee(r)$ is equidistant of distance q^{r-1}.

E.1.35. Find a check matrix of $Ham_7(2)$, the Hamming code over $\mathbb{F}_7$ of codimension 2, and use it to decode the message

$$3523410610521360.$$

Weight enumerator

The *weight enumerator* of a code C is defined as the polynomial

$$A(t) = \sum_{i=0}^{n} A_i t^i,$$

where A_i is the number of elements of C that have weight i.

It is clear that $A(t)$ can be expressed in the form

$$A(t) = \sum_{x \in C} t^{|x|},$$

for the term t^i appears in this sum as many times as the number of solutions of the equation $|x| = i$, $x \in C$.

Note that $A_0 = 1$, since 0_n is the unique element of C with weight 0. On the other hand $A_i = 0$ if $0 < i < d$, where d is the minimum distance of C. The determination of the other A_i is not easy in general and it is one of the basic problems of coding theory.

MacWilliams identities

A situation that might be favorable to the determination of the weight enumerator of a code is the case when it is constructed in some prescribed way from another code whose weight enumerator is known. Next theorem (7), which shows how to obtain the weight enumerator of $C^\perp$ from the weight enumerator of C (and conversely) is a positive illustration of this expectation.

Remark. In the proof of the theorem we need the notion of *character* of a group G, which by definition is a group homomorphism of G to $U(1) = \{z \in \mathbb{C} \,|\, |z| = 1\}$, the group of complex numbers of modulus 1. The constant map $g \mapsto 1$ is a character, called the *unit character*. A character different from the unit character is said to be *non-trivial* and the main fact we will need is that the additive group of finite field $\mathbb{F}_q$ has non-trivial characters. Actually any finite abelian group has non-trivial characters. Let us sketch how this can be established.

It is known that any finite abelian group G is isomorphic to a product of the form

$$\mathbb{Z}_{n_1} \times \cdots \times \mathbb{Z}_{n_k},$$

with k a positive integer and $n_1, \ldots, n_k$ integers greater than 1. For example, in the case of a finite field $\mathbb{F}_q$, we have (if $q = p^r$, p prime),

$$\mathbb{F}_q \simeq \mathbb{Z}_p^r$$

since $\mathbb{F}_q$ is vector space of dimension r over $\mathbb{Z}_p$. In any case, it is clear that if we know how to find a non-trivial character of $\mathbb{Z}_{n_1}$ then we also know how to find a non-trivial character of G (the composition of the non-trivial character of $\mathbb{Z}_{n_1}$ with the projection of $\mathbb{Z}_{n_1} \times \cdots \times \mathbb{Z}_{n_k}$ onto $\mathbb{Z}_{n_1}$ gives a non-trivial character of G). Finally note that if n is an integer greater than 1 and $\xi \neq 1$ is an n-th root of unity then the map $\chi : \mathbb{Z}_n \to U(1)$ such that $\chi(k) = \xi^k$ is well defined and is a non-trivial character of $\mathbb{Z}_n$.

E.1.36. Given a group G, the set $G^\vee$ of all characters of G has a natural group structure (the multiplication $\chi\chi'$ of two characters χ and χ' is defined by the relation $(\chi\chi')(g) = \chi(g)\chi'(g)$) and with this structure $G^\vee$ is called the *dual group* of G. Prove that if n is an integer greater than 1 and $\xi \in U(1)$ is a primitive n-th root of unity, then there is an isomorphism $\mathbb{Z}_n \simeq \mathbb{Z}_n^\vee$, $m \mapsto \chi_m$, where $\chi_m(k) = \xi^{mk}$. Use this to prove that for any finite abelian group G there exists an isomorphism $G \simeq G^\vee$.

1.44 Theorem (F. J. MacWilliams). *The weight enumerator $B(t)$ of the dual code $C^\perp$ of a code C of type $[n, k]$ can be determined from the weight enumerator $A(t)$ of C according to the following identity:*

$$q^k B(t) = (1 + q^*t)^n A\left(\frac{1-t}{1+q^*t}\right), \quad q^* = q - 1.$$

Proof: Let χ be a non-trivial character of the additive group of $\mathbb{F} = \mathbb{F}_q$ (see the preceeding Remark). Thus we have a map $\chi : \mathbb{F} \to U(1)$ such that $\chi(\alpha + \beta) = \chi(\alpha)\chi(\beta)$ for any $\alpha, \beta \in \mathbb{F}$ with $\chi(\gamma) \neq 1$ for some $\gamma \in \mathbb{F}$. Observe that $\sum_{\alpha \in \mathbb{F}} \chi(\alpha) = 0$, since

$$\chi(\gamma) \sum_{\alpha \in \mathbb{F}} \chi(\alpha) = \sum_{\alpha \in \mathbb{F}} \chi(\alpha + \gamma) = \sum_{\alpha \in \mathbb{F}} \chi(\alpha)$$

and $\chi(\gamma) \neq 1$.

Now consider the sum

$$S = \sum_{x \in C} \sum_{y \in \mathbb{F}^n} \chi(x, y) t^{|y|},$$

where $\chi(x, y) = \chi(\langle x|y \rangle)$. After reordering we have

$$S = \sum_{y \in \mathbb{F}^n} t^{|y|} \sum_{x \in C} \chi(x, y).$$

If $y \in C^\perp$, then $\langle x|y \rangle = 0$ for all $x \in C$ and so $\chi(x, y) = 1$ for all $x \in C$, and in this case $\sum_{x \in C} \chi(x, y) = |C|$. If $y \notin C^\perp$, then the map $C \to \mathbb{F}$ such that $x \mapsto \langle x|y \rangle$ takes each value of $\mathbb{F}$ the same number of times (because it is an $\mathbb{F}$-linear map) and hence, using that $\sum_{\alpha \in \mathbb{F}} \chi(\alpha) = 0$, we have that $\sum_{x \in C} \chi(x, y) = 0$. Putting the two cases together we have that

$$S = |C| \sum_{y \in C^\perp} t^{|y|} = q^k B(t).$$

On the other hand, for any given $x \in C$, and making the convention, for all $\alpha \in \mathbb{F}$, that $|\alpha| = 1$ if $\alpha \neq 0$ and $|\alpha| = 0$ if $\alpha = 0$, we have

$$\sum_{y \in \mathbb{F}^n} \chi(x,y)t^{|y|} = \sum_{y_1,\dots,y_n \in \mathbb{F}} \chi(x_1 y_1 + \dots + x_n y_n)t^{|y_1| + \dots + |y_n|}$$

$$= \sum_{y_1,\dots,y_n \in \mathbb{F}} \prod_{i=1}^{n} \chi(x_i y_i)t^{|y_i|}$$

$$= \prod_{i=1}^{n} \left(\sum_{\alpha \in \mathbb{F}} \chi(x_i \alpha)t^{|\alpha|} \right).$$

But it is clear that

$$\sum_{\alpha \in \mathbb{F}} \chi(x_i \alpha)t^{|\alpha|} = \begin{cases} 1 + q^* t & \text{if } x_i = 0 \\ 1 - t & \text{if } x_i \neq 0 \end{cases}$$

because $\sum_{\alpha \in \mathbb{F}^*} \chi(\alpha) = -1$. Consequently

$$\sum_{\alpha \in \mathbb{F}} \chi(x_i \alpha)t^{|\alpha|} = (1 + q^* t)^n \left(\frac{1 - t}{1 + q^* t} \right)^{|x|}$$

and summing with respect to $x \in C$ we see that

$$S = (1 + q^* t)^n A\left(\frac{1 - t}{1 + q^* t} \right),$$

as stated. $\square$

1.45 *Example* (Weight enumerator of the zero-parity code). The zero-parity code of length n is the dual of the repetition code C of length n. Since the weight enumerator of C is $A(t) = 1 + (q - 1)t^n$, the weight enumerator $B(t)$ of the zero-parity code is given by the relation

$$qB(t) = (1 + (q - 1)t)^n \left(1 + (q - 1)\left(\frac{1 - t}{1 + (q - 1)t} \right)^n \right)$$

$$= (1 + (q - 1)t)^n + (q - 1)(1 - t)^n.$$

In the binary case we have

$$2B(t) = (1 + t)^n + (1 - t)^n,$$

which yields

$$B(t) = \sum_{i=0}^{i \leq n/2} \binom{n}{2i} t^{2i}.$$

Note that this could have been written directly, for the binary zero-parity code of length n has only even-weight words and the number of those having weight $2i$ is $\binom{n}{2i}$.

1.46 Example (Weight enumerator of the Hamming codes). We know that the dual Hamming code $Ham_q^{\vee}(r)$ is *equidistant*, with minimum distance q^{r-1} (Proposition 1.43). This means that its weight enumerator, say $B(t)$, is the polynomial

$$B(t) = 1 + (q^r - 1)t^{q^{r-1}},$$

because $q^r - 1$ is the number of non-zero vectors in $Ham_q^{\vee}(r)$ and each of these has weight q^{r-1}. Now the MacWilliams identity allows us to determine the weight enumerator, say $A(t)$, of $Ham_q(r)$:

$$q^r A(t) = (1 + q^*t)^n B\left(\frac{1-t}{1+q^*t}\right)$$

$$= (1 + q^*t)^n + (q^r - 1)(1 + q^*t)^n \frac{(1-t)^{q^{r-1}}}{(1+q^*t)^{q^{r-1}}}$$

$$= (1 + q^*t)^n + (q^r - 1)(1 + q^*t)^{n-q^{r-1}}(1-t)^{q^{r-1}}$$

$$= (1 + q^*t)^{n-q^{r-1}}\left((1 + q^*t)^{q^{r-1}} + (q^r - 1)(1-t)^{q^{r-1}}\right)$$

$$= (1 + q^*t)^{\frac{q^{r-1}-1}{q-1}}\left((1 + q^*t)^{q^{r-1}} + (q^r - 1)(1-t)^{q^{r-1}}\right),$$

since $n = (q^r - 1)/(q - 1)$ and $n - q^{r-1} = (q^{r-1} - 1)/(q - 1)$.

In the binary case the previous formula yields that the weight enumerator $A(t)$ of $Ham(r)$ is

$$2^r A(t) = (1 + t)^{2^{r-1}-1}\left((1+t)^{2^{r-1}} + (2^r - 1)(1-t)^{2^{r-1}}\right).$$

For $r = 3$, $Ham(3)$ has type $[7, 4]$ and

$$8A(t) = (1 + t)^3\left((1+t)^4 + 7(1-t)^4\right),$$

or

$$A(t) = 1 + 7t^3 + 7t^4 + t^7.$$

This means that $A_0 = A_7 = 1$, $A_1 = A_2 = A_5 = A_6 = 0$ and $A_3 = A_4 = 7$. Actually it is easy to find, using the description of this code given in the example 1.40 (see also the example 1.9) that the weight 3 vectors of $Ham(3)$ are

$$[1, 1, 0, 1, 0, 0, 0], [0, 1, 1, 0, 1, 0, 0], [1, 0, 1, 0, 0, 1, 0], [0, 0, 1, 1, 0, 0, 1],$$
$$[1, 0, 0, 0, 1, 0, 1], [0, 1, 0, 0, 0, 1, 1], [0, 0, 0, 1, 1, 1, 0]$$

and the weight 4 vectors are

$$[1, 1, 1, 0, 0, 0, 1], [1, 0, 1, 1, 1, 0, 0], [0, 1, 1, 1, 0, 1, 0], [1, 1, 0, 0, 1, 1, 0],$$
$$[0, 1, 0, 1, 1, 0, 1], [1, 0, 0, 1, 0, 1, 1], [0, 0, 1, 0, 1, 1, 1]$$

Note that the latter are obtained from the former, in reverse order, by interchanging 0 and 1.

| ▲ | Library | Weight enumerator of the Hamming codes | ▲ |

hamming_weight_enumerator(q, r, T)
:= begin

  local $n=\dfrac{q^{r-1}-1}{q-1}$, $a=1+(q-1)\cdot T$, $e=q^{r-1}$

  $\dfrac{a^n \cdot (a^e + (q^r-1)\cdot (1-T)^e)}{q^r}$

  end

hamming_weight_enumerator(r,T) := hamming_weight_enumerator(2, r, T)

macwilliams(n,k,q,A,T) := $\dfrac{(1+(q-1)\cdot T)^n\cdot\left\{T\Rightarrow\dfrac{1-T}{1+(q-1)\cdot T}\right\}(A)}{q^k}$

macwilliams(n,k,A,T) := macwilliams(n,k,2,A,T)

---

hamming_weight_enumerator(3,T) → $T^7+7\cdot T^4+7\cdot T^3+1$
macwilliams(7,3,1+7·T^4,T) → $T^7+7\cdot T^4+7\cdot T^3+1$

Summary

- Basic notions: linear code, minimum weight, generating matrix, dual code, check matrix.

- Syndrome-leader decoding algorithm. It agrees with the minimum distance decoder g on the set D of g-decodable vectors.

- The Gilbert–Varshamov condition $vol_q(n-1, d-2) < q^{n-k}$ guarantees the existence of a linear code of type $[n, k, d]$ (n, k, d positive integers such that $k \leqslant n$ and $2 \leqslant d \leqslant n$). In particular the condition implies that $A_q(n, d) \geqslant q^k$, a fact that often can be used to improve the Gilbert lower bound.

- Definition of the Hamming codes

$$Ham_q(r) \sim [(q^r-1)/(q-1), (q^r-1)/(q-1)-r, 3]$$

by specifying a check matrix (a q-ary Hamming matrix of codimension r). They are perfect codes and the dual code $Ham_q^\vee(r)$ is equidistant with distance q^{r-1}.

- MacWilliams indentities (Theorem 7): If $A(t)$ and $B(t)$ are the weight enumerators of a code $C \sim [n, k]$ and its dual, then

$$q^k B(t) = (1+q^*t)^n A\left(\frac{1-t}{1+q^*t}\right), \qquad q^* = q-1.$$

The main example we have considered is the determination of the weight enumerator of the Hamming codes (example 1.46.

Problems

P.1.9 The ISBN code has been described in the Example 1.19. Prove that this code:

1. Detects single errors and transpositions of two digits.
2. Cannot correct a single error in general.
3. Can correct a single error if we know its position.

If instead of the ISBN weighted check digit relation we used the the relation

$$\sum_{i=1}^{10} x_i \equiv 0 \pmod{11},$$

which of the above properties would fail?

P.1.10 (The Selmer personal registration code, 1967). It is used in Norway to assign personal registration numbers and it can be defined as the set $R \subset (\mathbb{Z}_{11})^{11}$ formed with the vectors x such that $x_i \neq 10$ for $i = 1, \ldots, 9$ and $xH^T = 0$, where $H = \begin{pmatrix} 3 & 7 & 6 & 1 & 8 & 9 & 4 & 5 & 2 & 1 & 0 \\ 5 & 4 & 3 & 2 & 7 & 6 & 5 & 4 & 3 & 2 & 1 \end{pmatrix}$. Show that R can:

1) Correct one error if and only if its position is not 4 or 10.
2) Detect a double error if and only if it is not of the form $t(\varepsilon_4 - \varepsilon_{10})$, $t \in \mathbb{Z}_{11}^*$.

What can be said about the detection and correction of transpositions?

P.1.11 Encode pairs of information bits 00, 01, 10, 11 as 00000, 01101, 10111, 11010, respectively.

1. Show that this code is linear and express each of the five bits of the code words in terms of the information bits.
2. Find a generating matrix and a check matrix.
3. Write the coset classes of this code and choose a leaders' table.
4. Decode the word 11111.

P.1.12 The matrix

$$H = \begin{pmatrix} 1 & 0 & 0 & 0 & 1 & a \\ 1 & 1 & 0 & 0 & 0 & b \\ 1 & 0 & 1 & 0 & 0 & c \\ 0 & 1 & 1 & 1 & 0 & d \end{pmatrix}.$$

is the check matrix of a binary code C.

1. Find the list of code-words in the case $a = b = c = d = 1$.

2. Show that it is possible to choose a, b, c, d in such a way that C corrects simple errors and detects double errors. Is there a choice that would allow the correction of all double errors?

P.1.13 Let G_1 and G_2 be generating matrices of linear codes of types $[n_1, k, d_1]_q$ and $[n_2, k, d_2]_q$, respectively. Prove that the matrices

$$G_3 = \begin{pmatrix} G_1 & 0 \\ 0 & G_2 \end{pmatrix}, \qquad G_4 = G_1 | G_2$$

generate linear codes of types $[n_1 + n_2, 2k, d_3]_q$ and $[n_1 + n_2, k, d_4]_q$ with $d_3 = \min(d_1, d_2)$ and $d_4 \geqslant d_1 + d_2$.

P.1.14 Let $n = rs$, where r and s are positive integers. Let C be the binary code of length n formed with the words $x = a_1 a_2 \ldots a_n$ such that, when arranged in an $r \times s$ matrix

$$
\begin{array}{ccc}
a_1 & \cdots & a_r \\
a_{r+1} & \cdots & a_{2r} \\
\vdots & & \vdots \\
a_{(s-1)r+1} & \cdots & a_{sr}
\end{array}
$$

the sum of the elements of each row and of each column is zero.

1. Check that it is a linear code and find its dimension and minimum distance.

2. Devise a decoding scheme.

3. For $r = 3$ and $s = 4$, find a generating matrix and a check matrix of C.

P.1.15 A binary linear code C of length 8 is defined by the equations

$$x_5 = x_2 + x_3 + x_4$$
$$x_6 = x_1 + x_2 + x_3$$
$$x_7 = x_1 + x_2 + x_4$$
$$x_8 = x_1 + x_3 + x_4$$

Find a check matrix for C, show that $d_C = 4$ and calculate the weight enumerator of C.

P.1.16 (First order Reed–Muller codes). Let L_m be the vector space of polynomials of degree $\leqslant 1$ in m indeterminates and with coefficients in $\mathbb{F}$. The elements of L_m are expressions

$$a_0 + a_1 X_1 + \ldots + a_m X_m,$$

where $X_1, \ldots, X_m$ are indeterminates and $a_0, a_1, \ldots, a_m$ are arbitrary elements of $\mathbb{F}$. So $1, X_1, \ldots, X_m$ is a basis of L_m over $\mathbb{F}$ and, in particular, the dimension of L_m is $m + 1$.

Let n be an integer such that $q^{m-1} < n \leqslant q^m$ and pick distinct vectors $\boldsymbol{x} = x^1, \ldots, x^n \in \mathbb{F}^m$. Show that:

1) The linear map $\epsilon \colon L_m \to \mathbb{F}^n$ such that $\epsilon(f) = (f(x^1), \ldots, f(x^n))$ is injective.

2) The image of ϵ is a linear code of type $[n, m + 1, n - q^{m-1}]$.

Such codes are called (first order) *Reed–Muller* codes and will be denoted $RM_q^{\boldsymbol{x}}(m)$. In the case $n = q^m$, instead of $RM_q^{\boldsymbol{x}}(m)$ we will simply write $RM_q(m)$ and we will say that this is a *full* Reed–Muller code. Thus $RM_q(m) \sim (q^m, m + 1, q^{m-1}(q - 1))$.

1.47 Remark. There are many references that present results on the decoding of *RM* codes. The following two books, for example, include a lucid and elementary introduction: [1] (Chapter 9) and [27] (Chapter 4).

1.3 Hadamard codes

> The first error control code developed for a deep space appli-
> cation was the nonlinear $(32, 64)$ code used in the *Mariner '69*
> mission.
>
> S.B. Wicker, [28], p. 2124.

Essential points

- Hadamard matrices. The matrices $H^{(n)}$.
- The order of a Hadamard matrix is divisible by 4 (if it is not 1 or 2).
- Paley matrices.
- Uses of the Paley matrices for the construction of Hadamard matrices.
- Hadamard codes. Example: the Mariner code.
- The Hadamard decoder.
- Equivalence of some Hadamard codes with first order Reed–Muller codes.

Hadamard matrices

A *Hadamard matrix* of order n is a matrix of type $n \times n$ whose coefficients are 1 or -1 and such that

$$HH^T = nI.$$

Note that this relation is equivalent to say that the rows of H have norm $\sqrt{n}$ and that any two of them are orthogonal. Since H is invertible and $H^T = nH^{-1}$, we see that $H^T H = nH^{-1}H = nI$ and hence H^T is also a Hadamard matrix. Therefore H is a Hadamard matrix if and only if its columns have norm $\sqrt{n}$ and any two of them are orthogonal.

If H is a Hadamard matrix of order n, then

$$\det(H) = \pm n^{n/2}.$$

In particular we see that Hadamard matrices satisfy the equality in the Hadamard inequality

$$|\det(A)| \leqslant \prod_{i=1}^{n} (\sum_{i=1}^{n} a_{ij}^2)^{1/2},$$

which is valid for all $n \times n$ real matrices $A = (a_{ij})$.

E.1.37. Show that $n^{n/2}$ is an upper-bound for the value of $\det(A)$ when A runs through all real matrices such that $|a_{ij}| \leqslant 1$, and that the equality is attained only if A is a Hadamard matrix.   □

If the signs of a row (column) of a Hadamard matrix are changed, then the resulting matrix is clearly a Hadamard matrix again. If the rows (columns) of a Hadamard matrix are permuted in an arbitrary way, then the resulting matrix is a Hadamard matrix. Two Hadamard matrices are said to be *equivalent* (or *isomorphic*) if it is possible to go from one to the other using a sequence of these two operations. Observe that by changing the sign of all the columns whose first term is -1, and then of all the rows whose first term is -1, we obtain an equivalent Hadamard matrix whose first row and column are $\mathbf{1}_n$ and $\mathbf{1}_n^T$, respectively. Such Hadamard matrices are said to be *normalized* Hadamard matrices. For example,

$$H^{(1)} = \begin{pmatrix} 1 & 1 \\ 1 & -1 \end{pmatrix}$$

is the only normalized Hadamard matrix of order 2. More generally, if we define $H^{(n)}$ ($n \geqslant 2$) recursively by the formula

$$H^{(n)} = \begin{pmatrix} H^{(n-1)} & H^{(n-1)} \\ H^{(n-1)} & -H^{(n-1)} \end{pmatrix},$$

then $H^{(n)}$ is a normalized Hadamard matrix of order 2^n for all n (see the listing **Hadamard matrix H**$^{(n)}$, where we define the function hadamard_matrix_ that later will be superseded by hadamard_matrix).

It is also easy to check that if H and H' are Hadamard matrices of orders n and n', then the tensor product $H \otimes H'$ is a Hadamard matrix of order nn' (the matrix $H \otimes H'$, also called the *Kronecker product* of H and H', can be defined as the matrix obtained by replacing each entry a_{ij} of H by the matrix $a_{ij}H'$). For example, if H is a Hadamard matrix of order n, then

$$H^{(1)} \otimes H = \begin{pmatrix} H & H \\ H & -H \end{pmatrix}$$

is a Hadamard matrix of order $2n$. Note also that we can write

$$H^{(n)} = H^{(1)} \otimes H^{(n-1)}.$$

E.1.38. Check that the function tensor defined in the listing **Tensor product of two matrices** yields the tensor or Kronecker product of two matrices.

```
┌─┬─────────────────────────────────────────────────────────┬─┐
│▲│ Library │ Hadamard matrix H⁽ⁿ⁾                           │▲│
├─┴─────────────────────────────────────────────────────────┴─┤
```

$$\text{hadamard\_matrix\_(1)} := \begin{pmatrix} 1 & 1 \\ 1 & -1 \end{pmatrix}$$

```
hadamard_matrix_(n : ℤ)
check n>1
:= begin
     local H=hadamard_matrix_(n−1)
     ( H |  H )
     &
     ( H |−H )
   end
```

$$\text{hadamard\_matrix\_(1)} \longrightarrow \begin{pmatrix} 1 & 1 \\ 1 & -1 \end{pmatrix}$$

$$\text{hadamard\_matrix\_(2)} \longrightarrow \begin{pmatrix} 1 & 1 & 1 & 1 \\ 1 & -1 & 1 & -1 \\ 1 & 1 & -1 & -1 \\ 1 & -1 & -1 & 1 \end{pmatrix}$$

$|\text{hadamard\_matrix\_(3)}| \longrightarrow 4096$

$|\text{hadamard\_matrix\_(5)}| \longrightarrow 1208925819614629174706176$

$(2^5)^{\frac{2^5}{2}} \longrightarrow 1208925819614629174706176$

```
┌─┬─────────────────────────────────────────────────────────┬─┐
│▲│ Library │ Tensor or Kronecker product of two matrices    │▲│
├─┴─────────────────────────────────────────────────────────┴─┤
```

```
tensor(H : Matrix, H′ : Matrix) :=
begin
    local S, T=[ ]
    for h in H do
      S=[ ]
      for a in h do
        S = S | a·H′
      end
      T = T & S
    end
end
```

$$H=\text{hadamard\_matrix(1)} \longrightarrow \begin{pmatrix} 1 & 1 \\ 1 & -1 \end{pmatrix}$$

$$\text{tensor(H, H)} \longrightarrow \begin{pmatrix} 1 & 1 & 1 & 1 \\ 1 & -1 & 1 & -1 \\ 1 & 1 & -1 & -1 \\ 1 & -1 & -1 & 1 \end{pmatrix}$$

1.48 Proposition. *If H is a Hadamard matrix of order $n \geqslant 3$, then n is a multiple of 4.*

Proof: We can assume, without loss of generality, the H is normalized.

First let us see that n is even. Indeed, if r and r' are the number of times that 1 and -1 appear in the second row of H, then on one hand the scalar product of the first two rows of H is zero, by definition of Hadamard matrix, and on the other it is equal to $r - r'$, since H is normalized. Hence $r' = r$ and $n = r + r' = 2r$.

Now we can assume, after permuting the columns of H if necessary, that the second row of H has the form

$$(\mathbf{1}_r| - \mathbf{1}_r)$$

and that the third row (which exists because $n \geqslant 3$) has the form

$$(\mathbf{1}_s| - \mathbf{1}_t|\mathbf{1}_{s'}| - \mathbf{1}_{t'}),$$

with

$$s + t = s' + t' = r.$$

Since the third row contains r times 1 and r times -1, we also have that

$$s + s' = t + t' = r.$$

The equations $s + t = r$ and $s + s' = r$ yield $s' = t$. Similarly, $t' = s$ from $s + t = r$ and $t + t' = r$. Hence the third row of H has the form

$$(\mathbf{1}_s| - \mathbf{1}_t|\mathbf{1}_t| - \mathbf{1}_s).$$

Finally the condition that the second and third row are orthogonal yields that $2(s - t) = 0$ and hence $s = t$, $r = 2s$ and $n = 4s$. $\qquad\square$

E.1.39. For $n = 4$ and $n = 8$, prove that any two Hadamard matrices of order n are equivalent.

The Paley construction of Hadamard matrices

Assume that q, the cardinal of $\mathbb{F}$, is odd (that is, $q = p^r$, where p is a prime number greater than 2 and r is a positive integer). Let

$$\chi\colon \mathbb{F}^* \to \{\pm 1\}$$

be the *Legendre character* of $\mathbb{F}^*$, which, by definition, is the map such that

$$\chi(x) = \begin{cases} 1 & \text{if } x \text{ is a square} \\ -1 & \text{if } x \text{ is not a square} \end{cases}$$

We also extend χ to $\mathbb{F}$ with the convention that $\chi(0) = 0$.

1.49 Proposition. *If we set $r = (q - 1)/2$, then $\chi(x) = x^r$ for all $x \in \mathbb{F}$. Moreover, χ takes r times the value 1 and r times the value -1 and hence $\sum_{x \in \mathbb{F}} \chi(x) = 0$.*

Proof: Since $(x^r)^2 = x^{2r} = x^{(q-1)} = 1$ (because $\mathbb{F}^*$ has order $q - 1$; cf. Proposition 2.29), we see that $x^r = \pm 1$. If x is a square, say $x = w^2$, then $x^r = w^{2r} = 1$.

Note that there are exactly r squares in $\mathbb{F}^*$. Indeed, since $(-x)^2 = x^2$, the number of squares coincides with the number of distinct pairs $\{x, -x\}$ with $x \in \mathbb{F}^*$ and it is enough to observe that $x \neq -x$, for all x, because p is odd.

Since the equation $x^r = 1$ can have at most r solutions in $\mathbb{F}^*$, and the r squares are solutions, it turns out that $x^r \neq 1$ for the r elements of $\mathbb{F}^*$ that are not squares, hence $x^r = -1$ for all x that are not squares. Thus the value of x^r is 1 for the squares and -1 for the non-squares, as was to be seen. $\square$

The elements of $\mathbb{F}$ that are squares (non-squares) are also said to be *quadratic residues* (*quadratic non-residues*) of $\mathbb{F}$. The listing **Legendre character and the functions QR and NQR** includes the function Legendre(a,F), that computes $a^{(q-1)/2}$ ($q = |F|$) for $a \in F^*$, and the functions QR(F) and QNR(F) that return, respectively, the sets of quadratic residues and quadratic non-residues of any finite field $\mathbb{F}$.[3]

```
▲ Library │ Legendre character and the functions QR and NQR        ▲

  Legendre(a:Element(Field), F:GF)
  check subextension?(field(a), F )
  := if a=0  then
        0
     else if a^(cardinal(F)-1)/2=1 then
        1
     else
        -1
     end
  QR(F:GF)
  := begin
        local x, q=cardinal(F)
        x(j):= element(j, F)
        {x(j) with j in 1 .. (q-1) such_that Legendre(x(j), F)=1 }
     end
  QNR(F:GF):=
     begin
        local x, q=cardinal(F)
        x(j):= element(j,F)
        {x(j) with j in 2 .. (q-1) such_that Legendre(x(j), F)≠1 }
     end

  QR(Z_29) → {1, 4, 5, 6, 7, 9, 13, 16, 20, 22, 23, 24, 25, 28}
  QNR(Z_29) → {2, 3, 8, 10, 11, 12, 14, 15, 17, 18, 19, 21, 26, 27}
```

---

[3]For the convenience of the reader, the function Legendre is defined and used here instead of the equivalent built-in function legendre.

Paley matrix of a finite field of odd characteristic

Let $x_0 = 0, x_1, \ldots, x_{q-1}$ denote the elements of $\mathbb{F}$, in some order, and define the *Paley matrix* of $\mathbb{F}$,

$$S_q = (s_{ij}) \in M_q(\mathbb{F}),$$

by the formula

$$s_{ij} = \chi(x_i - x_j).$$

For its computation, see the listing **Paley matrix of a finite field**.

```
▲ Library │ Paley matrix of a finite field of char ≠ 2                          ▲

paley_matrix(F:GF)
check characteristic(F)≠2
:= begin
      local χ, x, n=cardinal(F)−1
      x(j) := element(j,F)
      χ(a):= legendre(a,F)
      [[χ(x(i)−x(j))with j in 0..n]with i in 0..n]
   end
```

$$\text{paley\_matrix}(\mathbb{Z}_5) \;\longrightarrow\; \begin{pmatrix} 0 & 1 & -1 & -1 & 1 \\ 1 & 0 & 1 & -1 & -1 \\ -1 & 1 & 0 & 1 & -1 \\ -1 & -1 & 1 & 0 & 1 \\ 1 & -1 & -1 & 1 & 0 \end{pmatrix}$$

1.50 Proposition. *If $U = U_q \in M_q(\mathbb{F})$ is the matrix whose entries are all 1, then $S = S_q$ has the following properties:*

$$\begin{array}{ll} 1) \ S1_q^T = 0_q^T & 2) \ 1_q S = 0_q \\ 3) \ SS^T = qI_q - U & 4) \ S^T = (-1)^{(q-1)/2}S. \end{array}$$

Proof: Indeed, 1 and 2 are direct consequences of the equality

$$\sum_{x \in \mathbb{F}} \chi(x) = 0,$$

which in turn was established in proposition 1.49.

The relation 3 is equivalent to say that

$$\sum_k \chi(a_i - a_k)\chi(a_j - a_k) = \begin{cases} q-1 & \text{if} \quad j = i \\ -1 & \text{if} \quad j \neq i \end{cases}$$

The first case is clear. To see the second, let $x = a_i - a_k$ and $y = a_j - a_i$, so that $a_j - a_k = x + y$. With these notations we want to show that the value of

$$s = \sum_x \chi(x)\chi(x + y)$$

is -1 for all $y \neq 0$. To this end, we can assume that the sum is extended to all elements x such that $x \neq 0, -y$. We can write $x + y = xz_x$, where $z_x = 1 + y/x$ runs through the elements $\neq 0, 1$ when x runs the elements $\neq 0, -y$. Therefore

$$s = \sum_{x \neq 0, -y} \chi(x)^2 \chi(z_x) = \sum_{z \neq 0, 1} \chi(z) = -1.$$

Finally the relation 4 is a consequence of the the formula $\chi(-1) = (-1)^{(q-1)/2}$ (proposition 1.49). □

Construction of Hadamard matrices using Paley matrices

1.51 Theorem. *If $n = q + 1 \equiv 0 \pmod 4$, then the matrix*

$$H_n = \begin{pmatrix} 1 & 1_q \\ 1_q^T & -I_q \pm S_q \end{pmatrix}$$

is a Hadamard matrix of order n.

Proof: The hypothesis $n = q + 1 \equiv 0 \pmod 4$ implies that $(q-1)/2$ is odd and hence $S = S_q$ is skew-symmetric ($S^T = -S$). Writing 1 instead of 1_q for simplicity, we have

$$H_n H_n^T = \begin{pmatrix} 1 & 1 \\ 1^T & -I_q + S \end{pmatrix} \begin{pmatrix} 1 & 1 \\ 1^T & -I_q - S \end{pmatrix}$$

$$= \begin{pmatrix} 1 + 1 \cdot 1^T & 1 + 1(-I_q - S) \\ 1^T + (-I_q + S)1^T & 1^T \cdot 1 + I_q + SS^T \end{pmatrix}$$

and the preceeding proposition implies that this matrix is equal to nI_n. Note that $1 \cdot 1^T = q$ and $1^T \cdot 1 = U_q$. We have shown the case $+S$; the case $-S$ is analogous. □

For the computation of the matrix H_n in the theorem, see the listing **Hadamard matrix of a finite field with q+1 mod 4 = 0**.

1.52 Remark. When $q + 1$ is not divisible by 4, in E.1.42 we will see that we can still associate a Hadamard matrix to $\mathbb{F}_q$, but its order will be $2q + 2$. Then we will be able to define hadamard_matrix(F:Field) for any finite field of odd characteristic by using the formula in the Theorem 1.51 when $q + 1$ is divisible by 4 (in this case it returns the same as hadamard_matrix_) and the formula in E.1.42 when $q + 1$ is not divisible by 4.

> The name conference matrix originates from an application to conference telephone circuits.
>
> J.H. van Lint and R.M. Wilson, [26], p. 173.

```
┌─────────────────────────────────────────────────────────────────┐
│ ▲ Library │ Hadamard matrix of a finite field with q+1 = 0 mod 4    ▲ │
├─────────────────────────────────────────────────────────────────┤
│ hadamard_matrix_(F:GF)                                            │
│ check cardinal(F)+1 mod 4 = 0                                     │
│ := begin                                                          │
│      local q=cardinal(F), u=constant_vector(q,1), A=I_q+paley_matrix(F) │
│      ( [ 1 ] │   u )                                              │
│      &                                                            │
│      ( u   │ -A )                                                 │
│    end                                                            │
└─────────────────────────────────────────────────────────────────┘
```

$$
\text{hadamard\_matrix\_}(\mathbb{Z}_7) \to
\begin{pmatrix}
1 & 1 & 1 & 1 & 1 & 1 & 1 & 1 \\
1 & -1 & 1 & 1 & -1 & 1 & -1 & -1 \\
1 & -1 & -1 & 1 & 1 & -1 & 1 & -1 \\
1 & -1 & -1 & -1 & 1 & 1 & -1 & 1 \\
1 & 1 & -1 & -1 & -1 & 1 & 1 & -1 \\
1 & -1 & 1 & -1 & -1 & -1 & 1 & 1 \\
1 & 1 & -1 & 1 & -1 & -1 & -1 & 1 \\
1 & 1 & 1 & -1 & 1 & -1 & -1 & -1
\end{pmatrix}
$$

Conference matrices and their use in contructing Hadamard matrices

E.1.40. A matrix C of order n is said to be a *conference matrix* if its diagonal is zero, the other entries are ± 1 and $CC^T = (n-1)I_n$. Prove that if a conference matrix of order $n > 1$ exists, then n is even. Prove also that by permuting rows and columns of C, and changing the sign of suitable rows and columns, we can obtain a symmetric (skew-symmetric) conference matrix of order n for $n \equiv 2 \pmod 4$ ($n \equiv 0 \pmod 4$).

E.1.41. Let S be the Paley matrix of a finite field $\mathbb{F}_q$, q odd, and let $\varepsilon = 1$ if S is symmetric and $\varepsilon = -1$ if S is skew-symmetric. Prove that the matrix

$$
C = \begin{pmatrix} 0 & \mathbf{1}_q \\ \varepsilon \mathbf{1}_q^T & S \end{pmatrix}
$$

is a conference matrix of order $q + 1$ (for the computation of this matrix, see the listing **Conference matrix of a finite field**). This matrix is symmetric (skew-symmetric) if $q \equiv 1 \pmod 4$ ($q \equiv 3 \pmod 4$). Note that in the case $q \equiv 3 \pmod 4$ the matrix $I_n + C$ is equivalent to the Hadamard matrix H_n of theorem 1.51.

E.1.42. If C is a skew-symmetric conference matrix of order n, show that $H = I_n + C$ is a Hadamard matrix (cf. the remark at the end of the previous exercise). If C is a symmetric conference matrix of order n, show that

$$
\overline{H} = \begin{pmatrix} I_n + C & -I_n + C \\ -I_n + C & -I_n - C \end{pmatrix}
$$

is a Hadamard matrix of order $2n$.   □

```
┌─┬───────┬──────────────────────────────────────┐
│▲│Library│ Conference matrix of a finite field  │▲│
├─┴───────┴──────────────────────────────────────┤
│ conference_matrix(F:GF):=                       │
│   begin                                         │
│     local S=paley_matrix(F), q=cardinal(F),     │
│         e=(-1)^(q-1)/2, u=constant_vector(q,1)   │
│     ([0]| u)                                     │
│     &                                            │
│     ( e·u | S)                                   │
│   end                                            │
└─────────────────────────────────────────────────┘
```

$$
\mathtt{\backslash caretconference\_matrix}(\mathbb{Z}_3) \ \longrightarrow \ \begin{pmatrix} 0 & 1 & 1 & 1 \\ -1 & 0 & -1 & 1 \\ -1 & 1 & 0 & -1 \\ -1 & -1 & 1 & 0 \end{pmatrix}
$$

The last exercise allows us, as announced in Remark 1.52, to associate to a finite field $\mathbb{F}_q$, q odd, a Hadamard matrix of order $q + 1$ or $2q + 2$ according to whether $q + 1$ is or is not divisible by 4, respectively. For the computational details, see the listing **Hadamard matrix of a finite field**. The function in this listing is called hadamard_matrix(F), and extends the definition of hadamard_matrix_ that could be applied only if $q + 1 \equiv 0$ mod 4.

```
┌─┬───────┬──────────────────────────────────────┐
│▲│Library│ Hadamard matrix of a finite field    │▲│
├─┴───────┴──────────────────────────────────────┤
│ hadamard_matrix(F:GF):=                         │
│   begin                                         │
│     local q=cardinal(F), r=(q-1)/2, e=(-1)^r,   │
│         I=I_{q+1}, C=conference_matrix(F)        │
│     if e=1  then                                 │
│       (  I+C | -I+C )                            │
│       &                                          │
│       (-I+C | -I-C )                             │
│     else                                         │
│        I+C                                       │
│     end                                          │
│   end                                            │
└─────────────────────────────────────────────────┘
```

```
Suggested examples (matrices of order 12 and 20
  which are not displayed)
hadamard_matrix(Z_5)
clear(x); F=Z_3[x]/(x^2+1);
hadamard_matrix(F)
```

1.53 Remark. The Hadamard matrices $H^{(2)}$, $H^{(3)}$, $H^{(4)}$, $H^{(5)}$ have orders $4, 8, 16, 32$. Theorem 1.51 applies for $q = 7, 11, 19, 23, 27, 31$ and yields Hadamard matrices H_8, H_{12}, H_{20}, H_{24}, H_{28} and H_{32}. The second part of the exercise E.1.42 yields Hadamard matrices $\overline{H}_{12}, \overline{H}_{20}, \overline{H}_{28}$ from the fields $\mathbb{F}_5$, $\mathbb{F}_9$ and $\mathbb{F}_{13}$.

It is conjectured that for all positive integers n divisible by 4 there is a Hadamard matrix of order n. The first integer n for which it is not known whether there exists a Hadamard matrix of order n is 428 (see [26] or [2]).

The matrices $H^{(3)}$ and H_8 are not equal, but we know that they are equivalent (see E.1.39). The matrices H_{12} and $\overline{H}_{12}$ are also equivalent (see P.1.19). In general, however, it is not true that all Hadamard matrices of the same order are equivalent (it can be seen, for example, that H_{20} and $\overline{H}_{20}$ are not equivalent).

For more information on the present state of knowledge about Hadamard matrices, see the web page

> http://www.research.att.com/˜njas/hadamard/

Hadamard codes

> One of the recent very interesting and successful applications
> of Hadamard matrices is their use as co-called *error-correcting*
> *codes*.
>
> J.H. van Lint and R.M. Wilson, [26], p. 181.

We associate a binary code $C = C_H$ of type $(n, 2n)$ to any Hadamard matrix H of order n as follows: $C = C' \sqcup C''$, where C' is obtained by replacing all occurrences of -1 in H by 0 and C'' is the result of complementing the vectors in C' (or replacing -1 by 0 in $-H$).

Since any two rows of H differ in exactly $n/2$ positions, if follows that the minimum distance of C_H is $n/2$ and so the type of C_H is $(n, 2n, n/2)$.

For the construction of C_H as a list of code-words, see the listing **Construction of Hadamard codes** (to save space in the examples, instead of the matrix of code-words we have written its transpose).

E.1.43. Are the Hadmanard codes equidistant? Can they have vectors of weight less than $n/2$?

E.1.44. Show that the code C_{H_8} is equivalent to the parity extension of the Hamming $[7, 4, 3]$ code. It is therefore a linear code (note that a Hadamard code of length n cannot be linear unless n is a power of 2.

1.54 *Example.* The code $C = C_{H^{(5)}}$, whose type is $(32, 64, 16)$, is the code used in the period 1969–1972 by the Mariner space-crafts to transmit pictures of Mars (this code turns out to be equivalent to the linear code $RM_2(5)$; see P.1.21). Each picture was digitalized by subdividing it into small squares (pixels) and assigning to each of them a number in the range $0..63$ (000000..111111 in binary) corresponding to the grey level. Finally the coding was done by assigning each grey level to a code-word of C in a

▲ Library	Construction of Hadamard codes	▲

```
hadamard_code(H:Matrix) := {-1⇒0}(H&-H)
hadamard_code(n:ℤ) check n>0 := hadamard_code(hadamard_matrix(n))
hadamard_code(F:GF) := hadamard_code(hadamard_matrix(F))
```

$$\text{hadamard\_code}(2)^\mathsf{T} \longrightarrow \begin{pmatrix} 1\,1\,1\,1\,0\,0\,0\,0 \\ 1\,0\,1\,0\,0\,1\,0\,1 \\ 1\,1\,0\,0\,0\,0\,1\,1 \\ 1\,0\,0\,1\,0\,1\,1\,0 \end{pmatrix}$$

$$\text{hadamard\_code}(\mathbb{Z}_7)^\mathsf{T} \longrightarrow \begin{pmatrix} 1\,0\,0\,0\,0\,0\,0\,0\,1\,1\,1\,1\,1\,1\,1 \\ 1\,1\,1\,1\,0\,1\,0\,0\,0\,0\,0\,0\,1\,0\,1\,1 \\ 1\,0\,1\,1\,1\,0\,1\,0\,0\,1\,0\,0\,0\,1\,0\,1 \\ 1\,0\,0\,1\,1\,1\,0\,1\,0\,1\,1\,0\,0\,0\,1\,0 \\ 1\,1\,0\,0\,1\,1\,1\,0\,0\,0\,1\,1\,0\,0\,0\,1 \\ 1\,0\,1\,0\,0\,1\,1\,1\,0\,1\,0\,1\,1\,0\,0\,0 \\ 1\,1\,0\,1\,0\,0\,1\,1\,0\,0\,1\,0\,1\,1\,0\,0 \\ 1\,1\,1\,0\,1\,0\,0\,1\,0\,0\,0\,1\,0\,1\,1\,0 \end{pmatrix}$$

one-to-one fashion. Hence the rate of this coding is 6/32 and its correcting capacity is 7. See figure 1.1 for an upper-bound of the error-reduction factor of this code. Decoding was done by the Hadamard decoder explained in the next subsection. For more information on the role and most relevant aspects of coding in the exploration of the solar system, see [28].

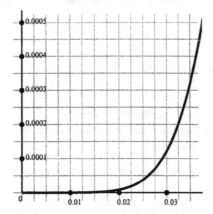

Figure 1.1: *The error-reduction factor of the Mariner code is given by* $\sum_{j=8}^{32} \binom{32}{j}(1-p)^{32-j}p^{j-1}$. *For* $p = 0.01$, *its value is* 8.49×10^{-8}, *and for* $p = 0.001$ *it is* 1.03×10^{-14}.

Decoding of Hadamard codes

Let $C = C_H$ be the Hadamard code associated to the Hadamard matrix H. Let n be the order of H and $t = \lfloor (n/2 - 1)/2 \rfloor = \lfloor (n-2)/4 \rfloor$. Since $d_C = n/2$, t is the correcting capacity of C.

Given a binary vector y of length n, let $\hat{y} \in \{1, -1\}^n$ be the result of replacing any occurrence of 0 in y by -1. For example, from the definition of C it follows that if $x \in C'$ (or $x \in C''$), then $\hat{x}$ (or $-\hat{x}$) is a row of H.

E.1.45. Let y be a binary vector of length n, z be the result of changing the sign of one component of y, and h be any row of H. Then $\langle \hat{z} | h \rangle = \langle \hat{y} | h \rangle \pm 2$.

1.55 Proposition. *Let $C = C_H$, where H is a Hadamard matrix of order n. Assume $x \in C$ is the sent vector, that e is the error vector and that $y = x + e$ is the received vector. Let $|e| = s$ and assume $s \leqslant t$. Then there is a unique row h_y of H such that $| \langle \hat{y} | h \rangle | > n/2$, and all other rows h' of H satisfy $| \langle \hat{y} | h \rangle | < n/2$. Moreover, $\hat{x} = h_y$ or $\hat{x} = -h_y$ according to whether $\langle \hat{y} | h \rangle > 0$ or $\langle \hat{y} | h \rangle < 0$.*

Proof: Assume $x \in C'$ (the case $x \in C''$ is similar and is left as an exercise). Then $h = \hat{x}$ is a row of H and $\langle \hat{x} | h \rangle = \langle h | h \rangle = n$. Since $\hat{y}$ is the result of changing the sign of s components of $\hat{x}$, E.1.45 implies that $\langle \hat{y} | h \rangle = n \pm 2r$, with $0 \leqslant r \leqslant s \leqslant t$, and so $\langle \hat{y} | h \rangle \geqslant n \pm 2t > n/2$. For any other row h' of H we have that $\langle \hat{x} | h' \rangle = \langle h | h' \rangle = 0$ and again by E.1.45 we obtain that $\langle \hat{y} | h' \rangle = \pm 2r$, with $0 \leqslant r \leqslant s \leqslant t$, and hence $| \langle \hat{y} | h' \rangle | \leqslant 2t < n/2$. $\square$

The preceding proposition suggests decoding the Hadamard code $C = C_H$ with the algorithm below (for its implementation, see **The Hadamard decoder**).

Hadamard decoder

    0) Let y be the received vector.

    1) Obtain $\hat{y}$.

    2) Compute $c = \langle \hat{y} | h \rangle$ for successive rows h of H, and stop as soon as $|c| > n/2$, or if there are no further rows left. In the latter case return a decoder error message.

    3) Otherwise, output the result of replacing -1 by 0 in h if $c > 0$ or in $-h$ if $c < 0$.

1.56 Corollary. *The Hadamard decoder corrects all error patterns of weight up to t.*

E.1.46 (Paley codes). Let $S = S_q$ be the Paley matrix of order q associated to $\mathbb{F}_q$, q odd. Let C_q be the binary code whose elements are the rows of the matrices $\frac{1}{2}(S + I + U)$ and $\frac{1}{2}(-S + I + U)$, together with the vectors $\mathbf{0}_n$ and $\mathbf{1}_n$. Show that C_q has type $(q, 2q + 2, (q - 1)/2)$ (for examples, see the listing **Construction of Paley codes**; to save space, the example has been written in columns rather than in rows). Note that for $q = 9$ we obtain a $(9, 20, 4)$ code, hence also an $(8, 20, 3)$ code (cf. E.1.4).

```
┌─────────────────────────────────────────────────────────────────┐
│ ▲ Library │ The Hadamard decoder                              ▲ │
├─────────────────────────────────────────────────────────────────┤
│ hadamard_decoder(y, H):=                                          │
│ begin                                                             │
│    local r=length(H)/2, c                                         │
│    y={0⇒-1}(y)                                                    │
│    for h in H do                                                  │
│       c=y·h                                                       │
│       if |c|>r then                                               │
│          if c>0 then                                              │
│             return({-1⇒0}(h))                                    │
│          else                                                     │
│             return({-1⇒0}(-h))                                   │
│          end                                                      │
│       end                                                         │
│    end                                                            │
│    "Decoder error"                                                │
│ end                                                               │
└─────────────────────────────────────────────────────────────────┘
```

```
\caretH=hadamard_matrix(Z₅); C=hadamard_code(H);
x=C₅; x, hadamard_decoder(x,H)
   ⟶ [1, 0, 0, 1, 1, 1, 1, 0, 0, 1, 0, 1],[1, 0, 0, 1, 1, 1, 1, 0, 0, 1, 0, 1]
x=C₁₅; y=flip(x,{3}); x, y
   ⟶ [0, 0, 0, 0, 1, 1, 0, 0, 1, 0, 1, 1],[0, 0, 1, 0, 1, 1, 0, 0, 1, 0, 1, 1]
hadamard_decoder(y,H) ⟶ [0, 0, 0, 0, 1, 1, 0, 0, 1, 0, 1, 1]
x=C₉; y=flip(x,{2,7}); x, y
   ⟶ [1, 1, 0, 1, 0, 0, 0, 0, 0, 0, 1, 1],[1, 0, 0, 1, 0, 0, 1, 0, 0, 0, 1, 1]
hadamard_decoder(y,H) ⟶ [1, 1, 0, 1, 0, 0, 0, 0, 0, 0, 1, 1]
x=C₁₈; y=flip(x,{7,12}); x, y
   ⟶ [0, 0, 1, 1, 0, 0, 0, 0, 1, 1, 0, 1],[0, 0, 1, 1, 0, 0, 1, 0, 1, 1, 0, 0]
hadamard_decoder(y,H) ⟶ [0, 0, 1, 1, 0, 0, 0, 0, 1, 1, 0, 1]
x=C₁₈; y=flip(x,{2,7,9}); x, y
   ⟶ [0, 0, 1, 1, 0, 0, 0, 0, 1, 1, 0, 1],[0, 1, 1, 1, 1, 0, 0, 1, 0, 0, 1, 1]
hadamard_decoder(y,H) ⟶ "Decoder error"
```

Summary

- The definition of Hadamard matrices (for example the matrices $H^{(n)}$, $n \geqslant 1$) and the fact that the order of a Hadamard matrix is divisible by 4 (or is equal to 1 or 2).

- The Paley matrix of a finite field of odd order q and their properties (Proposition 1.50).

- The Hadamard matrix associated to a finite field of odd order q such that $q + 1$ is divisible by 4 (Theorem 1.51).

- Conference matrices. Symmetric and skew-symmetric conference matrices of order n have an associated Hadamard matrix of order n and $2n$, respectively (E.1.42).

```
┌─────────────────────────────────────────────────────────────────────┐
│ ▲│ Library │ Construction of Paley codes                          ▲│
├─────────────────────────────────────────────────────────────────────┤
│   paley_code(F:GF)                                                  │
│   := begin                                                          │
│        local S=paley_matrix(F), q=cardinal(F), I=I_q, U=constant_matrix(q,1)  │
│                            S+I+U      -S+I+U                         │
│        constant_vector(q,0) &  ─────  &  ─────  & constant_vector(q,1)  │
│                              2          2                            │
│   end                                                               │
└─────────────────────────────────────────────────────────────────────┘
```

$\caret F = \mathbb{Z}_3[x]/(x^2+1);$

$$\text{paley\_code(F)}^{\mathsf{T}} \rightarrow \begin{pmatrix} 0\,1\,1\,1\,1\,0\,0\,1\,0\,0\,1\,0\,0\,0\,1\,1\,0\,1\,1\,1 \\ 0\,1\,1\,1\,0\,1\,0\,0\,1\,0\,0\,1\,0\,1\,0\,1\,1\,0\,1\,1 \\ 0\,1\,1\,1\,0\,0\,1\,0\,0\,1\,0\,0\,1\,1\,1\,0\,1\,1\,0\,1 \\ 0\,1\,0\,0\,1\,1\,1\,1\,0\,0\,0\,1\,1\,1\,0\,0\,0\,1\,1\,1 \\ 0\,0\,1\,0\,1\,1\,1\,0\,1\,0\,1\,0\,1\,0\,1\,0\,1\,0\,1\,1 \\ 0\,0\,0\,1\,1\,1\,1\,0\,0\,1\,1\,1\,0\,0\,0\,1\,1\,1\,0\,1 \\ 0\,1\,0\,0\,1\,0\,0\,1\,1\,1\,0\,1\,1\,0\,1\,1\,1\,0\,0\,1 \\ 0\,0\,1\,0\,0\,1\,0\,1\,1\,1\,1\,0\,1\,1\,0\,1\,0\,1\,0\,1 \\ 0\,0\,0\,1\,0\,0\,1\,1\,1\,1\,1\,1\,0\,1\,1\,0\,0\,0\,1\,1 \end{pmatrix}$$

- To each Hadamard matrix of order n there is associated a code

$$C_H \sim (n, 2n, n/2).$$

The codes C_H are the Hadamard codes.

- To each finite field of odd order q there is associated a Paley code $C_q \sim (q, 2q+2, (q-1)/2)$.

- The Hadamard code of the matrix $H^{(m)}$ is equivalnet to the (linear) Reed–Muller code $RM_2(m)$ (P.1.21).

Problems

P.1.17. Compare the error-reduction factor of the binary repetition code of length 5 with the Hadamard code $(32, 64, 16)$.

P.1.18. Prove that the results in this section allow us to construct a Hadamard matrix of order n for all $n \leqslant 100$ divisible by 4 except $n = 92$.

P.1.19. Prove that any two Hadamard matrices of order 12 are equivalent.

P.1.20. Fix a positive integer r and define binary vectors $u_0, \ldots, u_{r-1}$ so that

$$u_k = [b_0(k), ..., b_{r-1}(k)],$$

where $b_i(k)$ is the i-th bit in the binary representation of k. Show that $H^{(r)}$ coincides with the $2^r \times 2^r$ matrix $(-1)^{\langle u_i | u_j \rangle}$. For the computations related to this, and examples, see the listing **Bit-product of integers and Hadamard matrices**.

```
┌─────────────────────────────────────────────────────────────────┐
│ ▲ Library │ Bit−product of integers and Hadamard matrices    ▲ │
├─────────────────────────────────────────────────────────────────┤
```

$$\text{bit\_product}(a,b,r) := \sum_{i=0}^{r-1} \text{bit}(a,i) \cdot \text{bit}(b,i) \bmod 2$$

```
hadamard(r:ℤ)
check r>0
:= begin
     local n=2ʳ
     [[(−1)^bit_product(i,j,n) with j in 0..n−1] with i in 0..n−1]
   end
```

$$\text{hadamard}(2) \;\longrightarrow\; \begin{pmatrix} 1 & 1 & 1 & 1 \\ 1 & -1 & 1 & -1 \\ 1 & 1 & -1 & -1 \\ 1 & -1 & -1 & 1 \end{pmatrix}$$

P.1.21. As seen in P.1.16, the code $RM_2(m)$ has type $(2^m, 2^{m+1}, 2^{m-1})$. Prove that this code is equivalent to the Hadamard code corresponding to the Hadamard matrix $H^{(m)}$. Thus, in particular, $RM_2(5)$ is equivalent to the Mariner code. *Hint:* If H is the control matrix of $Ham_2(m)$, then the matrix $\widetilde{H}$ obtained from H by first adding the column 0_m to its left and then the row 1_{2^m} on top is a generating matrix of $RM_2(m)$.

1.4   Parameter bounds

> Probably the most basic problem in coding theory is to find the
> largest code of a given length and minimum distance.
>
> F. MacWilliams and N.J.A. Sloane, [11], p. 523

Essential points

- The Griesmer bound for linear codes.

- The Plotkin and Elias bounds (for general q-ary codes) and the Johnson bound (for general binary codes).

- The linear programming upper bound of Delsarte and the function LP(n,d) implementing it (follwing the the presentation in [11]).

- Asymptotic bounds and related computations: the lower bound of Gilbert–Varshamov and the upper bounds of Singleton, Hamming, Plotkin, Elias, and McEliece–Rodemich–Rumsey–Welch.

Introduction

The parameters n, k (or M) and d of a code satisfy a variety of relations. Aside from trivial relations, such as $0 \leqslant k \leqslant n$ and $1 \leqslant d \leqslant n$, we know the Singleton inequality

$$d \leqslant n - k + 1$$

(Proposition 1.11), the sphere upper bound

$$M \leqslant q^n / vol_q(n, t)$$

(Theorem 1.14) and the Gilbert lower bound

$$M \geqslant q^n / vol_q(n, d - 1)$$

(Theorem 1.17). We also recall that if q is a prime power and

$$vol_q(n - 1, d - 2) < q^{n-k},$$

then there exists a linear code of type $[n, k, d']$ with $d' \geqslant d$ and hence that $A(n, d) \geqslant q^k$ (Theorem 1.38). Moreover, it is not hard to see that this bound is always equal or greater than the Gilbert bound (see P.1.22).

In this section we will study a few other upper bounds that provide a better understanding of the function $A_q(n, d)$ and the problems surrounding its determination. We include many examples. Most of them have been chosen to illustrate how far the new resources help (or do not) in the determination of the values on Table 1.1.

> Redundancy adds to the cost of transmitting messages, and the art and science of error-correcting codes is to balance redundancy with efficiency.
>
> From the 2002/08/20 press release on the Fields Medal awards (L. Laforgue and V. Voevodski) and Nevanlinna Prize awards (Madhu Sudan). The quotation is from the description of Madhu Sudan's contributions to computer science, and in particular to the theory of error-correcting codes.

The Griesmer bound

Consider a linear code C of type $[n, k, d]$ and a vector $z \in C$ such that $|z| = d$. Let

$$S(z) = \{i \in 1..n \mid z_i \neq 0\}$$

(this set is called the *support* of z) and

$$\rho_z : \mathbb{F}^n \to \mathbb{F}^{n-d}$$

the linear map that extracts, from each vector in $\mathbb{F}^n$, the components whose index is not in $S(z)$. Then we have a linear code $\rho_z(C) \subseteq \mathbb{F}^{n-d}$ (we will say that it is the *residue of* C with respect to z).

If G is a generating matrix of C, then $\rho_z(C) = \langle \rho_z(G) \rangle$, where $\rho_z(G)$ denotes the matrix obtained by applying ρ_z to all the rows of G (we will also say that $\rho_z(G)$ is the *residue of* G with respect to z). Since $\rho_z(z) = 0$, it is clear that $\dim(\rho_z(C)) \leqslant k - 1$.

1.57 Lemma. *The dimension of $\rho_z(C)$ is $k - 1$ and its minimum distance is at least $\lceil d/q \rceil$.*

Proof: Choose a generating matrix G of C whose first row is z, and let G' be the matrix obtained from $\rho_z(G)$ by omiting the first row (which is $\rho_z(z) = 0$). It will suffice to show that the rows of G' are linearly independent.

To see this, let us assume, to argue by contradiction, that the rows of G' are linearly dependent. Then there is a non-trivial linear combination x of the rows of G other than the first such that $x_i = 0$ for $i \notin S(z)$ (in other words, $S(x) \subseteq S(z)$ and hence $|x| \leqslant |z| = d$). Since the rows of G are linearly independent, we have both $x \neq 0$ and $\lambda x \neq z$ for all $\lambda \in \mathbb{F}$. But the relation $x \neq 0$ allows us to find $\lambda \in \mathbb{F}^*$ such that $z_i = \lambda x_i$ for some $i \in S(z)$, and from this it follows that $|z - \lambda x| < d$, which is a contradiction because $z - \lambda x \in C - \{0\}$.

To bound below the minimum distance d' of $\rho_z(C)$, we may assume, after replacing C by a scalarly equivalent code if necessary, that $S(z) = \{1, \ldots, d\}$ and that $z_i = 1$ for all $i \in S(z)$. Then, given a non-zero vector $x' \in \rho_z(C)$ of minimum weight d', let $x \in C$ be any vector such that $\rho_z(x) = x'$. We can write $x = y|x'$, with $y \in \mathbb{F}^d$. We claim that we can choose y in such a way that $|y| \leqslant d - \lceil d/q \rceil$. Indeed, let $\alpha \in \mathbb{F}$ be an element chosen so that the number s of its ocurrences in y is maximum. With this condition it is clear that $sq \geqslant d$ (which is equivalent to $s \geqslant \lceil d/q \rceil$) and that $\bar{x} = x - \alpha z \in C$ has the form $\bar{x} = y'|x'$ with $|y'| \leqslant d - s \leqslant d - \lceil d/q \rceil$. Now we can bound d' below:

$$d' = |x'| = |\bar{x}| - |y'| \geqslant d - (d - \lceil d/q \rceil) = \lceil d/q \rceil,$$

which is what we wanted. $\square$

E.1.47. Let i be a positive integer. Check that

$$\left\lceil \frac{1}{q^{i-1}} \left\lceil \frac{d}{q} \right\rceil \right\rceil = \left\lceil \frac{d}{q^i} \right\rceil.$$

1.58 Proposition (Griesmer, 1960). *If $[n, k, d]$ are the parameters of a linear code, and we let*

$$G_q(k, d) = \sum_{i=0}^{k-1} \left\lceil \frac{d}{q^i} \right\rceil,$$

then

$$G_q(k, d) \leqslant n.$$

Proof: If we let $N_q(k, d)$ denote the minimum length n among the linear codes of dimension $k \geqslant 1$ and minimum distance d, it will be sufficient to show that $G_q(k, d) \leqslant N_q(k, d)$. Since this relation is true for $k = 1$, for the value of both $N_q(1, d)$ and $G_q(1, d)$ is d, we can assume that $k \geqslant 2$.

Suppose, as induction hypothesis, that $G_q(k - 1, d) \leqslant N_q(k - 1, d)$ for all d. Let C be a code of type $[n, k, d]_q$ with $n = N_q(k, d)$. Let C' be the residue of C with respect to a vector $z \in C$ such that $|z| = d$. We know that C' is a code of type $[n - d, k - 1, d']_q$ with $d' \geqslant \lceil d/q \rceil$ (Lemma 1.57). Therefore

$$N_q(k, d) - d = n - d \geqslant N_q(k - 1, d') \geqslant G_q(k - 1, d') \geqslant G_q(k - 1, \lceil d/q \rceil).$$

Taking into account E.1.47, the last term turns out to be $G_q(k, d) - d$ and from this the stated relation follows. $\square$

1.59 Remark. Let us see that the Griesmer bound is an improvement of the Singleton bound. The term corresponding to $i = 0$ in the summation expression of $G_q(k, d)$ in Proposition 1.58 is d and each of the other $k - 1$ terms is at least 1. Hence $G_q(k, n) \geqslant d + k - 1$ and so $G(k, n) \leqslant n$ implies that $d + k - 1 \leqslant n$, which is equivalent to the Singleton relation. $\quad\square$

1.60 Remark. In the proof of Proposition 1.58 we have introduced the function $N_q(k, d)$, which gives the minimum possible length of a q-ary linear code of dimension k and minimum distance d, and we have established that

$$N_q(k, d) \geqslant d + N_q(k - 1, \lceil d/q \rceil).$$

Using this inequality recursively, we have proved that $N_q(k, d) \geqslant G_q(k, d)$.

There are two main uses of this inequality. One is that if a code $[n, k, d]_q$ exists, then $G_q(d, k) \leqslant n$, and hence $G_q(d, k)$ yields a lower bound for the length n of codes of dimension k and miminum distance d. In other words, to search for codes $[n, k, d]_q$, with k and d fixed, the tentative dimension to start with is $n = G_q(k, d)$. For example, $G_2(4, 3) = 3 + \lceil 3/2 \rceil + \lceil 3/2^2 \rceil + \lceil 3/2^3 \rceil = 3 + 2 + 1 + 1 = 7$ and we know that there is a $[7, 4, 3]$ code, so $N_2(4, 3) = 7$, that is, 7 is the minimum length for a binary code of dimension 4 with correcting capacity 1. For more examples, see 1.61.

The other use of the Griesmer inequality is that the maximum k that satisfies $G_q(k, d) \leqslant n$, for fixed n and d, yields an upper bound q^k for the function $B_q(n, d)$ defined as $A_q(n, d)$, but using only linear codes instead of general codes. Note that $A_q(n, d) \geqslant B_q(n, d)$.

We refer to **Griesmer bound** for implementations and some examples. The function griesmer(k,d,q) computes $G_q(k, d)$, which we will call *Griesmer function*, and ub_griesmer(n,d,q) computes the upper bound on $B_q(n, d)$ explained above.

1.61 Examples. According to the listing **Griesmer bound**, and Remark 1.60, we have $B_2(8, 3) \leqslant 32$, $B_2(12, 5) \leqslant 32$, $B_2(13, 5) \leqslant 64$, $B_2(16, 6) \leqslant 256$. As we will see below, none of these bounds is sharp.

The first bound is worse than the Hamming bound ($A_2(8, 3) \leqslant 28$), and in fact we will see later in this section, using the linear programming bound, that $A_2(8, 3) = 20$ (which implies, by the way, that $B_2(8, 3) = 16$).

The bound $B_2(13, 5) \leqslant 64$ is not sharp, because if there were a $[13, 6, 5]$ code, then taking residue with respect to a weight 5 code vector would lead to a linear code $[8, 5, d]$ with $d \geqslant 3$ (as $\lceil d/2 \rceil = 3$) and $d \leqslant 4$ (by Singleton), but both $[8, 5, 4]$ and $[8, 5, 3]$ contradict the sphere bound.

Let us see now that a $(16, 256, 6)$ code cannot be linear. Indeed, if we had a linear code $[16, 8, 6]$, then there would also exist a $[15, 8, 5]$ code. Selecting now from this latter code all words that end with 00, we would be able to

```
┌─────────────────────────────────────────────────────────────────┐
│ ▲│Library │ Griesmer upper bound and Griesmer function        ▲│
├─────────────────────────────────────────────────────────────────┤
```

```
ub_griesmer(n,d,q)
:= begin
      local i=0, P=1
      while n>0 do
         n=n−⌈d/P⌉; P=P·q; i=i+1
      end
      P
   end
ub_griesmer(n,d) := ub_griesmer(n,d,2)
```

$$\text{griesmer(k,d,q)} := \sum_{i=0}^{k-1} \lceil d/q^i \rceil$$

```
griesmer(k,d) := griesmer(k,d,2)
```

```
ub_griesmer(8,3)   ⟶  32
ub_griesmer(12,5)  ⟶  32
ub_griesmer(13,5)  ⟶  64
ub_griesmer(16,6)  ⟶  256
ub_griesmer(14,9,3) ⟶  81
griesmer(4,3)   ⟶  7
griesmer(5,7)   ⟶  15
griesmer(12,7)  ⟶  22
griesmer(6,5,3) ⟶  11
```

construct a code $[13, 6, 5]$ that we have shown not to exist in the previous paragraph.

The fact that the bound $B_2(12,5) \leqslant 32$ is not sharp cannot be settled with the residue operation, because it would lead to a $[7, 4, 3]$ code, which is equivalent to the linear $Ham(3)$. The known arguments are much more involved and we omit them (the interested reader may consult the references indicated in [16], p. 124, or in [25], p. 52).

The last expresion in the same listing says that $B_3(14, 9) \leqslant 4$, but it is not hard to see that this is not sharp either (cf. [25], p. 53, or Problem P.1.23).

In the listing we also have that $G_2(5, 7) = 15$, $G_2(12, 7) = 22$ and $G(6, 5, 3) = 11$. The techniques developed in Chapter 3 will allow us to construct codes $[15, 5, 7]$ (a suitable BCH code) and $[11, 6, 5]_3$ (the ternary Golay code). This implies that $N_2(5, 7) = 15$ and $N_3(6, 5) = 11$. On the other hand, there is no $[22, 12, 7]$ code, because these parameters contradict the sphere bound, but in Chapter 3 we will construct a $[23, 12, 7]$ code (the binary Golay code) and therefore $N_2(12, 7) = 23$.

1.62 Remark. If we select all words of a code C whose i-th coordinate is a given symbol λ, and then delete this common i-th coordinate λ from the selected words, we get a code $C' \sim (n - 1, M', d)$, with $M' \leqslant M$. We say that this C' is a *shortening* of C (or the λ-shortening of C in the i-th coordinate). If C is linear, a 0-shortening is also linear, but a λ-shortening of a linear code is in general not linear for $\lambda \neq 0$. In the previous examples we have used this operation twice in the argument to rule out the existence of linear codes $[16, 8, 6]$.

The Plotkin bound

We will write β to denote the quotient $(q - 1)/q = 1 - q^{-1}$. Although the proof of the proposition below can be found in most of the books on error correcting codes (see, for example, [25] (5.2.4) or [20], Th. 4.5.6), we include it because it illustrates an interesting method that is relevant for subsequent exercises, examples and computations.

1.63 Proposition (Plotkin, 1951). *For any code C of type (n, M, d), we have*

$$d \leqslant \frac{\beta n M}{M - 1},$$

or, equivalently,

$$M(d - \beta n) \leqslant d.$$

Proof: The basic idea is that the average of the distances between pairs of distinct elements of C is at least d.

In order to calculate this average, arrange the elements of C in an $M \times n$ matrix. For a given column c of this matrix, let m_λ denote the number of times that $\lambda \in \mathbb{F}$ appears in that column, and note that

$$\sum_{\lambda \in \mathbb{F}} m_\lambda = M.$$

The contribution of c to the sum of the distances between pairs of distinct elements of C is

$$S_c = \sum_{\lambda \in \mathbb{F}} m_\lambda (M - m_\lambda) = M^2 - \sum_{\lambda \in \mathbb{F}} m_\lambda^2.$$

But

$$M^2 = \left(\sum_{\lambda \in \mathbb{F}} m_\lambda \right)^2 \leqslant q \sum_{\lambda \in \mathbb{F}} m_\lambda^2,$$

by the Cauchy–Schwartz inequality, and so $S_c \leqslant \beta M^2$. If we let S denote the sum of the S_c, where c runs through all columns of the matrix C, then $S \leqslant n\beta M^2$. Now it is clear that $M(M - 1)d \leqslant S$ and so

$$M(M - 1)d \leqslant n\beta M^2,$$

and this immediately yields the stated inequality.   □

1.64 Corollary (Plotkin upper bound). *If $d > \beta n$, then*

$$A_q(n, d) \leqslant \left\lfloor \frac{d}{d - \beta n} \right\rfloor.$$

E.1.48. Prove that the inequality $d \leqslant \beta n M/(M - 1)$ is an equality if and only if the code C is equidistant and the numbers m_λ are the same for all λ.

1.65 *Example*. For the dual Hamming codes the Plotkin inequality is an equality. Indeed, if $C = Ham_q^\vee(r)$, we know that C' has type $[n, r, d]$, with $d = q^{r-1}$ and $n = (q^r - 1)/(q - 1)$. Since $\beta n = q^{r-1} - q^{-1} < q^{r-1} = d$, Plotkin's upper bound can be applied. As $d/(d - \beta n) = q^r$, which is the cardinal of C,

$$A_q(\frac{q^r - 1}{q - 1}, q^{r-1}) = q^r.$$

This formula explains the entries $A_2(6, 3) = A_2(7, 4) = 8$ and $A_2(14, 7) = A_2(15, 8) = 16$ on the Table 1.1. Another example: with $q = 3$ and $r = 3$ we find $A_3(13, 9) = 27$.

E.1.49 (Improved Plotkin bound). For binary codes, $\beta = 1/2$, and the Plotkin upper bound, which is valid if $2d > n$, says that

$$A_2(n, d) \leqslant \lfloor 2d/(2d - n) \rfloor.$$

But this can be improved to

$$A_2(n, d) \leqslant 2 \lfloor d/(2d - n) \rfloor.$$

Moreover, if d is odd and $n < 2d + 1$, then $A_2(n, d) = A_2(n + 1, d + 1)$ and $n + 1 < 2(d + 1)$, so that

$$A_2(n, d) \leqslant 2 \lfloor (d + 1)/(2d + 1 - n) \rfloor.$$

Hint: With the notations of the proof of Proposition 1.63, we have $S_c = 2m_0(M - m_0)$ and so $S_c \leqslant M^2/2$ if M is even and $S_c \leqslant (M^2 - 1)/2$ if M is odd; it follows that $M(M - 1)d \leqslant S \leqslant nM^2/2$ if M is even and $M(M - 1)d \leqslant S \leqslant n(M^2 - 1)/2$ if M is odd.

E.1.50. Let C be a code of type $(n, M, d)_q$. Prove that there is a code of type $(n - 1, M', d)_q$ with $M' \geqslant M/q$. If C is optimal, we deduce that $A_q(n, d) \leqslant q A_q(n - 1, d)$ *Hint:* This can be achieved with the operation of *shortening* C: choose a symbol $\lambda \in \mathbb{F}$ whose occurrence in the last position is maximum, and let $C' = \{x \in \mathbb{F}^{n-1} \mid (x|\lambda) \in C\}$.

1.66 Remark. If C is a code of type (n, M, d) and $d \leqslant \beta n$, then we cannot apply the Plotkin bound as it is. But we can use the exercise E.1.50 to find a code C' of type $(n - m, M', d)$ such that $d > \beta(n - m)$ and with $M' \geqslant M/q^m$, whereupon we can bound M by $q^m M'$ and M' by the Plotkin bound corresponding to $(n - m, M', d)$. Note that $d > \beta(n - m)$ is equivalent to $m > n - d/\beta = (n\beta - d)/\beta$. These ideas, together with those of the previous exercise in the binary case, are used to optimize the function ub_plotkin defined in **Plotkin upper bound**.

```
▲ Library │ Plotkin upper bound                                    ▲
 ub_plotkin (n,d,q) :=
    begin
      local m, β=(q−1)/q
      if d>β·n  then
        if q=2  then
          if odd?(d)  then
              n=n+1; d=d+1
          end
          return 2·⌊ d / (2·d−n) ⌋
        else
            return ⌊ d / (d−β·n) ⌋
        end
      end
      m=(β·n−d)/β
      if m ∈ ℤ  then
         m=m+1
      else
         m=⌈m⌉
      end
      qᵐ·ub_plotkin (n−m, d, q)
    end
 Binary case:
 ub_plotkin (n,d) := ub_plotkin (n,d,2)
```

```
 ub_plotkin (13,5)  →  96
 ub_plotkin(32,16)  →  64
 {ub_plotkin(n,3) with n in 5..8}  →  {4, 8, 16, 32}
 {ub_plotkin(n,5) with n in 5..12}  →  {2, 2, 2, 4, 6, 12, 24, 48}
 {ub_plotkin(n,7) with n in 7..16}  →  {2, 2, 2, 2, 4, 4, 8, 16, 32, 64}
```

E.1.51. Show that for d even, $A_2(2d, d) \leqslant 4d$, and for d odd $A_2(2d+1, d) \leqslant 4d + 4$.

1.67 Remark. The second expression in the listing **Plotkin upper bound** shows that $A_2(32, 16) \leqslant 64$. But we know that there is a code (32,64,16), and hence $A_2(32, 16) = 64$. In other words, the Mariner code $RM(5)$ is optimal.

The same listing shows that the Plotkin bound gives the correct values

$4, 8, 16$ (for $d = 3$), $2, 2, 2, 4, 6, 12, 24$ (for $d = 5$) and $2, 2, 2, 2, 4, 4, 8, 16, 32$ (for $d = 7$) on the Table 1.1. Note that it gives good results when d is large with respect to n. When d is small with respect to n, the sphere bound is better. For example, the Plotkin bound gives $A_2(8, 3) \leqslant 32$, while the sphere bound is $A_2(8, 3) \leqslant 28$.

Elias upper bound

From the proof of the Plotkin bound it is clear that we cannot hope that it is sharp when the distances between pairs of distinct elements of the code are not all equal (cf. E.1.48). The idea behind the Elias bound (also called Bassalygo–Elias bound) is to take into account, in some way, the distribution of such distances in order to get a better upper estimate.

1.68 Proposition (Elias, 1954, Bassalygo, 1965). *Assume that q, n, d and r are positive integers with $q \geqslant 2$ and $r \leqslant \beta n$. Suppose, moreover, that $r^2 - 2\beta nr + \beta nd > 0$. Then*

$$A_q(n, d) \leqslant \frac{\beta nd}{r^2 - 2\beta nr + \beta nd} \cdot \frac{q^n}{\mathsf{vol}_q(n, r)}.$$

Proof: It can be found in many reference books, as for example [25] (5.2.11), or [20], Th. 4.5.24, and we omit it.   □

1.69 Remark. The methods used by Elias and Bassalygo are a refinement of those used in the proof of the Plotkin bound, but are a little more involved, and the result improves some of the previous bounds only for large values on n.

For example, for all values of n and d included in Table 1.1, the Elias bound is strictly greater than the corresponding value in the table, and the difference is often quite large. Thus the Elias bound for $A_2(13, 5)$ is 162, when the true value of this function is 64 (this will be establshed with the linear programming method later in this section).

The first value of n for which the Elias bound is better than the Hamming, Plotkin and Johnson bounds (the Johnson bound is studied below) is $n = 39$, and in this case only for $d = 15$. For $n = 50$ ($n = 100$), the Elias bound is better than the same three bounds for d in the range $15..20$ ($21..46$). For the precise asymptotic behaviour of this bound, see E.1.58.

Johnson's bound

Let us just mention another approach, for binary codes, that takes into account the distribution of distances. This is done by means of the auxiliary

```
┌─┬──────────────────────────────────────────────────────────────┬─┐
│▲│Library │ Elias upper bound                                    │▲│
├─┴──────────────────────────────────────────────────────────────┴─┤
│ ub_elias(n,d,q)                                                   │
│ := begin                                                          │
│      local f, R, M, A, β=1−1/q                                    │
│      f(r):= r²−2·β·n·r+β·n·d                                      │
│      R=0 .. ⌊β·n⌋                                                 │
│      M=qⁿ                                                         │
│      for r in R do                                                │
│        if f(r)>0  then                                            │
│                                                                   │
│          A=⌊ β·n·d    qⁿ            ⌋                             │
│            ⌊ ─────  · ───────────── ⌋                             │
│            ⌊  f(r)    volume(n,r,q) ⌋                             │
│                                                                   │
│          if A<M  then                                             │
│              M=A                                                  │
│          end                                                      │
│        end                                                        │
│      end                                                          │
│      if q=2 & odd?(d)  then                                       │
│        M=min(M, ub_elias(n+1,d+1))                                │
│      end                                                          │
│      M                                                            │
│    end                                                            │
│  Binary case:                                                     │
│  ub_elias(n,d):=ub_elias(n,d,2)                                   │
└───────────────────────────────────────────────────────────────────┘
  ub_elias(13,5)  →  162
  ub_elias(14,6)  →  162
```

function $A(n, d, w)$ that is defined as the greatest cardinal M of a binary code (n, M) with the condition that its minimum distance is $\geqslant d$ and *all* its vectors have weight w (we will say that such a code has type (n, M, d, w)).

1.70 Remark. Since the Hamming distance between two words of the same weight w is even, and not higher than $2w$, we have $A(n, 2h - 1, w) = A(n, 2h, w)$ and, if $w < h$, then $A(n, 2h, w) = 1$. So we may restrict the study of $A(n, d, w)$ to even d, say $d = 2h$, and $w \geqslant h$. Now we have $A(n, 2h, w) = A(n, 2h, n - w)$ (taking complements of the words of an (n, M, d, w) code gives an $(n, M, d, n - w)$ code and viceversa).

E.1.52. Show that $A(n, 2h, h) = \lfloor n/h \rfloor$.

1.71 Proposition (Johnson, 1962). *The function $A(n, d, w)$ satisfies the relation*

$$A(n, 2h, w) \leqslant \left\lfloor \frac{n}{w} A(n - 1, 2h, w - 1) \right\rfloor.$$

Hence, by induction,

$$A(n, 2h, w) \leqslant \left\lfloor \frac{n}{w} \left\lfloor \frac{n-1}{w-1} \left\lfloor \cdots \left\lfloor \frac{n-w+h}{h} \right\rfloor \cdots \right\rfloor \right\rfloor \right\rfloor.$$

Proof: Let C be a code $(n, M, 2h, w)$ with $M = A(n, 2h, w)$. Let $C' = \{x' \in \mathbb{Z}_2^{n-1} \mid (1|x') \in C\}$ and set $M' = |C'|$. Then C' is an $(n-1, M', 2h, w-1)$ code, and hence $M' \leqslant A(n-1, 2h, w-1)$. Arguing similarly with all other components, we see that there are at most $nA(n-1, 2h, w-1)$ ones in all words of C, and thus $wM \leqslant nA(n-1, 2h, w-1)$, which is equivalent to the claimed bound. $\square$

1.72 Theorem (Johnson, 1962). *For $d = 2t + 1$,*

$$A_2(n, d) \leqslant \cfrac{2^n}{\displaystyle\sum_{i=0}^{t} \binom{n}{i} + \cfrac{\dbinom{n}{t+1} - \dbinom{d}{t} A(n, d, d)}{\left\lfloor \dfrac{n}{t+1} \right\rfloor}}.$$

Note that this would be the binary sphere bound if the fraction in the denominator were omitted.

Proof: See, for example, [25] (5.2.15) or [20], Th. 4.5.14). $\square$

The bounds in Proposition 1.71 and Theorem 1.72 are computed in the listing **Johnson upper bound** with the function ub_johnson, the first bound with the arguments (n,d,w) and the second with the arguments (n,d). In the Johnson bound the generally unknown term $A(n, d, d)$ is replaced by its upper bound given in Proposition 1.71 (notice that this replacement does not decrease the right hand side of the Johnson bound). The case where d is even, say $d = 2t + 2$, can be reduced to the odd d case with the equality $A_2(n, 2t + 2) = A_2(n - 1, 2t + 1)$.

In the examples included in **Johnson upper bound** we can see that only a few values on the Table 1.1 are given correctly by the Johnson bound. But in fact the overall deviations are not very large and for a fixed d they become relatively small when n increases.

Linear programming bound

The best bounds have been obtained taking into account, for an (n, M, d) code C, the sequence $A_0, \ldots, A_n$ of non-negative integers defined as

$$A_i = |\{(x, y) \in C \times C \mid d(x, y) = i\}|.$$

```
┌─┬──────────┬────────────────────────────────────────────────┬─┐
│▲│ Library │ Johnson upper bound                            │▲│
└─┴──────────┴────────────────────────────────────────────────┴─┘
```

```
ub_johnson(n,d,w)
:= begin
     local h=⌊(d+1)/2⌋, b=1
     if h≤w  then
        for i in w−h .. 0 .. −1 do
```

$$b = \left\lfloor \frac{b \cdot (n-i)}{w-i} \right\rfloor$$

```
        end
     else
        1
     end
   end
ub_johnson(n,d)
:= begin
     local t, x
     if even?(d)  then
        n=n−1; d=d−1
     end
     t=⌊d/2⌋
```

$$x = \sum_{i=0}^{t} \binom{n}{i} + \frac{\left[\binom{n}{t+1} - \binom{d}{t}\right] \cdot \text{ub\_johnson}(n,d,d)}{\left\lfloor \dfrac{n}{t+1} \right\rfloor}$$

```
     ⌊2ⁿ/x⌋
   end
```

```
ub_johnson(13,5,5)  →  23
ub_johnson(13,5)  →  77
{ub_johnson(n,3) with n in 5..16}
   →  {4, 8, 16, 25, 51, 83, 167, 292, 585, 1024, 2048, 3615}
{ub_johnson(n,5) with n in 5..16}
   →  {2, 2, 3, 5, 10, 14, 28, 47, 77, 153, 256, 428}
{ub_johnson(n,7) with n in 7..16}  →  {2, 2, 2, 3, 7, 12, 18, 26, 51, 86}
```

We will set $a_i = A_i/M$ and we will say that $a_0, \ldots, a_n$ is the *distance distribution*, or *inner distribution*, of C. We have:

$$A_0 = M, \quad a_0 = 1; \tag{1.3}$$

$$A_i = a_i = 0 \text{ for } i = 1, \ldots, d-1; \tag{1.4}$$

$$A_0 + A_d + \cdots + A_n = M^2, \quad a_0 + a_d + \cdots + a_n = M. \tag{1.5}$$

1.73 Remark. For some authors the distance distribution of a code is the sequence $A_0, \ldots, A_n$.

E.1.53. Given $x \in C$, let $A_i(x)$ denote the number of words $x' \in C$ such that $hd(x, x') = i$. Check that $A_i = \sum_{x \in C} A_i(x)$. Show that $A_i(x) \leqslant A(n, d, i)$ and that $a_i \leqslant A(n, d, i)$.

E.1.54. Show that if C is linear then the distance distribution of C coincides with the weight enumerator.

The Krawtchouk polynomials

If C is a linear code with weight enumerator $a_0, \ldots, a_n$, and $b_0, \ldots, b_n$ is the weight enumerator of $C^\perp$, then we know (Theorem 7) that

$$\sum_{j=0}^{n} b_j z^j = \sum_{i=0}^{n} a_i (1-z)^i (1+(q-1)z)^{n-i}.$$

The coefficient of z^j in $(1-z)^i(1+(q-1)z)^{n-i}$ is

$$K_j(i) = \sum_{s=0}^{j} (-1)^s \binom{i}{s} \binom{n-i}{j-s} (q-1)^{j-s}$$

and hence

$$b_j = \sum_{i=0}^{n} a_i K_j(i) \qquad\qquad [1.6]$$

for $j = 0, \ldots, n$. Note that

$$K_j(x) = \sum_{s=0}^{j} (-1)^s \binom{x}{s} \binom{n-x}{j-s} (q-1)^{j-s} \qquad\qquad [1.7]$$

is a polynomial in x of degree j, which will be denoted $K_j(x, n, q)$ if we need to make the dependence on n and q explicit.

Since $K_j(0) = \binom{n}{j}$ and $a_i = 0$ for $i = 1, \ldots, d-1$, the formula [1.6] can be rewritten as

$$b_j = \binom{n}{j} + \sum_{i=d}^{n} a_i K_j(i). \qquad\qquad [1.8]$$

▲| Library | Krawtchouk polynomials |▲

The auxiliary Newton binomial function

$$N(x,j) := \frac{1}{j!} \prod_{i=1}^{j} (x-i+1)$$

The Krawtchouk polynomials

$$\text{krawtchouk}(x,k,n,q) := \sum_{j=0}^{k} (-1)^j \cdot N(x,j) \cdot N(n-x, k-j) \cdot (q-1)^{k-j}$$

krawtchouk$(x,k,n) := $ krawtchouk$(x,k,n,2)$

The next proposition yields another interpretation of the expressions $K_j(i)$.

1.74 Proposition (Delsarte's lemma). *Let $\omega \in \mathbb{C}$ be a primitive q-th root of unity and $T = \mathbb{Z}_q$. For $i = 0, \ldots, q$, let W_i be the subset of T^q whose elements are the vectors of weight i. Then, for any given $x \in W_i$,*

$$\sum_{y \in W_j} \omega^{\langle x | y \rangle} = K_j(i).$$

Proof: Passing to an equivalent code if necessary, we may assume that $x_l = 0$ for $l = i + 1, \ldots, n$. Let $\nu = \{\nu_1, \ldots, \nu_j\}$ be the support of y, with $\nu_1 < \cdots < \nu_j$, and let s in $0..j$ be the greatest subindex such that $\nu_s \leqslant i$ (with the convention $s = 0$ if $i < \nu_1$). If we restrict the sum in the statement to the set $W_j(\nu)$ formed with the vectors of weight j whose support is ν, we have (with $T' = T - \{0\}$)

$$\sum_{y \in W_j(\nu)} \omega^{\langle x | y \rangle} = \sum_{y_{\nu_1} \in T'} \cdots \sum_{y_{\nu_j} \in T'} \omega^{x_{\nu_1} y_{\nu_1} + \cdots + x_{\nu_j} y_{\nu_j}}$$

$$= (q-1)^{j-s} \sum_{y_{\nu_1} \in T'} \cdots \sum_{y_{\nu_s} \in T'} \omega^{x_{\nu_1} y_{\nu_1} + \cdots + x_{\nu_s} y_{\nu_s}}$$

$$= (q-1)^{j-s} \prod_{\ell=1}^{s} \sum_{y \in T'} \omega^{x_{\nu_\ell} y}$$

$$= (-1)^s (q-1)^{j-s}.$$

For the last equality note that the map $\chi_a : T \to \mathbb{C}$ such that $y \mapsto \omega^{ay}$, where $a \in T$ is fixed, is a character of the additive group of T, and hence, for $a \neq 0$, $\sum_{y \in T} \chi_a(y) = 0$ and $\sum_{y \in T'} \chi_a(y) = -\chi_a(0) = -1$ (cf. the beginning of the proof of Theorem 7). Now the result follows, because there are $\binom{i}{s}\binom{n-i}{j-s}$ ways to choose ν. □

Delsarte's theorem

With the notations above, notice that if $a_0, \ldots, a_n$ is the weight distribution of a linear code, then $\sum_{i=0}^{n} a_i K_j(i) \geqslant 0$ (since $b_j \geqslant 0$). The surprising fact is that if C is an unrestricted code and $a_0, \ldots, a_n$ is its inner distribution, then the inequality is still true:

1.75 Theorem (Delsarte, 1973). *$A_q(n,d)$ is bounded above by the maximum of*

$$1 + a_d + \cdots + a_n$$

subjet to the constraints

$$a_i \geqslant 0 \quad (d \leqslant i \leqslant n)$$

and

$$\binom{n}{j} + \sum_{i=d}^{n} a_i K_j(i) \geqslant 0 \quad for \quad j \in \{0, 1, \dots, n\},$$

where $K_j = K_j(x, n, q)$, $j = 0, \dots, n$, are the Krawtchouk polynomials.

Proof: We only have to prove the inequalities in the last display. Let $D_i \subseteq C \times C$ be the set of pairs (x, y) such that $hd(x, y) = i$. From the definition of the a_i, and using Delsarte's lemma, we see that

$$M \sum_{i=0}^{n} a_i K_j(i) = \sum_{i=0}^{n} \sum_{(x,y) \in D_i} \sum_{z \in W_j} \omega^{\langle x-y | z \rangle}$$

$$= \sum_{z \in W_j} \left| \sum_{x \in C} \omega^{\langle x | z \rangle} \right|^2,$$

which is clearly non-negative.   □

1.76 Remark. In the case of binary codes and d even, we may assume that $a_i = 0$ for odd i. Indeed, from a code (n, M, d) with d even we can get a code $(n - 1, M, d - 1)$ (as in E.1.12) and the parity extension of the latter is a code (n, M, d) which does not contain vectors of odd weight. The listing **Linear programming bound** contains an implementation of the linear programming bound for the binary case, basically following [11], Ch. 17, §4. Notice that in the case the constrained maximum is odd, then the function LP goes through a second linear programming computation that decreases the linear terms $\binom{n}{j}$ of the relations. For this clever optimization (due to Best, Brouwer, MacWilliams, Odlyzko and Sloane, 1977), see [11], p. 541.

1.77 Remark. The expression a\$\$i in the listing **Linear programming upper bound** constructs an identifier by appending to the identifier a the integer i regarded as a character string. For example, a\$\$3 is the identifier a3.

1.78 *Examples.* So we see that $A_2(13, 5) = A_2(14, 6) \leqslant 64$. Since there is a code Y such that $|Y| = 64$, as we will see in Chapter 3, it follows that $A_2(13, 5) = A_2(14, 6) = 64$.

The third and fourth expressions in the listing **Linear programming bound** show that we can pass a relation list as a third argument to LP. Such lists are meant to include any further knowledge we may gain on the numbers a_i, for particular values of n and d, beyond the generic constraints introduced in Theorem 1.75.

In the case $A_2(12, 5) = A_2(13, 6)$, it can be easily seen that $a_{12} \leqslant 1$, $a_{10} \leqslant 4$, with $a_{10} = 0$ when $a_{12} = 1$, and so we have an additional relation $a_{12} + 4a_{10} \leqslant 4$. With this we find that the linear programming bound gives

```
▲ Library │ Linear programming upper bound                                    ▲
  LP(n,d) := LP(n,d,{})
  LP(n,d,R'):=
  begin
     local K, A, α, a, R, m, b, β
     if odd?(d) then n=n+1; d=d+1 end
     K={krawtchouk(x,k,n) with k in 1.. n/2}
     A=[[evaluate(k,0)] │ [evaluate(k,j)with j in d..n..2]with k in K]
     α=[1] │ [a$$i with i in d..n..2]
     R={⟨α,a⟩⩾0 with a in A}
     m=constrained_maximum(sum(α),R│R')
     b=⌊m₁⌋
     if odd?(b) then
        β=[1−b⁻¹] │ tail(α)
        R={⟨β,a⟩⩾0 with a in A}
        m=constrained_maximum(sum(α),R│R')
     end
     {⌊m₁⌋, m₂}
  end
```

```
  LP(13,5)  →  {64, {a10⇒14,a12⇒0,a14⇒0,a6⇒42,a8⇒7}}
  LP(14,6)  →  {64, {a10⇒14,a12⇒0,a14⇒0,a6⇒42,a8⇒7}}
  LP(13,6,{a10+4·a12⩽4})  →  {32, {a10⇒4,a12⇒0,a6⇒24,a8⇒3}}
  LP(9,4,{a8⩽1, a6+4·a8⩽12})  →  {20, {a4⇒ 95/7 ,a6⇒ 106/21 ,a8⇒1}}
```

$A_2(12,5) \leqslant 32$. Since we will construct a code $(12,32,5)$ in Chapter 3 (the Nadler code), it turns out that $A_2(12,5) = 32$.

Finally we find that $A_2(8,3) = A_2(9,4) \leqslant 20$ using the extra relations $a_8 \leqslant 1$ and $a_6 + 4a_8 \leqslant 12$. Here $a_8 \leqslant A(9,4,8) = A(9,4,1) = 1$ (see Remark 1.70 and E.1.53), $a_6 \leqslant A(9,4,6) = A(9,4,3) \leqslant 3A(8,4,2) \leqslant 12$, and if $a_8 = 1$ then it is easy to see that $a_6 \leqslant A(8,3,5)$, so that $a_6 \leqslant A(8,3,5) = A(8,4,5) = A(8,4,3) \leqslant \lfloor \frac{8}{3}A(7,4,2) \rfloor \leqslant \lfloor \frac{8}{3} \lfloor \frac{7}{2} \rfloor \rfloor = 8$.

Since we know that there are codes $(8,20,3)$ (Example 1.4, or E.1.46), we conclude that $A_2(8,3) = 20$.

The linear programming bound also gives, without any additional relation, the values 128 and 256 (for $d = 5$) and 32 (for $d = 7$) on Table 1.1.

Asymptotic bounds

Let us consider an example first. By the Singleton bound we have $k \leqslant n - d + 1$. On dividing by n, we get

$$R = \frac{k}{n} \leqslant 1 - \frac{d}{n} + \frac{1}{n}.$$

If we now keep $\delta = \dfrac{d}{n}$ fixed and let n go to infinity, then we see that $\alpha(\delta) \leqslant 1 - \delta$, where $\alpha(\delta) = \limsup R$ is the *asymptotic rate* corresponding to δ. Thus we see that the Singleton bound gives rise to the bound $\alpha(\delta) \leqslant 1 - \delta$ of the asymptotic rate, and this inequality is called the *asymptotic Singleton bound*.

In general we may define the *asymptotic rate* $\alpha(\delta)$, $0 \leqslant \delta \leqslant 1$, as $\limsup_{n \to \infty} \log_q A_q(n, \lfloor \delta n \rfloor)/n$ and then there will be a bound of $\alpha(\delta)$, given by some function $\alpha'(\delta)$, corresponding to each of the known bounds of $A_q(n, d)$. These functions $\alpha'(\delta)$ are called *asymptotic bounds*. For example, $\alpha'(\delta) = 1 - \delta$ in the case of the Singleton bound.

Below we explain the more basic asymptotic bounds (for the corresponding graphs, see figure 1.2). We also provide details of how to compute them.

We will need the q-ary *entropy function* $H_q(x)$, which is defined as follows ($0 \leqslant x \leqslant 1$):

$$H_q(x) = x \log_q(q - 1) - x \log_q(x) - (1 - x) \log_q(1 - x).$$

The importance of this function in this context stems from the fact that for $0 \leqslant \delta \leqslant \beta$, $\beta = (q - 1)/q$, we have

$$\lim_{n \to \infty} n^{-1} \log_q(\mathit{vol}_q(n, \lfloor \delta n \rfloor)) = H_q(\delta) \qquad [1.9]$$

(see, for example, [25] (5.1.6), or [20], Cor. A.3.11). It is immediate to check that $H_q(x)$ is strictly increasing and $H_q(x) \geqslant x/\beta$ for $0 \leqslant x \leqslant \beta$, with equality if and only if $x = 0$ or $x = \beta$.

E.1.55 (Gilbert asymptotic bound). Use the formula [1.9] and the Gilbert bound (Theorem 1.17) to prove that

$$\alpha(\delta) \geqslant 1 - H_q(\delta),$$

for $0 \leqslant \delta \leqslant \beta$.

1.79 Remark. If $0 \leqslant \delta \leqslant \beta$, the Gilbert asymptotic bound shows that there exists a sequence of codes $[n_j, k_j, d_j]$, $j = 1, 2, ...$, that has the following property: given $\varepsilon > 0$, there exists a positive integer j_ε such that

$$\frac{d_j}{n_j} \geqslant \delta, \qquad \frac{k_j}{n_j} \geqslant 1 - H_q(\delta) - \varepsilon$$

for all $j > j_\varepsilon$. Any family of codes with this property is said to *meet the Gilbert–Varshamov bound*. The family is said to be *asymptotically good* if it satisfies the following weaker property: there exist positive real numbers R and δ such that $k_j/n_j \geqslant R$ and $d_j/n_j \geqslant \delta$ for all j. A family which is not asymptotically good is said to be *asymptotically bad*, and in this case either k_j/n_j or d_j/n_j approaches 0 when $j \to \infty$.

Long ago it was conjectured that $\alpha(\delta)$ would coincide with $1 - H_q(\delta)$, but this turned out to be incorrect. Indeed, in 1982 Tsfasman, Vladut and Zink proved, using modern methods of algebraic geometry, that the bound can be improved for $q \geqslant 49$. For lower values of q, however, and in particular for binary codes, the Gilbert bound remains the best asymptotic lower bound.

Asymptotic upper bounds

E.1.56 (Hamming asymptotic bound). Use the formula [1.9] and the sphere upper bound (Theorem 1.14) to show that

$$\alpha(\delta) \leqslant 1 - H_q(\delta/2),$$

for $0 \leqslant \delta \leqslant 1$.

E.1.57 (Plotkin asymptotic bound). Show that the Plotkin bound (Proposition 1.63) implies that

$$\alpha(\delta) \leqslant 1 - \delta/\beta \quad \text{if } 0 \leqslant \delta \leqslant \beta$$
$$\alpha(\delta) = 0 \quad \text{if } \beta \leqslant \delta \leqslant 1.$$

So in what follows we may restrict ourselves to the interval $0 \leqslant \delta \leqslant \beta$.

E.1.58 (Elias asymptotic bound). Use the formula [1.9] and the Elias bound (Proposition 1.68) to prove that

$$\alpha(\delta) \leqslant 1 - H_q(\beta - \sqrt{\beta(\beta - \delta)}) \quad \text{if } 0 \leqslant \delta < \beta$$
$$\alpha(\delta) = 0 \quad \text{if } \beta \leqslant \delta \leqslant 1.$$

Note that the second part is known from the Plotkin asymptotic bound.

McEliece *et al.* asymptotic bound

This result was obtained in 1977 by McEliece, Rodemich, Rumsey and Welsh (MRRW) with a delicate analysis of the asymptotic behaviour of the linear programming bound for binary codes and it remains the best of the known asymptotic upper bounds for such codes (cf. [6], chapter 9). The result is the following:

$$\alpha(\delta) \leqslant \min_{0 \leqslant u \leqslant 1 - 2\delta} (1 + g(u^2) - g(u^2 + 2\delta u + 2\delta)),$$

where $g(x) = H_2((1 - (1 - x)^{1/2})/2)$.

Van Lint asymptotic bound

The authors of the previous bound also showed that for $0.273 < \delta \leqslant 0.5$ the minimum is attained at $u = 1 - 2\delta$, in which case the bound takes the following form:

$$\alpha(\delta) \leqslant H_2(1/2 - \delta^{1/2}(1 - \delta)^{1/2}).$$

This bound can be considered in the interval $0 \leqslant \delta \leqslant 0.273$ as well, but it is slightly above the MRRW bound in this range, and near 0 it even becomes worse than the Hamming bound. In any case, this bound was established directly by van Lint using fairly elementary methods of linear programming (see [25] (5.3.6)).

For a further discussion on "perspectives on asymptotics", the reader can consult Section 4, Lecture 9, of [22].

▲ Library │ The entropy function and the asymptotic bounds   ▲

```
entropy(x:IR,q:Z)
check 0⩽x & x⩽1 & q⩾2
:= if x=0 then 1 elif x=1 then log_q(q−1)
    else x·log_q(q−1)−x·log_q(x)−(1−x)·log_q(1−x) end
entropy(x) := entropy(x,2)
gilbert(x:IR) check 0⩽x & x⩽1 := if x⩽1/2 then 1−entropy(x) else 0 end
hamming(x:IR) check 0⩽x & x⩽1 := 1−entropy(x/2)
```

$$\text{elias}(x:\mathbb{R}) \text{ check } 0 \leqslant x \ \& \ x \leqslant \frac{1}{2} := 1 - \text{entropy}((1 - \sqrt{1 - 2 \cdot x})/2)$$

$$\text{vanlint}(x:\mathbb{R}) \text{ check } 0 \leqslant x \ \& \ x \leqslant 1 := \text{entropy}\left(\frac{1}{2} - \sqrt{x \cdot (1 - x)}\right)$$

```
mceliece(x) check 0⩽x & x⩽1
:= begin
    local g, ε=0.005, δ=0.04
```

$$g(t) := \text{entropy}\left(\frac{1 - \sqrt{1 - t}}{2}\right)$$

```
    infimum(u↦1+g(u²)−g(u²+2·x·u+2·x), ε .. (1−2·x−ε) .. δ·(1−2·x))
end
```

```
entropy(0.2) → 0.72193
mceliece(0.2) → 0.46137
```

The results ... somehow bound the possible values of parameters. Still the problem of actually finding out what is possible and what is not is very difficult

Tsfasman–Vladut, [24], p. 35

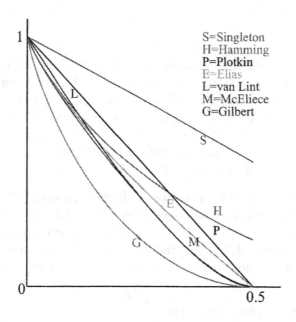

Figure 1.2: *Asymptotic bounds*

Summary

- Griesmer bound: if $[n, k, d]$ are the parameters of a linear code, then

$$\sum_{i=0}^{k-1} \left\lceil \frac{d}{q^i} \right\rceil \leqslant n.$$

This gives an upper bound for $B_q(n, d)$, the highest cardinal q^k for a linear code $[n, k, d]_q$.

- Plotkin bound: for any code C of type (n, M, d), $d \leqslant \dfrac{\beta n M}{M - 1}$, which is equivalent to $M(d - \beta n) \leqslant d$. As a consequence, if $d > \beta n$, then $A_q(n, d) \leqslant \left\lfloor \dfrac{d}{d - \beta n} \right\rfloor$. For binary codes this can be improved to $A_2(n, d) \leqslant 2 \lfloor d/(2d - n) \rfloor$.

- Computations and examples related to the Elias, Johnson and linear programming upper bounds.

- Computations related to the asymptotic bounds.

Problems

P.1.22. Let $C \subseteq \mathbb{F}_q^n$ be a linear code $[n, k, d]$ and assume k is maximal (for n and d fixed). Show that $|C| \geqslant q^n/vol_q(n, d - 1)$, which proves that the Gilbert lower bound can always be achieved with linear codes when q is a primer power. Show also that the Gilbert–Varshamov lower bound is never worse than the Gilbert bound. *Hint:* If it were $|C| < q^n/vol_q(n, d - 1)$, then there would exist $y \in \mathbb{F}_q^n$ such that $hd(y, x) \geqslant d$ for all $x \in C$, and the linear code $C + \langle y \rangle$ would have type $[n, k + 1, d]$.

P.1.23. Show that there does not exist a binary linear code $[14, 4, 5]_3$ ($3^4 = 81$ is the value of the Griesmer bound for $n = 14$ and $d = 5$). *Hint:* Forming the residue with respect to a word of weight 9, and using the Singleton bound, we would have a $[5, 3, 3]_3$ code, and this cannot exist as can be easily seen by studying a generating matrix of the form $I_3|P$, where P is a ternary 3×2 matrix.

P.1.24 (Johnson, 1962). Prove that

$$A(n, 2h, w) \leqslant \left\lfloor \frac{hn}{hn - w(n - w)} \right\rfloor,$$

provided $hn > w(n - w)$. *Hint:* First notice that if $x, y \in \mathbb{Z}_2^n$ satisfy $|x| = |y| = w$ and $hd(x, y) \geqslant 2h$ then $|x \cdot y| \leqslant w - h$; next, given a code $C \sim (n, M, 2h, w)$ with $M = A(n, 2h, w)$, bound above by $(w - h)M(M - 1)$ the sum S of all $|x \cdot x'|$ for $x, x' \in C$, $x \neq x'$; finally note that if m_i is the number of 1s in position i for all words of C, then $S = \sum_{i=1}^n m_i^2 - wM$ and $S \geqslant wM(wM/n - 1)$.

P.1.25 (The Best code, 1978). We know that $A_2(8, 3) = 20$, hence $A_2(9, 3) \leqslant 2A_2(8, 3) = 40$. In what follows we will construct a code $(10, 40, 4)$. Consider the matrix $G = M|M$, where $M = I_3|R^T$ and $R = \begin{pmatrix} 0 & 1 & 1 \\ 1 & 1 & 0 \end{pmatrix}$. Let $C' = \langle G \rangle \cup \{10000\,00100\}$. If we let $\sigma = (1, 2, 3, 4, 5)(6, 7, 8, 9, 10) \in S_{10}$, let $C_i = \sigma^i(C')$, $i = 0, 1, 2, 3, 4$, and set $C = \bigcup_{i=0}^4 C_i$. Show that C has type $(10, 40, 4)$.

P.1.26. Write a function that returns the λ-shortening C' of a code C with respect to a position. If C is given as a list, C' should be given as a list. If C is linear with control matrix H (generating matrix G), C' should be given as a list in the case $\lambda \neq 0$ and by a generating matrix if $\lambda = 0$.

P.1.27. Find the linear programming bounds of $A_2(n, 3)$ for $n = 9, 10, 11$ and compare the results with the corresponding values in Table 1.1.

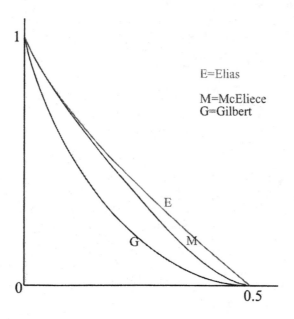

Figure 1.3: *Asymptotic bounds: MBBW and Elias*

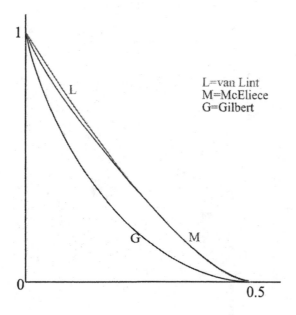

Figure 1.4: *Asymptotic bounds: MBBW and van Lint*

2 Finite Fields

> During the last two decades more and more abstract algebraic
> tools such as the theory of finite fields and the theory of poly-
> nomials over finite fields have influenced coding.
>
> R. Lidl and H. Niederreiter, [10], p. 305.

This chapter is devoted to the presentation of some of the basic ideas and results of the theory of *finite fields* that are used in the theory of error-correcting
codes. It is a self-contained exposition, up to a few elementary ideas on rings
and polynomials (for the convenience of the reader, the latter are summarized
below). On the computational side, we include a good deal of details and examples of how finite fields and related objects can be constructed and used.

Notations and conventions

A *ring* is an abelian group $(A, +)$ endowed with a product $A \times A \to A$,
$(x, y) \mapsto x \cdot y$, that is *associative, distributive* with respect to the sum on
both sides, and with a unit element 1_A which we always assume to be different from the zero element 0_A of $(A, +)$. If $x \cdot y = y \cdot x$ for all $x, y \in A$
we say that the ring is *commutative*. A *field* is a commutative ring for which
all nonzero elements have a multiplicative inverse. Examples: $\mathbb{Z}$ (the ring of
integers); $\mathbb{Q}, \mathbb{R}, \mathbb{C}$ (the fields of rational, real and complex numbers); if A is
a commutative ring, the polynomials in one variable with coefficients in A
form a commutative ring, denoted $A[X]$, with the usual addition and multiplication of polynomials, and the square matrices of order n with coefficients
in A form a ring, denoted $M_n(A)$, with the usual addition and multiplication
of matrices, and which is non-commutative for $n > 1$.

An additive subgroup B of a ring A is said to be a *subring* if it is closed
with respect to the product and $1_A \in B$. If A is a field, a subring B is said
to be a *subfield* if $b^{-1} \in B$ for any nonzero element b of B. For example,
$\mathbb{Z}$ is a subring of $\mathbb{Q}$; $\mathbb{Q}$ is a subfield of $\mathbb{R}$; if A is a commutative ring, A is

a subring of the polynomial ring $A[X]$; if B is a subring of a commutative ring A, then $B[X]$ is a subring of $A[X]$ and $M_n(B)$ is a subring of $M_n(A)$.

An additive subgroup I of a commutative ring A is said to be an *ideal* if $a \cdot x \in I$ for all $a \in A$ and $x \in I$. For example, if $a \in A$, then the set $(a) = \{b \cdot a \mid b \in A\}$ is an ideal of A, and it is called the *principal ideal* generated by a. In general there are ideals that are not principal, but there are rings (called *principal rings*) in which any ideal is principal, as in the case of the ring of integers $\mathbb{Z}$. If I and J are ideals of A, then $I \cap J$ is an ideal of A. The set $I + J$ of the sums $x + y$ formed with $x \in I$ and $y \in J$ is also an ideal of A. For example, if m and n are integers, then $(m) \cap (n) = (l)$, where $l = \mathrm{lcm}(m, n)$, and $(m) + (n) = (d)$, where $d = \gcd(m, n)$. A field K only has two ideals, $\{0\}$ and K, and conversely, any ring with this property is a field.

Let A and A' be rings and $f : A \to A'$ a group homomorphism. We say that f is a *ring homomorphism* if $f(x \cdot y) = f(x) \cdot f(y)$ for all $x, y \in A$ and $f(1_A) = 1_{A'}$. For example, if A is a commutative ring and I is an ideal of A, $I \neq A$, then the quotient group A/I has a unique ring structure such that the natural quotient map $A \to A/I$ is a homomorphism. The *kernel* $\ker(f)$ of a ring homomorphism $f : A \to A'$ is defined as $\{a \in A \mid f(a) = 0_{A'}\}$ and it is an ideal of A. One of the fundamental results on ring homomorphisms is that if I is an ideal of A and $I \subseteq \ker(f)$, then there is a unique homomorphism $\bar{f} : A/I \to A'$ such that $\bar{f}(\pi(a)) = f(a)$ for all $a \in A$, where $\pi : A \to A/I$ is the quotient homomorphism. The homomorphism $\bar{f}$ is injective if and only if $I = \ker(f)$ and in this case, therefore, $\bar{f}$ induces an isomorphism between $A/\ker(f)$ and $f(A)$, the image of A by f. Note that any homomorphism of a field K into a ring must be injective, for the kernel of this homomorphism is an ideal of K that does not conatain 1_K and hence it must be the ideal $\{0\}$.

If $f : A \to A'$ is a ring homomorphism, and B is a subring of both A and A', then we say that f is a B-homomorphism if $f(b) = b$ for all $b \in B$. We also say that A and A' are *extensions* of B, and write A/B and A'/B to denote this fact, and a B-homomorphism from A to A' is also called a homomorphism of the extension A/B to the extension A'/B. The homomorphisms $A/B \to A/B$ that are bijective are said to be *automorphisms* of the extension A/B. The set of automorphisms of an extension A/B is a group with the composition of maps.

If we copy the definition of K-vector space using a commutative ring A instead of a field K, we get the notion of A-*module*. For example, just as $K[X]$ is a K-vector space, $A[X]$ is and A-module. An A-module M is *free* if it has a *basis*, that is, if there exists a subset of M such that any element of M can be expressed in a unique way as a linear combination of elements of that subset with coefficients of A. For example, $A[X]$ is a free A-module, because $\{1, X, X^2, \dots\}$ is a basis.

2.1 $\mathbb{Z}_n$ and $\mathbb{F}_p$

> The real world is full of nasty numbers like 0.79134989..., while
> the world of computers and digital communication deals with
> nice numbers like 0 and 1.
>
> Conway–Sloane, [6], p. 56.

Essential points

- Construction of the ring $\mathbb{Z}_n$ and computations with its ring operations.

- $\mathbb{Z}_p$ is a field if and only if p is prime ($\mathbb{Z}_p$ is also denoted $\mathbb{F}_p$).

- The group $\mathbb{Z}_n^*$ of invertible elements of $\mathbb{Z}_n$ and Euler's function $\varphi(n)$, which gives the cardinal of $\mathbb{Z}_n^*$.

Let n be a positive integer. Let $\mathbb{Z}_n$ denote $\mathbb{Z}/(n)$, the ring of integers modulo n. The elements of this ring are usually represented by the integers a such that $0 \leqslant a \leqslant n - 1$. In this case the natural quotient map $\mathbb{Z} \to \mathbb{Z}_n$ is given by $x \mapsto [x]_n$, where $[x]_n$ is the remainder of the Euclidean division of x by n. This remainder is also denoted $x \bmod n$ and it is called the reduction of x modulo n. The equality $[x]_n = [y]_n$ is also written $x \equiv y \pmod{n}$ and it is equivalent to say that $y - x$ is divisible by n. The operations of sum and product in $\mathbb{Z}_n$ are the ordinary operations of sum and product of integers, but reduced modulo n.

For example, if $n = 11$ we have $3+7 \equiv 10, 3 \cdot 7 \equiv 10, 5^2 \equiv 3 \pmod{11}$. See **Examples of ring and field computations**. For the notations see A.6 and A.5 (p. 223). The function geometric_series(a,r) returns the vector $[1, a, \ldots, a^{r-1}]$.

The decision of whether an integer x does or does not have an inverse modulo n, and its computation in case such an inverse exists, can be done efficiently with a suitable variation of the Euclidean division algorithm. This algorithm is implemented in the **WIRIS** function bezout(m,n), which returns, given integers m and n, a vector $[d, a, b]$ of integers such that $d = \gcd(m,n)$ and $d = am + bn$. In case $d = 1$, a is the inverse of $m \pmod{n}$, and this integer is the value returned by inverse(m,n). The function inverse(m), where $m \in \mathbb{Z}_n$, returns the same value, but considered as an element of $\mathbb{Z}_n$, rather than as an integer. See **More examples on ring computations**. The function invertible?(a) returns true or false according to whether a is or is not

> Examples of ring and field computations
> A=$\mathbb{Z}_{35}$ $\rightarrow$ Z35
> 43 mod 35 $\rightarrow$ 8
> a=43:A $\rightarrow$ 8
> geometric_series(a,12) $\rightarrow$ [1, 8, 29, 22, 1, 8, 29, 22, 1, 8, 29, 22]
> a^3+5·a^7 $\rightarrow$ 27
> x=25:$\mathbb{Z}_{17}$ $\rightarrow$ 8
> y=$\dfrac{1}{x}$ $\rightarrow$ 15
> x·y $\rightarrow$ 1
> geometric_series(x,17) $\rightarrow$ [1, 8, 13, 2, 16, 9, 4, 15, 1, 8, 13, 2, 16, 9, 4, 15, 1]

an invertible element (in the ring to which it belongs). Note that if we ask for $1/a$ and a is not invertible, than the expression is returned unevaluated (as 1/(7:A) is the listing).

> More examples on ring computations
> A=$\mathbb{Z}_{35}$; a=8:A;
> invertible?(a) $\rightarrow$ true
> inverse(a) $\rightarrow$ 22
> 1/a $\rightarrow$ 22
> inverse(8,35) $\rightarrow$ 22
> invertible?(7:A) $\rightarrow$ false
> 1/(7:A)
> n=2345; m=1234;
> [d,a,b] = bezout(n,m) $\rightarrow$ [1, 311, −591]
> a·n+b·m $\rightarrow$ 1

The invertible elements of $\mathbb{Z}_n$ form a group, $\mathbb{Z}_n^*$. Since an $[x]_n \in \mathbb{Z}_n$ is invertible if and only if $gcd(x, n) = 1$, the cardinal of $\mathbb{Z}_n^*$ is given by Euler's function $\varphi(n)$, as this function counts, by definition, the number of integers x in the range $1..(n-1)$ such that $gcd(x, n) = 1$. In particular we see that $\mathbb{Z}_n$ is a field if and only if n is prime. For example, in $\mathbb{Z}_{10}$ the invertible elements are $\{1, 3, 7, 9\}$ (hence $\varphi(10) = 4$) and $3^{-1} \equiv 7 \pmod{10}$, $7^{-1} \equiv 3 \pmod{10}$, $9^{-1} \equiv 9 \pmod{10}$. On the other hand, in $\mathbb{Z}_{11}$ all non-zero elements are invertible (hence $\varphi(11) = 10$) and $\mathbb{Z}_{11}$ is a field.

If $p > 0$ is a prime integer, the field $\mathbb{Z}_p$ is also denoted $\mathbb{F}_p$.

2.1 Proposition. *The value of $\varphi(n)$ is determined by the following rules:*

1) *If $gcd(n_1, n_2) = 1$, then $\varphi(n_1 n_2) = \varphi(n_1)\varphi(n_2)$.*

2) *If p is a prime number and r a positive integer, then*

$$\varphi(p^r) = p^{r-1}(p - 1).$$

Proof: The map $\pi : \mathbb{Z} \to \mathbb{Z}_{n_1} \times \mathbb{Z}_{n_2}$ such that $k \mapsto ([k]_{n_1}, [k]_{n_2})$ is a ring homomorphism and its kernel is $(n_1) \cap (n_2) = (m)$, where $m = lcm(n_1, n_2)$. Since n_1 and n_2 are coprime, we have $m = n_1 n_2$ and hence there exists an injective homomorphism $\mathbb{Z}/(m) = \mathbb{Z}_{n_1 n_2} \to \mathbb{Z}_{n_1} \times \mathbb{Z}_{n_2}$. But this homomorphism must be an isomorphism, as the cardinals of $\mathbb{Z}_{n_1 n_2}$ and $\mathbb{Z}_{n_1} \times \mathbb{Z}_{n_2}$ are both equal to $n_1 n_2$.

For part 2, observe that the elements which are *not* invertible in $\mathbb{Z}_{p^r}$ are

$$0, p, 2p, ..., (p^{r-1} - 1)p,$$

hence $\varphi(p^r) = p^r - p^{r-1} = p^{r-1}(p - 1)$. $\qquad\qquad\square$

For example, $\varphi(10) = \varphi(2 \cdot 5) = \varphi(2)\varphi(5) = (2 - 1)(4 - 1) = 4$, in agreement with what we already checked above, and $\varphi(72) = \varphi(8 \cdot 9) = \varphi(8)\varphi(9) = (8 - 4)(9 - 3) = 24$. In WIRIS the function $\varphi(n)$ is given by phi_euler(n).

```
φ=phi_euler;
φ(105) → 48
φ(123456789) → 82260072
φ(1234567890123456789) → 814197031044480000
```

2.2 Remark. Since $\varphi(p) = p - 1$ for p prime, and since there are infinitely many primes, the quotient $\varphi(n)/n$ can take values as close to 1 as wanted. On the other hand, $\varphi(n)/n$ can be as close to 0 as wanted (see P.2.1). For a computational illustration, see **Small values of $\varphi(n)/n$.**

▲ Library	Small values of φ(n)/n	▲
The primes in 2..N		
P(N) := {n in 2..N such_that prime?(n)}		
m(N)=φ(p_N)/p_N, p_N the product of the primes in 2..N		
$m(N) := \displaystyle\prod_{p \text{ in } P(N)} \frac{p-1}{p}$:Float		

```
m(10) → 0.22857
m(100) → 0.12032
m(1000) → 0.080965
```

2.3 Proposition. *For each positive integer n, $\sum_{d|n}\varphi(d) = n$ (the sum is extended to all positive divisors of n).*

Proof: It is enough to notice that the denominators of the n fractions obtained by reducing $1/n, 2/n, \ldots, n/n$ to irreducible terms are precisely the divisors d of n and that the denominator d appears $\varphi(d)$ times. $\qquad\square$

```
φ=phi_euler;
n=9·5·7;
D=divisors(n)  →  {1, 3, 9, 5, 15, 45, 7, 21, 63, 35, 105, 315}

 ∑   φ(d)  →  315
d in D
```

E.2.1. Show that $\varphi(n)$ is even for all $n > 2$. In particular we have that $\varphi(n) = 1$ only happens for $n = 1, 2$.

E.2.2. Show that $\{n \in \mathbb{Z}^+ \mid \varphi(n) = 2\} = \{3, 4, 6\}$ and $\{n \in \mathbb{Z}^+ \mid \varphi(n) = 4\} = \{5, 8, 10, 12\}$.

E.2.3. The groups $\mathbb{Z}_5^*$, $\mathbb{Z}_8^*$, $\mathbb{Z}_{10}^*$ and $\mathbb{Z}_{12}^*$ have order 4. Determine which of these are cyclic and which are not (recall that a group of order four is either cyclic, in which case it is isomorphic to $\mathbb{Z}_4$, or isomorphic to $\mathbb{Z}_2^2$).

E.2.4. Given a finite group G with identity e, the *order* of an element a is the smallest positive integer r such that $a^r \in \{e = a^0, \cdots, a^{r-1}\}$. Show that actually $a^r = e$ and that r divides $|G|$. *Hint:* For the latter use that $H = \{e = a^0, \cdots, a^{r-1}\}$ is a subgroup of G and remember that the cardinal of any subgroup divides the cardinal of G (Lagrange theorem).

E.2.5. Fix a positive integer n and let a be any integer such that $\gcd(a, n) = 1$. Show that $a^{\varphi(n)} \equiv 1 \pmod{n}$.

E.2.6. Let F be a finite field and $q = |F|$. Show that $x^{q-1} = 1$ for all $x \in F^*$ and that $x^q - x = 0$ for all $x \in F$.

Summary

- Integer arithmetic modulo n (the ring $\mathbb{Z}/(n)$, which we denote $\mathbb{Z}_n$). The ring $\mathbb{Z}_n$ is a field if and only if n is prime and in this case is also denoted $\mathbb{F}_n$.

- The group $\mathbb{Z}_n^*$ of invertible elements of $\mathbb{Z}_n$. Its cardinal is given by Euler's function $\varphi(n)$, whose main properties are given in Propositions 2.1 and 2.3.

- Computation of the inverse of an element in $\mathbb{Z}_n^*$ by means of Bezout's theorem.

Problems

P.2.1. For $k = 1, 2, 3, ...$, let p_k denote the k-th prime number and let

$$N_k = (p_1 - 1)(p_2 - 1) \cdots (p_k - 1), \quad P_k = p_1 p_2 \cdots p_k.$$

Prove that the minimum of $\varphi(n)/n$ in the range $(P_k) \cdots (P_{k+1} - 1)$ is N_k/P_k. Deduce from this that in the range $1..(2 \times 10^{11})$ we have $\varphi(n) > 0.1579n$.

P.2.2. Prove that $\lim\inf_{n \to \infty} \varphi(n)/n = 0$.

P.2.3. Show that the values of n for which $\mathbb{Z}_n^*$ has 6 elements are $7, 9, 14, 18$. Since any abelian group of order 6 is isomorphic to $\mathbb{Z}_6$, the groups $\mathbb{Z}_7^*$, $\mathbb{Z}_9^*$, $\mathbb{Z}_{14}^*$, $\mathbb{Z}_{18}^*$ are isomorphic. Find an isomorphism between $\mathbb{Z}_7^*$ and $\mathbb{Z}_9^*$.

P.2.4. Show that the values of n for which $\mathbb{Z}_n^*$ has 8 elements are $15, 16, 20,$ $24, 30$. Now any abelian group of order 8 is isomorphic to $\mathbb{Z}_8$, $\mathbb{Z}_2 \times \mathbb{Z}_4$ or $\mathbb{Z}_2^3$. Show that $\mathbb{Z}_{24}^* \simeq \mathbb{Z}_2^3$ and that the other four groups are isomorphic to $\mathbb{Z}_2 \times \mathbb{Z}_4$.

2.2   Construction of finite fields

> The theory of finite fields is a branch of modern algebra that has
> come to the fore in the last 50 years because of its diverse ap-
> plications in combinatorics, coding theory, cryptology, and the
> mathematical theory of switching circuits, among others.
>
>   R. Lidl and H. Niederreiter, [10], p. ix.

Essential points

- Characteristic of a finite field.

- The cardinal of a finite field is a power of the characteristic.

- Frobenius automorphism (absolute and relative).

- Construction of a field of q^r elements if we know a field of q elements
  and an irreducible polynomial of degree r over it (basic construction).

- Splitting field of a polynomial with coefficients in a finite field and
  construction of a field of p^r elements for any prime number p and any
  positive integer r.

- Existence of irreducible polynomials over $\mathbb{Z}_p$ (or over any given finite
  field) of any degree. This guarantees the construction of a field of
  p^r elements that is more explicit and effective than using the splitting
  field method.

Ruffini's rule

Let us begin with two elementary propositions and some examples. Recall
that if K is a field, X an indeterminate and $f \in K[X]$ a polynomial of
degree $r > 0$, then f is said to be *irreducible* if it cannot be expressed as a
product $f = gh$ with $g, h \in K[X]$ and $deg\,(g), deg\,(h) < r$ (otherwise it is
said to be *reducible*).

2.4 Proposition (Ruffini's rule). *Let K be a field. Given a polynomial $f \in$
$K[X]$, an element $\alpha \in K$ is a root of f (that is, $f(\alpha) = 0$) if and only if f is
divisible by $X - \alpha$. In particular we see that if $f \in K[X]$ is irreducible and
$deg\,(f) > 1$, then f has no roots in K.*

Proof: For any $\alpha \in K$, the Euclidean division of $f \in K[X]$ by $X - \alpha$ tells us that there is $g \in K[X]$ and $c \in K$ such that $f = (X - \alpha)g + c$. Replacing X by α yields $c = f(\alpha)$. Thus we see that α is a root of f if and only if $c = 0$, that is, if and only if f is divisible by $(X - \alpha)$.

If f is irreducible and $deg\,(f) > 1$, then it cannot have a root in K, for otherwise it would have the form $(X - \alpha)g$ with $\alpha \in K$ and $g \in K[X]$ of positive degree, and this would contradict the irreducibility of f. $\square$

Thus the condition that $f \in K[X]$ (K a field, $deg\,(f) > 1$) has no root in K is necessary for f to be irreducible, but in general it is not sufficient. We will see, however, that it is sufficient when the degree of f is 2 or 3.

2.5 Proposition. *Let $f \in K[X]$ be a polynomial of degree 2 or 3. If f is reducible, then f has a root in K. Thus f is irreducible if (and only if) it has no root in K.*

Proof: A reducible polynomial of degree 2 is the product of two linear factors, and a reducible polynomial of degree 3 is the product of a linear and a quadratic factor. In both cases f has a linear factor. But if $aX + b \in K[X]$ is a linear factor of f (this means $a \neq 0$), then so is $X - \alpha$, where $\alpha = -b/a$, and hence α is a root of f. $\square$

2.6 Examples. If $a \in K$, the polynomial $X^2 - a$ is irreducible over K if and only if a is not the square of an element of K. For example, the squares in $\mathbb{Z}_7$ are $0, 1, 4, 2$. Therefore the polynomials $X^2 - 3 = X^2 + 4$, $X^2 - 5 = X^2 + 2$ and $X^2 - 6 = X^2 + 1$ are irreducible over $\mathbb{Z}_7$. Analogously, the polynomial $X^3 - a$ is irreducible over K if and only if a is not a cube in K. Since the cubes in $\mathbb{Z}_7$ are $0, 1, 6$, the polynomials $X^3 - 2 = X^3 + 5$, $X^3 - 3 = X^3 + 4$, $X^3 - 4 = X^3 + 3$ and $X^3 - 5 = X^3 + 2$ are irreducible over $\mathbb{Z}_7$.

E.2.7. Show that in a finite field of characteristic 2 every element is a square (recall that in a field of odd characteristic, as established in Proposition 1.49, half of the nonzero elements are squares and the other half are not). Show also that every polynomial of the form $X^2 + a$ is itself a square, hence reducible. *Hint:* If F is a field of characteristic 2, then the map $F \to F$ such that $x \mapsto x^2$ is a homomorphism.

E.2.8. Let K be a field of odd characteristic and $f = X^2 + aX + b \in K[X]$. Prove that f is irreducible if and only if the discriminant of f, namely $a^2 - 4b$, is not a square in K.

Characteristic of a finite field

We will use the *characteristic homomorphism* $\phi_A : \mathbb{Z} \to A$ of any ring A, that is, the map such that $\phi_A(n) = n \cdot 1_A$ for all $n \in \mathbb{Z}$. Note that ϕ_A is the only ring homomorphism of $\mathbb{Z}$ to A. Moreover, the image of ϕ_A is the smallest subring of A, in the sense that it is contained in any subring of A, and it is called the *prime ring* of A.

The kernel of ϕ_A is an ideal of $\mathbb{Z}$ and hence there is a uniquely determined non-negative integer n_A such that $ker\,(\phi_A) = (n_A)$. We say that n_A is the *characteristic* of A. For example, the characteristic of any domain is 0 (like $\mathbb{Z}$, $\mathbb{Q}$, $\mathbb{R}$ and $\mathbb{C}$), while $\mathbb{Z}_n$ has characteristic n for any integer $n > 1$.

```
characteristic(Z)    ⟶  0
characteristic(Z₇)   ⟶  7
characteristic(Z₈)   ⟶  8
cardinal(Q)          ⟶  +∞
cardinal(Z₁₇)        ⟶  17
```

In the case of rings of characteristic 0, ϕ_A is injective and hence an isomorphism of $\mathbb{Z}$ with the prime ring of A. Thus we may identify $\mathbb{Z}$ with the prime ring of A (the integer n is identified with $n \cdot 1_A \in A$). For example, if $A = M_r(\mathbb{Q})$ (the ring of rational matrices of order r), then A has characteristic 0 and $n \in \mathbb{Z}$ is identified with the diagonal matrix nI_r.

In the case when $n_A > 0$, then $n_A > 1$ (for 1_A is assumed to be different from 0_A) and by definition we have that the relation $n \cdot 1_A = 0_A$, $n \in \mathbb{Z}$, is equivalent to $n_A | n$. Note also that in this case we have a unique homomorphism (necessarily injective) $\mathbb{Z}_{n_A} \hookrightarrow A$ whose image is the prime ring of A, and so we can identify $\mathbb{Z}_{n_A}$ with the prime ring of A.

2.7 Proposition. *The characteristic n_A of a ring A has the following properties:*

1) *If A is finite, then $n_A > 0$.*

2) *If A is a domain, then either $n_A = 0$ or n_A is a prime number, and it is a prime number if A is finite.*

3) *The characteristic of a finite field F is a prime number p, and so in this case there is a unique homomorphism* (necessarily injective) $\mathbb{Z}_p \hookrightarrow F$ *whose image is the prime ring of F* (in this case a subfield).

Proof: If A is finite, then $\phi_A(\mathbb{Z})$ is finite. Since $\mathbb{Z}$ is infinite, ϕ_A cannot be injective. Hence $ker\,(\phi_A) \neq 0$ and therefore $n_A \neq 0$.

If A is a domain and $n = n_A > 0$, then $\mathbb{Z}_n$ is also a domain, because it is isomorphic to a subring of A. But $\mathbb{Z}_n$ is not a domain unless n is prime (and if n is prime then we know that $\mathbb{Z}_n$ is a field).

Since a finite field F is a finite domain, its characteristic $p = n_F$ is a prime number. The other statements are clear by what we have said so far. $\square$

2.8 Proposition. *If F is a finite field, then its cardinal has the form p^r, where p is a prime number and r is a positive integer. Moreover, p coincides with the characteristic of F.*

Proof: If p is the characteristic of F, then p is a prime number and F has a subfield isomorphic to $\mathbb{Z}_p$. Thus we can regard F as $\mathbb{Z}_p$-vector space. Since F is finite, the dimension of F over $\mathbb{Z}_p$, say r, is also finite and we have $F \simeq \mathbb{Z}_p^r$ as $\mathbb{Z}_p$-vector spaces. Hence $|F| = |\mathbb{Z}_p^r| = p^r$. $\square$

E.2.9. Let L be a finite field and K a subfield of L. If $|K| = q$, show that $|L|$ has the form q^r for some positive integer r. Deduce from this that the cardinal of any subfield F of L such that $K \subseteq F$ has the form q^s, with $s|r$. What does this say when $K = \mathbb{Z}_p$, p the characteristic of L?

E.2.10 (Absolute Frobenius automorphism). Show that in a field K of characteristic p the map

$$F : K \to K \quad \text{such that} \quad x \mapsto x^p$$

satisfies

$$(x + y)^p = x^p + y^p \quad \text{and} \quad (xy)^p = x^p y^p$$

for all $x, y \in K$ and

$$\lambda^p = \lambda \quad \text{for all} \quad \lambda \in \mathbb{Z}_p$$

(for this relation use E.2.6). If K is finite, then F is an automorphism of $K/\mathbb{Z}_p$ (it is called the *Frobenius automorphism* of K).

E.2.11 (Relative Frobenius automorphism). Let L be a field of characteristic p and $K \subseteq L$ a finite subfield. If $q = |K|$, show that the map

$$F_K : L \to L \quad \text{such that} \quad x \mapsto x^q$$

satisfies

$$(x + y)^q = x^q + y^q \quad \text{and} \quad (xy)^q = x^q y^q$$

for all $x, y \in L$ and

$$\lambda^q = \lambda \quad \text{for all} \quad \lambda \in K$$

(for this relation use E.2.6). If L is finite, then F_K is an automorphism of L/K (it is called the *Frobenius automorphism* of L relative to K). $\square$

If K has cardinal q, the function frobenius(K) yields the map $x \mapsto x^q$, that is, the Frobenius map relative to K. When applied to an element x, frobenius(x) is equivalent to the Frobenius map relative to the prime field of the field of x. For examples, see the listing **Frobenius automorphisms**.

Frobenius automorphisms
$K=\mathbb{Z}_2[x]/(x^2+x+1); L=K[y]/(y^2+x\cdot y+1);$
f is the Frobenius relative to the prime field
f=frobenius $\rightarrow$ frobenius
f_K is the Frobenius relative to K
f_K=frobenius(K) $\rightarrow$ x$\mapsto$x^4
f(x) $\rightarrow$ x+1
f(y) $\rightarrow$ x·y+1
f_K(x) $\rightarrow$ x
f_K(y) $\rightarrow$ y+x

The basic construction

We will use the general ideas from polynomial algebra summarized in the points 1-5 below. If K is any field and $f \in K[X]$ is any polynomial over K of degree $r > 0$, then the quotient ring $K' = K[X]/(f)$ has the following properties:

1) The natural homomorphism $K \rightarrow K'$ is injective and so *we may iden-tify K with a subfield of K'.*

2) If $x = [X]$ is the class of X modulo f, then

$$\{1, x, \ldots, x^{r-1}\}$$

   is a basis of K' considered as a vector space over K. This means that every element of K' can be uniquely expressed in the form $\alpha_0 + \alpha_1 x + \cdots + \alpha_{r-1}x^{r-1}$, $\alpha_0, \ldots, \alpha_{r-1} \in K$. So the elements of K' can be represented as polynomials in x of degree $< r$ with coefficients in K or, in other words, K' can be identified with the vector space $K[x]_r$ of polynomials in x of degree $< r$ with coefficients in K.

3) In terms of the representation in the previous paragraph, the *sum* of two elements of K' agrees with the sum of the corresponding polynomials, and its *product* can be obtained by reducing modulo f the ordinary product of polynomials. This reduction can be done as follows. If $f = a_0 + a_1 X + \cdots + a_{r-1}X^{r-1} + X^r$, then we have $x^r = -(a_0 + a_1 x + \cdots + a_{r-1}x^{r-1})$, and this relation can be used recursively to express x^j, for all $j \geqslant r$, as a linear combination of $1, x, \ldots, x^{r-1}$.

4) *The ring K' is a field if and only if f is irreducible over K.* In this case, the inverse of $\alpha = \alpha_0 + \alpha_1 x + \cdots + \alpha_{r-1}x^{r-1}$ can be obtained, as in the case of modular arithmetic, with a suitable variation of the Euclidean division algorithm applied to f and α. The field K' is usually denoted $K(x)$, because it coincides with the minimum field that contains K and x, but note that it coincides with $K[x]$, the ring of polynomial expressions in x with coefficients in K.

5) If K is finite and $q = |K|$, then $|K'| = q^r$, where r is the degree of f.

If p is a prime number and we apply 5 to $K = \mathbb{Z}_p$, we see that if we know an irreducible polynomial $f \in \mathbb{Z}_p[X]$, then we also have a finite field K' of cardinal p^r. For example, $f = X^3 - X + 1$ is an irreducible polynomial over $\mathbb{Z}_3$ (use Proposition 2.5) and so $K' = \mathbb{Z}_3[X]/(f)$ is a field and if x is the class of X mod f then the 27 elements of K' have the form $a + bx + cx^2$, $a, b, c \in \mathbb{Z}_3$. At the end of this section we include several other examples.

For the computational implementation of the basic construction, see A.7, and for sample computations related to the preceeding example, see the listing **The function extension(K,f)**. Note that extension(K,f) is equivalent to the expression K[x]/(f), where x is the variable of the polynomial f. The name of this object is still K[x], but the x here is now bound to the class $[x]$ modulo f of the initial variable x of f.

```
The function extension(K,f)
let K=Z₃;
F=extension(Z₃, x³+2 x+1)  ⟶  K[x]
characteristic(F)  ⟶  3
q=cardinal(F)  ⟶  27
geometric_series(2x+1,4)  ⟶  [1, 2·x+1, x²+x+1, 2·x+2]
L={xʲ→j with j in 0..(q−1)};
x¹¹, L(x²+x+2)  ⟶  x²+x+2,11
F=Z₃[t]/(t³+2·t+1)  ⟶  K[t]
cardinal(F)  ⟶  27
geometric_series(2t+1,4)  ⟶  [1, 2·t+1, t²+t+1, 2·t+2]
```

2.9 Remark. In the points 1-3 above, K can be replaced by any commutative ring and f by a monic polynomial (a polynomial f is *monic* if its the coefficient of its highest degree monomial is 1). Then K' is a ring, and a *free module* over K with basis $\{1, x, \ldots, x^{r-1}\}$. In other words, assuming that the ring K has already been constructed, that x is a free identifier (this can be achieved, if in doubt, with the sentence clear x) and that $f(x)$ is a monic polynomial over K, then K' can be constructed as shown in **The function extension(K,f) also works for rings**.

We end this subsection with a result that establishes a useful relation between the field $K' = K[X]/(f)$, $f \in K[X]$ irreducible, and any root of f.

2.10 Proposition. *Let L be a field and K a subfield. Let $f \in K[X]$ be a monic irreducible polynomial, and set $r = \deg(f)$, $K' = K[X]/(f)$. If $\beta \in L$ is a root of f, and x is the class of X modulo f, then there exists a*

```
The function extension(K,f) also works for rings
let A=ℤ₁₂ ⟶ A
B=extension(A, x³+x+5) ⟶ A[x]
b=x¹⁰+x³ ⟶ 10·x²+2·x+10
b+b⁴ ⟶ 6·x²+6·x+6
x⁻¹ ⟶ 7·x²+7
B=A[t]/(t³+t+5) ⟶ A[t]
1/t ⟶ 7·t²+7
```

unique K-homomorphism $K' \to K[\beta]$ *such that* $x \mapsto \beta$ *(K-homomorphism means that it is the identity on K). This homomorphism is necessarily an isomorphism. Hence* $K[\beta]$ *is a field and therefore* $K[\beta] = K(\beta)$.

Proof: Since the image of $\alpha_0 + \alpha_1 x + \cdots + \alpha_{r-1} x^{r-1} \in K'$ by a K-homomorphism such that $x \mapsto \beta$ is $\alpha_0 + \alpha_1 \beta + \cdots + \alpha_{r-1} \beta^{r-1}$, uniqueness is clear. For the existence, consider the map $K[X] \to L$ such that $h \mapsto h(\beta)$. It is a ring homomorphism with image $K[\beta]$ and its kernel contains the principal ideal (f), because $f(\beta) = 0$ by hypothesis. Hence there is an induced homomorphism $K' \to L$ such that

$$\alpha_0 + \alpha_1 x + \cdots + \alpha_{r-1} x^{r-1} \mapsto \alpha_0 + \alpha_1 \beta + \cdots + \alpha_{r-1} \beta^{r-1}.$$

Since K' is a field, this homomorphism is necessarily injective, hence an isomorphism between K' and the image $K[\beta]$.   □

E.2.12. Show that the kernel of the ring homomorphism $h \mapsto h(\beta)$ defined in the preceeding proposition coincides with the principal ideal (f).

2.11 *Example.* The quadratic polynomial $X^2 + X + 1$ has no roots in $\mathbb{Z}_2$, hence it is irreducible. The field $\mathbb{Z}_2(x) = \mathbb{Z}_2[X]/(X^2 + X + 1)$, where we write x for the class of X, has 4 elements: $0, 1, x, x + 1$. The multiplication table is easy to find using the basic relation $x^2 = x+1$: we have $x(x+1) = 1$ and $(x + 1)^2 = x$. Note that $\{x^0, x^1, x^2\} = \{1, x, x + 1\}$, so that the multiplicative group of $\mathbb{Z}_2(x)$ is cyclic.

```
Construction of F₄
clear(x);
f=x²+x+1 ⟶ x²+x+1
let F4= ℤ₂[x]/(f) ⟶ F4
geometric_series(x,3) ⟶ [1, x, x+1]
```

2.12 *Example.* The polynomial $T^3 + T + 1 \in \mathbb{Z}_2[T]$ is irreducible, because it has degree 3 and does not vanish for $T = 0$ or for $T = 1$. Hence $\mathbb{Z}_2(t) =$

$\mathbb{Z}_2[T]/(T^3 + T + 1)$ is a field of 8 elements, where we let t denote the class of T. The elements of $\mathbb{Z}_2(t)$ have the form $a + bt + ct^2$, with $a, b, c \in \mathbb{Z}_2$, and the multiplication table is determined by the basic relation $t^3 = t + 1$. Note that $t^4 = (t + 1)t = t^2 + t$. Since the cardinal of the multiplicative group of this field has 7 elements, it must be cyclic. In fact, as the following listing shows, it is a cyclic group generated by t.

```
Construction of F₈
clear(x);
f=x³+x+1  →  x³+x+1
let F8= Z₂[x]/(f)  →  F8
geometric_series(x,7)  →  [1, x, x², x+1, x²+x, x²+x+1, x²+1]
```

2.13 Example. Let $K = \mathbb{Z}_2(x)$ be the field introduced in Example 2.11. Consider the polynomial $f = Y^2 + xY + 1 \in K[Y]$. Since $f(a) \neq 0$ for all $a \in K$, f is irreducible over K (see 2.5). Therefore $K(y) = K[Y]/(f)$ is a field of $4^2 = 16$ elements. For later examples, this field will be denoted $\mathbb{Z}_2(x, y)$. Its elements have the form $a + by$, with $a, b \in K$, and its multiplication table is determined by the multiplication table of K and the basic relation $y^2 = xy + 1$. The listing below shows that the multiplicative group of this field is also cyclic: $y + 1$ is a generator, but note that y is not.

```
Construction of F₁₆ in two steps
f=x²+x+1  →  x²+x+1
let F4=Z₂[x]/(f)  →  F4
g=y²+x·y+1  →  y²+x·y+1
let F16=F4[y]/(g)  →  F16
geometric_series(y,15)
   →  [1, y, x·y+1, x·y+x, y+x, 1, y, x·y+1, x·y+x, y+x, 1, y, x·y+1, x·y+x, y+x]
geometric_series(y+1,10)
   →  [1, y+1, x·y, y+x, y+(x+1), x, x·y+x, (x+1)·y, x·y+(x+1), x·y+1]
```

2.14 Example. Since -1 is not a square in $\mathbb{Z}_3$, $\mathbb{Z}_3(x) = \mathbb{Z}_3[X]/(X^2 + 1)$ is a field of 9 elements, where x denotes the class of X. Its elements have the form $a + bx$, with $a, b \in \mathbb{Z}_3$ and the basic relation for the multiplication table is $x^2 = 2$. The listing below shows that the multiplicative group of this field is also cyclic: $x + 1$ is a generator, but note that x is not.

2.15 Example. The polynomial $f = X^3 + 5$ is irreducible over $\mathbb{Z}_7$. Therefore $\mathbb{Z}_7(x) = \mathbb{Z}_7[X]/(f)$ is a field of $7^3 = 343$ elements, where x denotes the class of X. The elements of $\mathbb{Z}_7(x)$ have the form $a + bx + cx^2$, with

⎡ Construction of F$_9$
⎢ f=x^2+1 → x^2+1
⎢ let F9=Z$_3$[x]/(f) → F9
⎢ geometric_series(x,8) → [1, x, 2, 2·x, 1, x, 2, 2·x]
⎣ geometric_series(x+1,8) → [1, x+1, 2·x, 2·x+1, 2, 2·x+2, x, x+2]

$a, b, c \in \mathbb{Z}_7$. The multiplication table of $\mathbb{Z}_7(x)$ is determined by the relations $x^3 = 2$, which is the form it takes the relation $f(x) = 0$, and $x^4 = 2x$.

⎡ Construction of F$_{343}$
⎢ let F343=Z$_7$[x]/(x^3+5) → F343
⎢ {element(j,F343) with j in 60..65}
⎣ → {x^2+x+4, x^2+x+5, x^2+x+6, x^2+2·x, x^2+2·x+1, x^2+2·x+2}

E.2.13. Find an irreducible polynomial of $\mathbb{Z}_3[T]$ of the form $T^3 + a$, and use it to construct a field $\mathbb{Z}_3(t)$ of 27 elements.

E.2.14. Show that the polynomial $S^4 + S + 1 \in \mathbb{Z}[S]$ is irreducible and use it to construct a field $\mathbb{Z}_2(s)$ of 16 elements. Later we will see that this field must be isomorphic to the field $\mathbb{Z}_2(x, y)$ of example 2.13. For a method to find isomorphisms between two finite fields with the same cardinal, see P.2.5.

Splitting fields

Given a prime number p and a positive integer r, the basic construction would yield a field of p^r elements if we knew an irreducible polynomial f of degree r over $\mathbb{Z}_p$. We have used this in the examples in the previous subsection for a few p and r, by choosing f explicitly, but at this point we do not know whether an f exists for every p and r.

In this subsection we will show that we can use the basic construction in another way (the *method of the splitting field*) that leads quickly to a proof of the existence of a field whose cardinal is p^r for any p and r. From this construction, which is difficult to use in explicit computations, we will deduce in the next subsection a formula for the number of monic irreducible polynomials of any degree over a field of q elements, and procedures for finding these polynomials. We will then be able to construct a field of p^r elements for any p and r which can be handled explicitly and effectively.

2.16 Theorem (Splitting field of a polynomial). *Given a field K and a monic* *polynomial $f \in K[X]$, there is a field extension L/K and elements*

$$\alpha_1, \ldots, \alpha_r \in L$$

such that

$$f = \prod_{j=1}^{r} (X - \alpha_j) \ and \ L = K(\alpha_1, \ldots, \alpha_r).$$

Proof: Let r be the degree of f. If $r = 1$, it is enough to take $L = K$. So assume $r > 1$ and that the theorem is true for polynomials of degree $r - 1$. If all irreducible factors of f have degree 1, then f has r roots in K and we can take $L = K$. Thus we can assume that f has at least one irreducible factor g of degree greater than 1. By the fundamental construction applied to g, there is an extension K'/K and an element $\alpha \in K'$ such that $K' = K(\alpha)$ and $g(\alpha) = 0$, hence also $f(\alpha) = 0$. Now the theorem follows by induction applied to the polynomial $f' = f/(X - \alpha) \in K'[X]$ (since α is a root of f, f is divisible by $X - \alpha$, by Proposition 2.4). $\qquad \square$

2.17 Theorem (Splitting field of $X^{q^r} - X$). *Let K be a finite field and $q = |K|$. Let L be the splitting field of $h = X^{q^r} - X \in K[X]$. Then $|L| = q^r$.*

Proof: By definition of splitting field, there are elements $\alpha_i \in L$, $i = 1, \ldots, q^r$, such that $h = \prod_{i=1}^{q^r} (X - \alpha_i)$ and $L = K(\alpha_1, \ldots, \alpha_{q^r})$. Now the elements α_i must be different, for otherwise h and its derivative h' would have a common root, which is impossible because h' is the constant -1. On the other hand the set $\{\alpha_1, \ldots, \alpha_{q^r}\}$ of roots of h in L form a subfield of L, because if α and β are roots, then $(\alpha - \beta)^{q^r} = \alpha^{q^r} - \beta^{q^r} = \alpha - \beta$ and $(\alpha\beta)^{q^r} = \alpha^{q^r} \beta^{q^r} = \alpha\beta$ (hence $\alpha - \beta$ and $\alpha\beta$ are also roots of h) and if α is a nonzero root then $(1/\alpha)^{q^r} = 1/\alpha^{q^r} = 1/\alpha$ (hence $1/\alpha$ is also a root of h). Since $\lambda^q = \lambda$ for all $\lambda \in K$, the elements of K are also roots of h. It follows that $L = K(\alpha_1, \ldots, \alpha_{q^r}) = \{\alpha_1, \ldots, \alpha_{q^r}\}$ and consequently $|L| = q^r$. $\quad \square$

2.18 Corollary (Existence of finite fields). *Let p be a primer number and r a nonnegative integer. Then there exists a finite field whose cardinal is p^r.*

Proof: By the theorem, the splitting field of $X^{p^r} - X$ over $\mathbb{Z}_p$ has cardinal p^r. $\qquad \square$

E.2.15. Given a finite field L such that $|L| = p^r$, the cardinal of any subfield is p^s, where $s|r$ (E.2.9). Prove that the converse is also true: given a divisor s of r, there exists a unique subfield of L whose cardinal is p^s.

Existence of irreducible polynomials over $\mathbb{Z}_p$

In this section we will establish a formula for the the number $I_q(n)$ of monic irreducible polynomials of degree n over a finite field K of q elements.

2.19 Proposition. *Let f be a monic irreducible polynomial of degree r over K. Then f divides $X^{q^n} - X$ in $K[X]$ if and only if $r|n$.*

Proof: If L is the splitting field of $X^{q^n} - X$ over K, then Theorem 2.17 tells us that $|L| = q^n$ and $X^{q^n} - X = \prod_{\alpha \in L}(X - \alpha)$. Let $f \in K[X]$ be monic and irreducible, set $K' = K[X]/(f)$ and write $x \in K'$ for the class of X mod f.

If we assume that f divides $X^{q^n} - X$, then there exists $\beta \in L$ such that $f(\beta) = 0$ and by Propostion 2.10 we have that $K[\beta]$ is isomorphic to $K' = K[X]/(f)$. Thus L has a subfield with cardinal q^r, which implies that $r|n$.

Assume, for the converse, that $r|n$. Then $X^{q^r} - X$ divides $X^{q^n} - X$ (because $q^r - 1$ divides $q^n - 1$ and hence $X^{q^r-1} - 1$ divides $X^{q^n-1} - 1$). To see that f divides $X^{q^n} - X$ it is thus enough to prove that f divides $X^{q^r} - X$. Let h be the remainder of the Euclidean division of $X^{q^r} - X$ by f. Since K' has q^r elements, x is a root of $X^{q^r} - X$ and hence also a root of h. This implies that h is zero, because otherwise $1, x, \ldots, x^{r-1}$ would not be linearly independent, and hence f divides $X^{q^r} - X$.   □

2.20 Corollary. *We have the following decomposition:*

$$X^{q^n} - X = \prod_{d|n} \prod_{f \in Irr_K(d)} f,$$

where $Irr_K(d)$ is the set of monic irreducible polynomials of degree d over K. Consequently,

$$q^n = \sum_{d|n} dI_q(d).$$

2.21 *Example.* Since $2^3 = 8$, the irreducible factors $X^8 - X$ over $\mathbb{Z}_2$ coincide with the irreducible polynomials of degrees 1 and 3 over $\mathbb{Z}_2$. Similarly, the irreducible factors $X^{16} - X$ over $\mathbb{Z}_2$ coincide with the irreducible polynomials of degrees 1, 2 and 4 over $\mathbb{Z}_2$. Now the factoring of $f \in K[X]$ into irreducible factors can be obtained with the **WIRIS** function factor(f,K). We can see the results of this example in the listing below.

> Irreducible polynomials of degree 1 and 3 over $\mathbb{Z}_2$
>
> k=$\mathbb{Z}_2$;
>
> factor(X^8−X, k) $\longrightarrow$ X·(X+1)·(X^3+X+1)·(X^3+X^2+1)
>
> Irreducible polynomials of degree 1, 2 and 4 over $\mathbb{Z}_2$
>
> factor(X^{16}−X,k)
>
> $\longrightarrow$ X·(X+1)·(X^2+X+1)·(X^4+X+1)·(X^4+X^3+1)·(X^4+X^3+X^2+X+1)

To obtain an expression for $I_q(n)$ we need the *Möbius μ function*, which is defined as follows: if d is a positive integer, then

$$\mu(d) = \begin{cases} 1 & \text{if } n = 1 \\ (-1)^k & \text{if } d = p_1 \cdots p_k, \text{ with the } p_i \text{ different primes} \\ 0 & \text{otherwise.} \end{cases}$$

The function μ is implemente in the WIRIS function mu_moebius.

```
Some values of μ
μ=mu_moebius;
μ(1) → 1
μ(101) → −1
μ(35) → 1
μ(101101101) → −1
factor(101101101) → 3·101·333667
```

E.2.16. Show that if $gcd\,(d, d') = 1$, then $\mu(dd') = \mu(d)\mu(d')$.

E.2.17. For all integers $n > 1$, prove that $\sum_{d|n} \mu(d) = 0$.

E.2.18 (Möbius inversion formula). Let A be an abelian additive group and $h : \mathbb{N} \to A$ and $h' : \mathbb{N} \to A$ maps such that

$$h'(n) = \sum_{d|n} h(d)$$

for all n. Prove that then

$$h(n) = \sum_{d|n} \mu(d)h'(n/d)$$

for all n, and conversely. There is a multiplicative version of Möbius inversion formula: Let A be an abelian multiplicative group and $h : \mathbb{N} \to A$ and $h' : \mathbb{N} \to A$ maps such that

$$h'(n) = \prod_{d|n} h(d)$$

for all n. Prove that then

$$h(n) = \prod_{d|n} h'(n/d)^{\mu(d)}$$

for all n, and conversely.

2.22 Example. Since for any positive integer n we have $n = \sum_{d|n} \varphi(d)$ (Proposition 2.3), the Möbius inversion formula tells us that we also have

$$\varphi(n) = \sum_{d|n} \mu(d)\frac{n}{d}.$$

```
n=9·5·7;
D=divisors(n);
phi_euler(n)  →  144
∑      mu_moebius(d)· n/d  →  144
d in D
```

If we apply the Möbius inversion formula to the relation $q^n = \sum_{d|n} dI_q(d)$ in Corollary 2.20, we get:

2.23 Proposition. *The number $I_q(n)$ of irreducible polynomials of degree n over a field K of q elements is given by the formula*

$$I_q(n) = \frac{1}{n}\sum_{d|n} \mu(d)q^{n/d}.$$

2.24 Corollary. *For all $n > 0$, $I_q(n) > 0$. In other words, if K is a finite field and $q = |K|$, then there are irreducible polynomials over K of degree n for all $n \geqslant 1$.*

Proof: Since for $d > 1$ we have $\mu(d) \geqslant -1$, it is easy to see that the formula in Proposition 2.23 implies that $nI_q(n) \geqslant q^n - (q^n - 1)/(q - 1) > 0$.   □

```
▲ Library | Computing I_q(n)                                              ▲

  μ=mu_moebius;
  Irr(q,n):= 1/n ·   ∑       μ(d)·q^{n/d}
                   d in divisors(n)
  Irr(n):=Irr(2,n)
```

```
Irr(q,35)  →  1/35·q^35 − 1/35·q^7 − 1/35·q^5 + 1/35·q
[Irr(n) with n in 1..12]  →  [2, 1, 2, 3, 6, 9, 18, 30, 56, 99, 186, 335]
```

If we apply the Möbius multiplicative inversion formula to the relation $X^{q^n} - X = \prod_{d|n}\prod_{f \in Irr_K(d)} f$ in Corollary 2.20, we get:

2.25 Proposition. *The product $P_K(n)$ of all monic irreducible polynomials of degree n over K is given by the formula*

$$P_K(f) = \prod_{d|n}(X^{q^{n/d}} - X)^{\mu(d)}.$$

▲ Library │ Product of the monic irreducible polynomials of degree n over F_q	▲

μ=mu_moebius

$$P(q,n,X):= \prod_{d \text{ in divisors}(n)} (X^{q^{n/d}}-X)^{\mu(d)}$$

$P(n,X):=P(2,n,X)$

$P4=P(4,X) \rightarrow X^{12}+X^9+X^6+X^3+1$

factor($P4,\mathbb{Z}_2$) $\rightarrow (X^4+X+1)\cdot(X^4+X^3+1)\cdot(X^4+X^3+X^2+X+1)$

$P(3,4,X) \rightarrow X^{72}+X^{64}+X^{56}+X^{48}+X^{40}+X^{32}+X^{24}+X^{16}+X^8+1$

2.26 Remark. Given a finite field K with $q = |K|$ and a positive integer r, to construct a field with q^r elements we can pick a monic irreducible polynomial $f \in K[X]$, which exists because $I_q(r) > 0$, and form $K[X]/(f)$. Usually this construction is more effective than the splitting field method introduced in Corollary 2.18.

To find irreducible polynomials, WIRIS has, in addition to the possibility of factoring $X^{q^n} - X$ and which has already been illustrated above, two main functions that are very useful for dealing with irreducible polynomials: the predicate irreducible?(f,K), that returns true or false according to whether $f \in K[x]$ is or is not irreducible over K, and irreducible_polynomial(K,r,X), that returns a monic polynomial of degree r in the variable X that is irreducible over K. The latter function generates monic polynomials of degree r in the variable X and coefficients in K and returns the first that is found to be irreducible. This works well because the density of monic irreducible polynomials of degree r among all monic polynomials of degree r is, by the formula in Proposition 2.23, of the order of $1/r$.

irreducible?($x^3+x^2+1,\mathbb{Z}_2$) $\rightarrow$ true

f=sum(geometric_series(x,7)) $\rightarrow x^6+x^5+x^4+x^3+x^2+x+1$

irreducible?($f,\mathbb{Z}_2$) $\rightarrow$ false

irreducible_polynomial($\mathbb{Z}_2,5,X$) $\rightarrow X^5+X^4+X^2+X+1$

irreducible_polynomial($\mathbb{Z}_2,6,X$) $\rightarrow X^6+X+1$

irreducible_polynomial($\mathbb{Z}_2,7,X$) $\rightarrow X^7+X+1$

irreducible_polynomial($\mathbb{Z}_2,8,X$) $\rightarrow X^8+X^7+X^3+X^2+1$

2.27 Remark. The function irreducible_polynomial(K,r,X) may give a different results (for the same K, r and X) for different calls, because it is based (unless r and K are rather small) on choosing a random polynomial and checking whether it is irreducible or not.

Summary

- The characteristic of a finite field K is the minimum positive number p such that $p1_K = 0_K$. It is a primer number and there is a unique isomorphism between $\mathbb{Z}_p$ and the minimum subfield of K (which is called the prime field of K). From this it follows $|K| = p^r$, where r is the dimension of K as a $\mathbb{Z}_p$ vector space.

- The (absolute) Frobenius automorphism of a finite field K is the map $K \to K$ such that $x \mapsto x^p$, where p is the characteristic of K. If L is a finite field and K is a subfield with $q = |K|$, then the Frobenius automorphism of L over K (or relative to K) is the map $x \mapsto x^q$.

- Basic construction: if K is a finite field of cardinal q and $f \in K[X]$ is a monic irreducible of degree r, then $K[X]/(f)$ is a field that contains K as a subfield and its cardinal is q^r.

- For any finite field K and any positive integer r, there exist monic irreducible polynomials $f \in K[X]$ of degree r, and hence there are fields of cardinal q^r ($q = |K|$) for all $r \geqslant 1$. In fact, we have derived an expression for the number $I_r(q)$ of monic irreducible polynomials of degree r over a field of q elements (Proposition 2.23). From this it follows that a positive integer can be the cardinal of a finite field if and only if it is a positive power of a prime number.

- Splitting field of a polynomial with coefficients in a finite field (Theorem 2.16).

- Using the splitting field of $X^{q^n} - X \in K[X]$, where K is a finite field of q elements, we have seen that $X^{q^n} - X \in K[X]$ is the product of all monic irreducible polynomials with coefficients in K whose degree is a divisor of n. We also have found a formula for the product $P_K(n)$ of all monic irreducible polynomials in $K[X]$ of n.

Problems

P.2.5 (Finding isomorphisms between finite fields). Let L be a finite field and $K \subseteq L$ a subfield. Set $q = |K|$ and let r be the positive integer such that $|L| = q^r$. Assume that we have a monic irreducible polynomial $f \in K[X]$ and that we can find $\beta \in L$ such that $f(\beta) = 0$. Show that there is a unique K-isomorphism $K[X]/(f) \simeq L$ such that $x \mapsto \beta$, where x is the class of X modulo f. Prove also that the K-isomorphisms between $K[X]/(f)$ and L are in one-to-one correspondence with the roots of f in L.

P.2.6. The polynomials $f = X^3 + X + 1, g = X^3 + X^2 + 1 \in \mathbb{Z}_2[X]$ are irreducible (as shown in Example 2.21 they are the only irreducible polynomials of degree 3 over $\mathbb{Z}_2$). Use the previous problem to find all the isomorphisms between $\mathbb{Z}_2[X]/(f)$ and $\mathbb{Z}_2[X]/(g)$.

P.2.7. Use P.2.5 to find all the isomorphisms between the field described in E.2.14 and the field $\mathbb{Z}_2(x, y)$ defined in Example 2.13.

P.2.8. A *partition* of a positive integer n is a set of pairs of positive integers $\lambda = \{(r_1, m_1), \ldots, (r_k, m_k)\}$ such that $r_1, \ldots, r_k$ are distinct and $n = m_1 r_1 + \ldots + m_k r_k$. The partition $\{(n, 1)\}$ is said to be *improper*, all others are said to be *proper*. Let now K be a finite field of q elements. Use the uniqueness (up to order) of the decomposition of a monic $f \in K[X]$ as a product of monic irreducible factors to prove that

$$P_q(\lambda) = \prod_{i=1}^{k} \binom{I_q(r_i) + m_i - 1}{m_i}$$

gives the number of monic polynomials in $K[X]$ that are the product of m_i irreducible monic polynomials of degree r_i, $i = 1, \ldots, k$. Deduce from this that $R_q(n) = \sum_\lambda P_q(\lambda)$, where the sum ranges over all proper partitions of n, is the number of polynomials in $K[X]$ that are reducible, monic and of degree n. Finally show that $R_q(n) + I_q(n) = q^n$ and use this to check the formula for $I_q(n)$ obtained in Proposition 2.23 for n in the range 1..5.

2.3   Structure of the multiplicative group of a finite field

In the examples of finite fields presented in the preceeding section we have seen that the multiplicative group was cyclic in all cases. In this section we will see that this is in fact the case for all finite fields. We will also consider consequences and applications of this fact.

Essential points

- Order of a nonzero element.
- Primitive elements (they exist for any finite field).
- Discrete logarithms.
- Period (or exponent) of a polynomial over a finite field.

Order of an element

If K is a finite field and α is a nonzero element of K, the *order* of α, $ord(\alpha)$, is the least positive integer r such that $\alpha^r = 1$ (in E.2.4 we saw that this r exists). Note that $ord(\alpha) = 1$ if and only if $\alpha = 1$.

2.28 *Example.* in $\mathbb{Z}_5$ we have $ord(2) = 4$, because $2 \neq 1$, $2^2 = 4 \neq 1$, $2^3 = 3 \neq 1$ and $2^4 = 1$. Similarly we have $ord(3) = 4$ and $ord(4) = 2$.

2.29 Proposition. *If the cardinal of K is q and $\alpha \in K^*$ has order r, then* $r|(q-1)$.

Proof: It is a special case of E.2.4.   □

2.30 Remark. This proposition implies that $\alpha^{q-1} = 1$ for all nonzero elements $\alpha \in K$, a fact that was already established in E.2.6. On the other hand, if $q - 1$ happens to be prime and $\alpha \neq 1$, then necessarily $ord(\alpha) = q - 1$. More generally, *r is the least divisor of $q - 1$ such that $\alpha^r = 1$.*

2.31 *Examples.* 1) In the field $\mathbb{Z}_2(x)$ of example 2.11 we have, since $q-1 = 3$ is prime, $ord(1) = 1$, $ord(x) = 3$ and $ord(x+1) = 3$. Note that in 2.11 we had already checked this from another perspective.

2) In the field $\mathbb{Z}_7(x)$ of the example 2.15, $q - 1 = 342 = 2 \cdot 3^2 \cdot 19$. Since $x^3 = 2$ and $2^3 = 1$, we have $x^9 = 1$ and hence $ord(x)|9$. Finally $ord(x) = 9$, as $x^3 \neq 1$.

3) In the field $\mathbb{Z}_2(x, y)$ of the example 2.13, which has cardinal 16, $ord(y) = 5$, because $ord(y)$ has to be a divisor of $16 - 1 = 15$ greater than 1 and it is easy to check that $y^3 = xy + x \neq 1$ and $y^5 = (xy+1)(xy+x) = 1$.

Computationally, $ord(\alpha)$ is returned by the function order(α). It implements the idea that the order r of α is the least divisor of $q - 1$ such that $\alpha^r = 1$. See the listing **Order of field elements** for examples.

```
Order of field elements
irreducible?(t⁵−t+1, Z₅) → true
F=Z₅[t]/(t⁵−t+1); q=cardinal(F) → 3125
r=order(t) → 1562
q−1
───  → 2
 r
{x,x'}=square_roots(t) → {2·t⁴+t³+2·t²+3·t+4, 3·t⁴+4·t³+3·t²+2·t+1}
order(x) → 3124
order(2·t) → 3124
```

E.2.19. In a finite field K every element α satisfies the relation $\alpha^q = \alpha$, where $q = |K|$ (E.2.6). Hence the elements of K are roots of the polynomial $X^q - X$. Show that

$$X^q - X = \prod_{a \in K} (X - a) \quad \text{and} \quad X^{q-1} - 1 = \prod_{a \in K^*} (X - a).$$

Note finally that these formulae imply that $\sum_{a \in K} a = 0$ if $K \neq \mathbb{Z}_2$ (as $X^q - X$ has no term of degree $q - 1$ if $q \neq 2$) and that $\prod_{a \in K^*} a = -1$ (See also Example 1.30, where this fact was proved in a different way).

2.32 Proposition. *Let K be a finite field, α a nonzero element of K and $r = ord(\alpha)$.*

1) *If x is a nonzero element such that $x^r = 1$, then $x = \alpha^k$ for some integer k.*

2) *For any integer k, $ord(\alpha^k) = r/\gcd(k, r)$.*

3) *The elements of order r of K have the form α^k, with $\gcd(k, r) = 1$. In particular we see that if there is an element of order r, then there are precisely $\varphi(r)$ elements of order r.*

Proof: Consider the polynomial $f = X^r - 1 \in K[X]$. Since f has degree r and K is a field, f has at most r roots in K. At the same time, since r is the order of α, all the r elements of the subgroup $R = \{1, \alpha, ..., \alpha^{r-1}\}$ are roots of f and so f has no roots other than the elements of R. Now by hypothesis x is a root of f, and hence $x \in R$. This proves 1.

To establish 2, let $d = gcd(r, k)$ and $s = r/d$. We want to show that α^k has order s. If $(\alpha^k)^m = 1$, then $\alpha^{mk} = 1$ and therefore $r|mk$. Dividing by d we get that $s|mk/d$. Since $(s, k/d) = 1$, we also get that $s|m$. Finally it is clear that $(\alpha^k)^s = \alpha^{ks} = (\alpha^r)^{k/d} = 1$, and this ends the proof.

Part 3 is a direct consequence of 1, 2 and the definition of $\varphi(r)$. $\qquad\square$

Primitive elements

A nonzero element of a finite field K of cardinal $q = p^r$ is said to be a *primitive element* of K if $ord(\alpha) = q - 1$. In this case it is clear that

$$K^* = \{1, \alpha, \alpha^2, ..., \alpha^{q-2}\}.$$

Of this representation of the elements of K we say that it is the *exponential representation* (or also the *polar representation*) relative to the primitive element α. With such a representation, the product of elements of K is computed very easily: $\alpha^i \alpha^j = \alpha^k$, where $k = i + j \mod q - 1$.

2.33 Examples. The elements 2 and 3 are the only two primitive elements of $\mathbb{Z}_5$ and x and $x + 1$ are the only two primitive elements of the field $\mathbb{Z}_2(x)$ introduced in Example 2.11. The element x of the field $\mathbb{Z}_7(x)$ introduced in Example 2.15 is not primitive, since we know that $ord(x) = 9$ (see Example 2.31). Finally, the element y of the field $\mathbb{Z}_2(x, y)$ considered in Example 2.13 is not primitive, since we saw that $ord(y) = 5$ (Example 2.31.3).

2.34 Theorem. *Let K be a finite field of cardinal q and d is a positive integer. If $d|(q - 1)$, then there are exactly $\varphi(d)$ elements of K that have order d. In particular, exactly $\varphi(q - 1)$ of its elements are primitive. In particular, any finite field has a primitive element.*

Proof: Let $p(d)$ be the number of elements of K that have order d. It is clear that

$$\sum_{d|(q-1)} p(d) = q - 1,$$

since the order of any nonzero element is a divisor of $q - 1$. Now observe that if $p(d) \neq 0$, then $p(d) = \varphi(d)$ (see Proposition 2.32.3). Since we also have $\sum_{d|(q-1)} \varphi(d) = q - 1$ (Proposition 2.3), it is clear that we must have $p(d) = \varphi(d)$ for any divisor d of $q - 1$. $\qquad\square$

The **WIRIS** function primitive_element(K) returns a primitive element of K. Here are some examples:

```
|| irreducible?(t⁷−t+1,Z₇) ⟶ true
|| let K=Z₇[t]/(t⁷−t+1); q=cardinal(K);
|| α = primitive_element(K) ⟶ 2·t
|| order(α), q−1 ⟶ 823542, 823542
|| order(t) ⟶ 274514
```

The algorithm on which the function primitive_element(K) is based is called *Gauss' algorithm* and it is explained and analyzed in P.2.10.

2.35 Proposition. *Let L be a finite field, K a subfield of L and $q = |K|$. Let r be the positive integer such that $|L| = q^r$. If α is a primitive element of L, then $1, \alpha, \ldots, \alpha^{r-1}$ is a basis of L as a K-vector space.*

Proof: If $1, \alpha, \ldots, \alpha^{r-1}$ are linearly dependent over K, there exist

$$a_0, a_1, \ldots, a_{r-1} \in K,$$

not all zero, such that

$$a_0 + a_1\alpha + \cdots + a_{r-1}\alpha^{r-1} = 0.$$

Letting $h = a_0 + a_1 X + \cdots + a_{r-1}X^{r-1} \in K[X]$, we have a polynomial of positive degree such that $h(\alpha) = 0$. From this it follows that there exists a monic irreducible polynomial $f \in K[X]$ of degree $< r$ such that $f(\alpha) = 0$. Using now Proposition 2.10, we get an isomorphism from $K' = K[X]/(f)$ onto $K[\alpha]$ such that $x \mapsto \alpha$. But this implies that $\text{ord}(\alpha) = \text{ord}(x) \leqslant q^{r-1} - 1$, which contradicts the fact that $\text{ord}(\alpha) = q^r - 1$. ☐

E.2.20. Let K be a finite field different from $\mathbb{F}_2$. Use the existence of primitive element to prove again that $\sum_{x \in K} x = 0$. ☐

Primitive polynomials

If f is an irreducible polynomial of degree r over $\mathbb{Z}_p$, p prime, then $\mathbb{Z}_p(x) = \mathbb{Z}_p[X]/(f)$ is a field of cardinal p^r, where x is the class of X modulo f. We know that x may be a primitive element, as in the case of the field $\mathbb{Z}_2(x)$ of example 2.11, or it may not, as in the case of the field $\mathbb{Z}_7(x)$ studied in the example 2.15 (cf. the examples 2.33). If x turns out to be a primitive element, we say that f is *primitive* over $\mathbb{Z}_p$. It is interesting, therefore, to explore how to detect whether or not a given polynomial is primitive. First see **Example of a primitive polynomial** (of degree 4 over $\mathbb{Z}_3$).

E.2.21. Let K be a finite field and $f \in K[X]$ be a monic irreducible polynomial, $f \neq X$. Let x be the class of X in $K[X]/(f)$. If $m = \deg(f)$, show that the order of x is the minimum divisor d of $q^m - 1$ such that $f | X^d - 1$. For

```
Example of a primitive polynomial
f=X⁴+X−1;
F81=extension(ℤ₃,x, f);
order(x) → 80
primitive_element(F81) → x
```

the computation of this order, which is also called the *period* (or *exponent*) of f, see **The function** period(f,q) .

▲ Library	The function period(f,q)	▲

```
period(f,q):=
   begin
      local r=degree(f), D=set(divisors(qʳ−1)), X=variable(f)
      for d in D do
         if remainder(Xᵈ−1, f)=0  then
            return d
         end
      end
   end
```

```
p=5;
f = xᵖ−x+1 : ℤₚ[x] → x⁵+4·x+1
pᵖ−1, period(f,p) → 3124, 1562
```

The discrete logarithm

Suppose L is a finite field and $\alpha \in L$ a primitive element of L. Let K be a subfield of L and let $q = |K|$ and r the dimension of L over K as a K-vector space. We know that $1, \alpha, \dots, \alpha^{r-1}$ form a basis of L as a K-vector space (Proposition 2.35), and thereby every element of L can be written in a unique way in the form $a_0 + a_1\alpha + \dots + a_{r-1}\alpha^{r-1}$, $a_0, \dots, a_{r-1} \in K$. We say that this is the *additive representation* over K of the elements of L, relative to the primitive element α.

With the additive representation, addition is easy, as it reduces the sum of two elements of L to r sums of elements of K. In order to compute products, however, it is more convenient to use the *exponential representation* with respect to a primitive element α. More precisely, if $x, y \in L^*$, and we know exponents i, j such that $x = \alpha^i$ and $y = \alpha^j$, then $xy = \alpha^{i+j}$.

In order to be able to use the additive and exponential representations cooperatively, it is convenient to compute a table with the powers α^i ($r \leq i \leq q - 2$, $q = |L|$) in additive form, say

$$\alpha^i = a_{i0} + a_{i1}\alpha + \dots + a_{i,r-1}\alpha^{r-1},$$

since this allows us to transfer, at our convenience, from the exponential to the additive representation and vice versa. If we let $ind(x)$ (or $ind_\alpha(x)$ if we want to declare α explicitely) to denote the exponent i such that $x = \alpha^i$, then the product xy is equal to α^k, where $k = ind(x) + ind(y) \pmod{q-1}$.

The listing **Computation of products by means of an index table** indicates a possible computational arrangement of this approach, and the example below presents in detail $\mathbb{F}_{16}$ done 'by hand'. Note that in the listing the function ind_table(x) constructs the table $\{x^j \to j\}_{j=0,\cdots,r-1}$, $r = ord(x)$, augmented with $\{0 \to \_\}$ in order to be able to include 0_K (in the definition of the function we use $0 \cdot x$ because this is the zero element of the field where x lives, whereas 0 alone would be an integer). We have used the underscore character to denote the index of 0_K, but in the literature it is customary to use the symbol ∞ instead.

▲ Library | Computation of products by means of an index table	▲
‖ ind_table(x) := {0·x→_} | {x^j→j with j in 0..(order(x)−1)}	

```
K=ℤ₃[x]/(x³−x+1); q=cardinal(K);
Set up an index table L and an exponential table E (inverse of L)
α=primitive_element(K); L=ind_table(α); E=inverse(L);
The product p(a,b) of a and b using table lookup
p(a,b) := E( L(a)+L(b) mod q−1 );
Example
a=α⁶; b=α¹⁵;
L(a), L(b) → 6, 15
a·b, p(a,b) → x²+1, x²+1
```

By analogy with the logarithms, $ind_\alpha(x)$ is also denoted $log_\alpha(x)$, or $log(x)$ if α can be understood. We say that it is the *discrete logarithm* of x with respect to α.

2.36 Example. Consider again the field $\mathbb{Z}_2(t) = \mathbb{Z}_2[T]/(f)$, $f = T^4 + T + 1$, studied in E.2.14. The element t is primitive, as $ord(t)|15$ and $t^3 \neq 1$, $t^5 = t^2 + t \neq 1$. So we have an exponential representation

$$\mathbb{Z}_2(t)^* = \{1, t, ..., t^{14}\}.$$

Consider now Table 2.1. The column on the right contains the coefficients of t^k, $k = 0, 1, ..., 14$, with respect to the basis $1, t, t^2, t^3$. For example, $t^8 \equiv t^2 + 1 \equiv 0101$. The first column contains the coefficients with respect to $1, t, t^2, t^3$ of the nonzero elements x of $\mathbb{Z}_2(t)$ ordered as the integers $1, ..., 15$ written in binary, together with the corresponding $log(x)$. For example, $5 \equiv 0101 \equiv t^2 + 1$, and its index is 8.

Table 2.1: *Discrete logarithm of* $\mathbb{Z}_2(t)$, $t^4 = t + 1$

x	$log(x)$	k	t^k		x	$log(x)$	k	t^k
0001	0	0	0001		1001	14	8	0101
0010	1	1	0010		1010	9	9	1010
0011	4	2	0100		1011	7	10	0111
0100	2	3	1000		1100	6	11	1110
0101	8	4	0011		1101	13	12	1111
0110	5	5	0110		1110	11	13	1101
0111	10	6	1100		1111	12	14	1001
1000	3	7	1011					

2.37 Remark. Index tables are unsuitable for large fields because the corresponding tables take up too much space. They are also unsuitable for small fields if we have the right computing setup because the field operations of a good implementation are already quite efficient. As an illustration of the contrast between pencil and machine computations, see **Examples of Lidl–Niederreiter** (cf. [10], p. 375).

> Examples of Lidl–Niederreiter
> $f = x^6 + x^5 + x^3 + x^2 + 1$;
> F64=extension($\mathbb{Z}_2$,b,f); L=ind_table(b);
> $s = (b^6 + b^{25} + b^{44}) \cdot (1 + b^{35})^{-1} + b^{28} \;\rightarrow\; b^5 + 1$
> L(s) $\rightarrow$ 38
> (in agreement with the result $s = b^{38}$)
>
> Zech logarithm for F$_{64}$
> Z={j→L(1+b^j) with j in 0..62};
> Z(4), Z(19), Z(35), Z(50) $\rightarrow$ 32, 34, 31, 60
> (thus, for example, $1 + b^{19} = b^{34}$)

Note in particular how to set up a table of *Jacobi logarithms*,

$$Z = \{j \mapsto ind(1 + \alpha^j)\}$$

(also called *Zech logarithms*), α a primitive element. Since we have the identity $1 + \alpha^j = \alpha^{Z(j)}$, they are useful to transform from the additive form to the exponential form. Here is a typical case:

$$\alpha^i + \alpha^j = \alpha^i(1 + \alpha^{j-i}) = \alpha^{i+Z(j-i)}$$

(assuming $i \leqslant j$).

Nevertheless, the problem of *computing ind$_\alpha(x)$* for any x is very interesting and seemingly difficult. Part of the interest comes from its uses in some cryptographic systems that exploit the fact that the discrete exponential $k \mapsto \alpha^k$ is easy to compute, while computing the inverse function $x \mapsto ind_\alpha(x)$ is hard. See, for example, [10], Chapter 9.

Summary

- Order of a nonzero element of a field of q elements, and the fact that it always divides $q - 1$. If d is a divisor of $q - 1$, there are precisely $\varphi(d)$ elements of order d.

- Primitive elements (they exist for any finite field). Since they are the elements of order $q - 1$, there are exactly $\varphi(q - 1)$ primitive elements.

- Period (or exponent) of an irreducible polynomial over a finite field. Primitive polynomials.

- Discrete and Jacobi (or Zech) logarithms.

Problems

P.2.9. Let K be a finite field, $\alpha, \beta \in K^*$, $r = ord(\alpha)$, $s = ord(\beta)$. Let $t = ord(\alpha\beta)$, $d = gcd(r, s)$, $m = lcm(r, s)$ and $m' = m/d$. Prove that $m' | t$ and $t | m$. Hence $ord(\alpha\beta) = rs$ if $d = 1$. Give examples in which $d > 1$ and $t = m'$ (repectively $t = m$).

> We now present an algorithm for finding primitive roots in an arbitrary finite field. It is due to Gauss himself.
>
> R.J. McEliece, [13], p. 38.

P.2.10 (Gauss algorithm). This is a fast procedure for constructing a primitive element of a finite field K of q elements.

1) Let a be any nonzero element of K and $r = ord(a)$. If $r = q - 1$, return a. Otherwise we have $r < q - 1$.

2) Let b an element which is not in the group $\langle a \rangle = \{1, a, \ldots, a^{r-1}\}$ (choose a nonzero b at random in K and compute b^r; if $b^r \neq 1$, then $b \notin \langle a \rangle$, by proposition 2.32, and otherwise we can try another b; since r is a proper divisor of $q - 1$, there are at least half of the nonzero elements of K that meet the requirement).

3) Let s be the order of b. If $m = q - 1$, return b. Otherwise we have $s < q - 1$.

4) Compute numbers d and e as follows. Start with $d = 1$ and $e = 1$ and look successively at each common prime divisor p of r and s. Let m be the minimum of the exponents with which p appears in r and s and set $d = dp^m$ if m occurs in r (in this case p appears in s with exponent m or greater) and $e = ep^m$ otherwise (in this case p appears in s with exponent m and in r with exponent strictly greater than m).

5) Replace a by $a^d b^e$, and reset $r = ord(a)$. If $r = q - 1$, then return a. Otherwise go to 2.

To see that this algorithm terminates, show that the order of $a^d b^e$ in step 5 is $lcm(r, s)$ (use P.2.9) and that $lcm(r, s) > \max(r, s)$.

P.2.11. Let K be a finite field, $q = |K|$, and $f \in K[X]$ a monic polynomial of degree $m \geq 1$ with $f(0) \neq 0$. Prove that there exists a positive integer $e \leq q^m - 1$ such that $f | X^e - 1$. The least e satisfying this condition is the *period*, or *exponent*, of f. If f is reducible, e need not be a divisor of $q^m - 1$. See **The function coarse_period(f,q)** , which calls period(f,q) if f is irreducible and otherwise proceeds by sluggishly trying the condition $f | (X^e - 1)$ for all e in the range $deg\,(f)..(q^m - 1))$. *Hint:* If x is the class of X in $K' = K[X]/(f)$, the q^m classes x^j $(j = 0, \ldots, q^m - 1)$ are nonzero and K' has only $q^m - 1$ nonzero elements.

```
◄ Library | The function coarse_period(f,q)                          ▲

  coarse_period (f,q) :=
    begin
       local r=degree(f), X=variable(f), E=1..qʳ-1
       if irreducible?(f,coefficient_ring(f)) then
          return period(f,q)
       end
       for e in E do
          if remainder(Xᵉ-1,f)=0  then
             return e
          end
       end
    end
```

```
  u=1:ℤ₂;

  coarse_period((X²+X+u)², 2)

  coarse_period(X¹⁰+X⁹+X³+X²+u, 2)
```

```
  f=x⁴+x-1 : ℤ₃[x];
  irreducible?(f,ℤ₃) → true
  period(f,3) → 80
  coarse_period(f,3) → 80
```

P.2.12. Find all primitive irreducible polynomials of degree $r = 2..6$ over $\mathbb{Z}_2$.

2.4 Minimum polynomial

> The definitions, theorems, and examples in this chapter can be
> found in nearly any of the introductory books on modern (or
> abstract) algebra.
>
> R. Lidl and H. Niederreiter, [10], p. 37.

Essential points

- Minimum polynomial of an element of a field with respect to a subfield.

- The roots of the minimum polynomial of an element are the conjugates of that element.

- Between two finite fields of the same order p^r there exist precisely r isomorphisms.

- Norm and trace of an element of a field with respect to a subfield.

Let L be a finite field and K a subfield of L. If the cardinal of K is q, we know that the cardinal of L has the form q^m, where m is the dimension of L as a K-vector space.

Let $\alpha \in L$. Then the $m + 1$ elements $1, \alpha, ..., \alpha^m$ are linearly dependent over K and so there exists elements $a_0, a_1, \ldots, a_m \in K$, not all zero, such that $a_0 + a_1 \alpha + \ldots + a_m \alpha^m = 0$. This means that if we set $f = a_0 + a_1 X + \ldots + a_m X^m$, then f is a nonzero polynomial such that $f(\alpha) = 0$. With this observation we can now prove the next proposition.

2.38 Proposition. *There is a unique monic polynomial $p \in K[X]$ that satisfies the following two conditions:*

1) $p(\alpha) = 0.$

2) *If $f \in K[X]$ is a polynomial such that $f(\alpha) = 0$, then $p|f$.*

The polynomial p is irreducible and satisfies that

3) $\deg(p) \leqslant m.$

Proof: Among all the nonzero polynomials f such that $f(\alpha) = 0$, choose one of minimum degree, p. The degree of p cannot be greater than m, as we know that there are nonzero polynomials f of degree m or less such that $f(\alpha) = 0$.

To see 2, assume $f(\alpha) = 0$ and let r be the residue of the Euclidean division of f by p, say $f = gp + r$ and with either $r = 0$ or $deg\,(r) < deg\,(p)$. Since $p(\alpha) = 0$, we also have $r(\alpha) = 0$. Hence the only possibility is $r = 0$, as otherwise we would contradict the way p was chosen.

To prove the uniqueness of p, let p' be another monic polynomial that satisfies 1 and 2. Then we have $p|p'$ and $p'|p$, and so p and p' are proportional by a nonzero constant factor. But this factor must be 1, because both p and p' are monic.

Finally if $p = fg$ is a factorization of p, then $f(\alpha) = 0$ or $g(\alpha) = 0$, and so $p|f$ or $p|g$. If $p|f$, then $f = pf'$, for some $f' \in K[X]$, and so $p = fg = pf'g$. Thus $1 = f'g$, f' and g are constants and $p = fg$ is not a proper factorization of p. The case $p|g$ leads with a similar argument to the same conclusion. Therefore p is irreducible. $\square$

The polynomial p defined in Proposition 2.38 is called the *minimum polynomial* of α over K, and is usually denoted

p_α. The degree of p_α is also known as the *degree of* α over K and is denoted $r = deg\,(\alpha)$.

E.2.22. Show that there exits a unique K-isomorphism $K[X]/(p_\alpha) \simeq K[\alpha]$ such that $x \mapsto \alpha$, where x is the class of X.

E.2.23. Show that if α is a primitive element of L, then α has degree m over K.

E.2.24. If f is an irreducible polynomial over a field K, and α is a root of f in an extension L of K, show that f is the minimum polynomial of α over K (note in particular that if $K(x) = K[X]/(f)$, then f is the minimum polynomial of x over K).

E.2.25. With the same notations as in the Proposition 2.38, prove that $r = deg\,(p)$ is the first positive integer r such that $\alpha^r \in \langle 1, \alpha, ..., \alpha^{r-1}\rangle_K$. As a result, r is also the first positive integer such that $K[\alpha] = \langle 1, \alpha, ..., \alpha^{r-1}\rangle_K$.

2.39 *Example.* Let us find the minimum polynomial, p, over $\mathbb{Z}_2$ of the element y of $\mathbb{Z}_2(x, y)$ introduced in the example 2.13 (cf. **The function minimum_polynomial**). We have $y^2 = xy + 1$. Hence $y^2 \in \langle 1, y \rangle_{\mathbb{Z}_2(x)}$, a fact that is nothing but a rediscovery that the minimum polynomial of y over $\mathbb{Z}_2(x)$ is $Y^2 + xY + 1$. But $y^2 \notin \langle 1, y \rangle_{\mathbb{Z}_2}$, and so p has degree higher than 2. We have $y^3 = xy^2 + y = x^2 y + x + y = xy + x$ and, as it is easy to see, $y^3 \notin \langle 1, y, y^2 \rangle_{\mathbb{Z}_2}$. Now $y^4 = xy^2 + xy = x^2 y + x + xy = y + x = y^3 + y^2 + y + 1$, and therefore $p = Y^4 + Y^3 + Y^2 + Y + 1$. As $ord(y) = 5$, p is not a primitive polynomial. $\square$

> The function minimum_polynomial
> $k=\mathbb{Z}_2$; $K=k[x]/(x^2+x+1)$; let $L=K[y]/(y^2+x\cdot y+1)$;
> minimum_polynomial(y,k,X) ⟶ $X^4+X^3+X^2+X+1$
> minimum_polynomial(y,T) ⟶ $T^4+T^3+T^2+T+1$
> minimum_polynomial(y,K,Y) ⟶ $Y^2+x\cdot Y+1$
>
> $K=\mathbb{Z}_3[\alpha]/(\alpha^3-\alpha+1)$; $\beta=\alpha^2+1$;
> minimum_polynomial(β,X) ⟶ $X^3+X^2+2\cdot X+1$

The set of *conjugates* over K of an element $\alpha \in L$ is defined to be

$$C_\alpha = \{\alpha, \alpha^q, \alpha^{q^2}, \ldots, \alpha^{q^{r-1}}\},$$

where r is the least positive integer such that $\alpha^{q^r} = \alpha$.

> The calls conjugates(x) and conjugates(x,k)
> k, x, K, y, L: same meaning as in the previous listing
> $k=\mathbb{Z}_2$; $K=k[x]/(x^2+x+1)$; $L=K[y]/(y^2+x\cdot y+1)$;
> conjugates(x,k) ⟶ $\{x, x+1\}$
> conjugates(x) ⟶ $\{x, x+1\}$
> conjugates(y,k) ⟶ $\{y, x\cdot y+1, y+x, x\cdot y+x\}$
> conjugates(y) ⟶ $\{y, x\cdot y+1, y+x, x\cdot y+x\}$
> conjugates(y,K) ⟶ $\{y, y+x\}$
> conjugates(x,K) ⟶ $\{x\}$

2.40 Proposition. *Let K be a finite field and $q = |K|$. Let L be a finite field extension of K and $\alpha \in L$. If C_α is the set of conjugates of α over K, then the minimum polynomial p_α of α over K is given by the expression*

$$p_\alpha = \prod_{\beta \in C_\alpha} (X - \beta).$$

Proof: Extend the Frobenius automorphism of L/K to a ring automorphism of $L[X]$ by mapping the polynomial $\sum_{i=0}^n a_i X^i$ to $\sum_{i=0}^n a_i^q X^i$. Then the polynomial $f = \prod_{\beta \in C_\alpha}(X - \beta)$ is invariant by this automorphism, because $\beta^q \in C_\alpha$ if $\beta \in C_\alpha$. Therefore $f \in K[X]$. If $\beta \in L$ is a root of p_α, then β^q is also a root of p_α, as it is easily seen by mapping the Frobenius automorphism to the relation $p_\alpha(\beta) = 0$. Using this observation repeatedly it follows, since α is a root of p_α, that $p_\alpha(\beta) = 0$ for all $\beta \in C_\alpha$ and so $f|p_\alpha$. But p_α is irreducible and f has positive degree, so $f = p_\alpha$, as both polynomials are monic. □

Uniqueness of finite fields

Next statement shows that two finite fields of the same cardinal are isomorphic.

2.41 Theorem. *Let K and K' be two finite fields with the same cardinal, q. Then there exists an isomorphism $\varphi : K \to K'$.*

Proof: Consider the polynomial $X^q - X \in \mathbb{Z}_p[X]$. Regarded as a polynomial with coefficients in K, we have $X^q - X = \prod_{\alpha \in K}(X - \alpha)$ (see E.2.19). Similarly we have $X^q - X = \prod_{\alpha' \in K'}(X - \alpha')$.

Let α be a primitive element of K and let $f \in \mathbb{Z}_p[X]$ be its minimum polynomial. As $|K| = |K'| = p^r$, f has degree r (Proposition 2.35), and since all the roots of f are in K (Proposition 2.40), we also have that

$$f \,|\, (T^{p^r-1} - 1)$$

as polynomials with coefficients in K. But since these polynomials are monic with coefficients in $\mathbb{Z}_p$, this relation is in fact valid as polynomials with coefficients in $\mathbb{Z}_p$. Now $(T^{p^r-1} - 1)$ also decomposes completely in K' and hence f has a root $\alpha' \in K'$. Now Proposition 2.10 says that there is a unique isomorphism $\mathbb{Z}_p[X]/(f) \simeq \mathbb{Z}_p[\alpha'] = K'$ such that $x \mapsto \alpha'$, where x is the class of X modulo f. But there is also a unique isomorphism $\mathbb{Z}_p[X]/(f) \simeq \mathbb{Z}_p[\alpha] = K$ and so there exists a unique isomorphism $K \simeq K'$ such that $\alpha \mapsto \alpha'$. $\square$

2.42 Remark. The proof above actually shows that there are precisely r distinct isomorphisms between K and K'.

E.2.26. The polynomials $X^3 + X + 1, X^3 + X^2 + 1 \in \mathbb{Z}_2[X]$ are the two irreducible polynomials of degree 3 over $\mathbb{Z}_2$. Construct an explicit isomorphism between the fields $\mathbb{Z}_2[X]/(X^3 + X + 1)$ and $\mathbb{Z}_2[X]/(X^3 + X^2 + 1)$. (cf. the listing below).

```
K=Z₂[β]/(β³+β²+1);
roots(X³+X+1, K)  →  {β+1, β²+1, β²+β}
```

E.2.27. Prove that $x^2 + x + 4$ and $x^2 + 1$ are irreducible polynomials over the field $\mathbb{F}_{11}$ and construct an explicit isomorphism between $\mathbb{F}_{11}[x]/(x^2 + x + 4)$ and $\mathbb{F}_{11}[x]/(x^2 + 1)$.

Trace and norm

Let L be a finite field and K a subfield. If $|K| = q$, then we know that $|L| = q^m$, where $m = [L : K]$, the dimension of L as a K-vector space.

Definitions

Let $\alpha \in L$. Then we can consider the map $m_\alpha : L \to L$ such that $x \mapsto \alpha x$. This map is clearly K-linear and we define the *trace* and the *norm* of α with respect to K, denoted $Tr_{L/K}(\alpha)$ and $Nm_{L/K}(\alpha)$ respectively, by the formulae

$$Tr_{L/K}(\alpha) = \text{trace}(m_\alpha), \quad Nm_{L/K}(\alpha) = \det(m_\alpha).$$

Let us recall that the right hand sides are defined as the sum of the diagonal elements and the determinant, respectively, of the matrix of m_α with respect to any basis of L as a K-vector space.

E.2.28. Check that $Tr_{L/K} : L \to K$ is a K-linear map and that $Tr_{L/K}(\lambda) = m\lambda$ for all $\lambda \in K$.

E.2.29. Similarly, check that $Nm_{L/K} : L^* \to K^*$ is a homomorphism of groups and that $Nm_{L/K}(\lambda) = \lambda^m$ for all $\lambda \in K$.

Formulas in terms of the conjugates of α

Let $p_\alpha = X^r + c_1 X^{r-1} + \cdots + c_r$ be the minimum polynomial of α with respect to K. If we put $\{\alpha_1, \ldots, \alpha_r\}$ to denote the set of conjugates of α with respect to K ($\alpha_1 = \alpha$), then we know that $p_\alpha = \prod_{i=1}^r (X - \alpha_i)$. In particular we have that

$$c_1 = -\sigma(\alpha) \quad \text{and} \quad c_r = (-1)^n \pi(\alpha),$$

where $\sigma(\alpha) = \alpha_1 + \cdots + \alpha_r$ and $\pi(\alpha) = \alpha_1 \cdots \cdots \alpha_r$.

2.43 Theorem. *With the notations above, and defining $s = m/r$, we have:*

$$Tr_{L/K}(\alpha) = s\sigma(\alpha) \quad \text{and} \quad Nm_{L/K}(\alpha) = (\pi(\alpha))^s.$$

Proof: The set $a = \{1, \alpha, \cdots, \alpha^{r-1}\}$ is a basis of $K(\alpha)$ as a K-vector space. If we let $\beta_1, \ldots, \beta_s$, $s = m/r$, denote a basis of L as a $K(\alpha)$-vector space, then $L = \oplus_{j=1}^s K(\alpha)\beta_j$ (as $K(\alpha)$-vector spaces). It follows that m_α has a matrix that is the direct sum of s identical blocks, each of which is the matrix A of $m_\alpha : K(\alpha) \to K(\alpha)$ with respect to the basis a. Since $m_\alpha(\alpha^i) = \alpha^{i+1}$, which for $i = r - 1$ is equal to

$$-(c_r + c_{r-1}\alpha + \cdots + c_1\alpha^{r-1}),$$

we have that

$$A = \begin{pmatrix} 0 & & & & -c_r \\ 1 & \ddots & & & -c_{r-1} \\ & \ddots & \ddots & & \vdots \\ & & 1 & 0 & -c_2 \\ & & & 1 & -c_1 \end{pmatrix}$$

and so

$$\mathrm{trace}(m_\alpha) = s\,\mathrm{trace}(A) = s(-c_1) = s\sigma(\alpha),$$

and

$$\det(m_\alpha) = (\det(A))^s = ((-1)^r c_r)^s = \pi(\alpha)^s,$$

as stated.   □

If L and K are constructed fields, then $Tr_{L/K}(\alpha)$ is returned by Tr(α,L,K), or trace(α,L,K). If K is not specified, it is taken as the prime field of L. And if neither K nor L are specified, it is assumed that L is the field of α and K its prime field. Similar notations and conventions hold for $Nm_{L/K}(\alpha)$ and Nm(α,L,K), or norm(α,L,K).

```
Trace examples
k=Z₃; K=k[α]/(α²+1); L=K[β]/(β³−β+1);
u=1 :k  →  1
Tr(u), Tr(α), Tr(β), Tr(α·β), Tr(β²), Tr(α·β²)  →  1,0,0,0,2,0
Tr(u,K), Tr(α,K)  →  2,0
Tr(u,K,k), Tr(α,K,k)  →  2,0
Tr(u,L), Tr(α,L), Tr(β,L), Tr(α·β,L), Tr(β²,L), Tr(α·β²,L)  →  0,0,0,0,1,0
Tr(u,L,k), Tr(α,L,k), Tr(β,L,k), Tr(α·β,L,k), Tr(β²,L,k), Tr(α·β²,L,k)  →  0,0,0,0,1,0
Tr(u,L,K), Tr(α,L,K), Tr(β,L,K), Tr(α·β,L,K), Tr(β²,L,K), Tr(α·β²,L,K)
    →  0,0,0,0,2,2·α
```

```
Norm examples
k=Z₃; K=k[α]/(α²+1); L=K[β]/(β³−β+1);
u=1 :k  →  1
Nm(u), Nm(α), Nm(β), Nm(α·β), Nm(β²), Nm(α·β²)  →  1,1,2,1,1,1
Nm(u,K), Nm(α,K)  →  1,1
Nm(u,K,k), Nm(α,K,k)  →  1,1
Nm(u,L), Nm(α,L), Nm(β,L), Nm(α·β,L), Nm(β²,L), Nm(α·β²,L)  →  1,1,1,1,1,1
Nm(u,L,k), Nm(α,L,k), Nm(β,L,k), Nm(α·β,L,k), Nm(β²,L,k), Nm(α·β²,L,k)
    →  1,1,1,1,1,1
Nm(u,L,K), Nm(α,L,K), Nm(β,L,K), Nm(α·β,L,K), Nm(β²,L,K), Nm(α·β²,L,K)
    →  1, 2·α, 2,α, 1, 2·α
```

Summary

- Minimum polynomial of an element of a field with respect to a subfield.

- The roots of the minimum polynomial of an element are the conjugates of that element with respect to the subfield.

- Between two finite fields of the same order p^r, there exist precisely r isomorphisms.

- Trace and norm of an element of a field L with $|L| = q^m$ with respect to a subfield K with $|K| = p^r$: if we set $s = m/r$, then the trace is the sum of the conjugates of α multiplied by s and the norm is the s-th power of the product of the conjugates of α.

Problems

P.2.13. Let K be a finite field and $q = |K|$. Let $f \in K[X]$ be monic and irreducible and $r = deg\,(f)$. Show that $K' = K[X]/(f)$ is the splitting field of f.

P.2.14. Find the trace of $a_0 + a_1 t + a_2 t^2 + a_3 t^3 + a_4 t^4 \in \mathbb{Z}_2[T]/(T^5 + T^2 + 1)$, where t is the class of T.

P.2.15. Let $K = \mathbb{Z}_2[X]/(X^4 + X + 1)$ and $\alpha = [X]$. Compute the matrix $M = Tr(\alpha^i \alpha^j)$ for $0 \leqslant i, j \leqslant 3$ and show that $\det(M) = 1$. Deduce that there exists a unique $\mathbb{Z}_2$-linear basis $\beta_0, \beta_1, \beta_2, \beta_3$ of K such that $Tr(\alpha^i \beta_j) = \delta_{ij}$ $(0 \leqslant i, j \leqslant 3)$.

P.2.16. Let $K = \mathbb{Z}_3$, $L = K[X]/(X^2 + 1)$ and $\beta = \alpha + 1$, where α is the class of X. Find $Nm(\beta^i)$ for $0 \leqslant i \leqslant 7$.

3 Cyclic Codes

The codes discussed in this chapter can be implemented relatively easily and possess a great deal of well-understood mathematical structure.

W. W. Peterson and E. J. Weldon, Jr., [16], p. 206.

Cyclic codes are linear codes that are invariant under cyclic permutations
of the components of its vectors. These codes have a nice algebraic structure
(after reinterpreting vectors as univariate polynomials) which favors its study
and use in a particularly effective way.

More specifically, we will see that cyclic codes of length n over a finite
field K are in one-to-one correspondence with the monic divisors g of

$$X^n - 1 \in K[X],$$

and we will establish how to construct generating and control matrices in
terms of g. Then we will investigate, in order to generate all factors g, how
to effectively factor $X^n - 1 \in K[X]$ into irreducible factors.

Not surprisingly, several of the better known codes (like the Golay codes),
or families of codes (like the *BCH* codes), are cyclic.

In the last section we will present the Meggitt decoding algorithm and
its computational implementation. Moreover, we will illustrate in detail its
application in the case of the Golay codes.

3.1 Generalities

> During the last two decades more and more algebraic tools
> such as the theory of finite fields and the theory of polynomi-
> als over finite fields have influenced coding.
>
> Lidl–Niederreiter, [10], p. 470.

Essential points

- Cyclic ring structure of $\mathbb{F}^n$.
- Generating and control polynomials of a cyclic code.
- Parameters of a cyclic code.
- Generating matrices and systematic encoding.
- Control matrices.

We will say that a linear code $C \subseteq \mathbb{F}^n$ is *cyclic* if $\sigma(a) \in C$ for all $a \in C$, where

$$\sigma(a) = (a_n, a_1, a_2, \ldots, a_{n-1}) \quad \text{if} \quad a = (a_1, \ldots, a_n). \qquad [3.1]$$

E.3.1. Assume that $\sigma(a) \in C$ for all rows a of a generating matrix G of C. Show that C is cyclic.

Cyclic ring structure of $\mathbb{F}^n$

In order to interpret the condition [3.1] in a more convenient form, we will identify $\mathbb{F}^n$ with $\mathbb{F}[x]_n$ (the vector space of univariate polynomials of degree $< n$ with coefficients in $\mathbb{F}$ in the indeterminate x) by means of the isomorphism that maps the vector $a = (a_1, \ldots, a_n) \in \mathbb{F}^n$ to the polynomial $a(x) = a_1 + a_2 x + \ldots + a_n x^{n-1}$ (note that this is like the interpretation of 357 as $3 \times 10^2 + 5 \times 10 + 7$). See**The isomorphism** $\mathbb{F}^n \xrightarrow{\sim} F[x]_n$ for how we can deal with this in computational terms.

Moreover, we will endow $\mathbb{F}[x]_n$ with the ring product obtained by identifying $\mathbb{F}[x]_n$ with the underlying vector space of the ring $\mathbb{F}[X]/(X^n - 1)$ via the unique $\mathbb{F}$-linear isomorphism that maps x to the class $[X]$ of X. If $f \in \mathbb{F}[X]$, its image $\bar{f} = f(x)$ in $\mathbb{F}[x]_n$, which we will cal the *n-th cyclic reduction of* f, is obtained by replacing, for all non-negative integers j, the monomial X^j

```
The isomorphism Fⁿ → F[x]ₙ
reverse_print(true);
a=[1, 2, 3];
a=polynomial(a,X)  →  1+2·X+3·X²
vector(a)  →  [1, 2, 3]
reverse_print(false);
```

by the monomial $x^{[j]_n}$, where $[j]_n$ is the remainder of the Euclidean division of j by n. We will denote by $\pi : \mathbb{F}[X] \to \mathbb{F}[x]_n$ the map such that $f \mapsto \bar{f}$ (see **Cyclic reduction of two polynomials**).

▲ Library | Cyclic reduction of polynomials   ▲

```
Cyclic reduction of a polynomial with respect to a variable x
cyclic_reduction(f, n:ℤ, x:Identifier)
check n>0 & collect(f,x) ∈ Polynomial
:= begin
      local c
      for j in n..degree(f,x) do
          c=coefficient(f,x,j)
          if c=0 then
              continue
          end
          f=f−c·xʲ ; f=f+c·xʲ mod n
      end
      collect(f,x)
  end
Cyclic reduction of a univariate polynomial f
cyclic_reduction(f, n:ℤ)
check n>0 & length(variables(f))⩽1
:= if variables(f)={ } then
      f
  else
      cyclic_reduction(f,n,variable(f))
  end
```

```
cyclic_reduction(1+2·x+3·x²+4·x³,3)  →  3·x²+2·x+5
cyclic_reduction(sum(geometric_series(x,25)),10)
  →  2·x⁹+2·x⁸+2·x⁷+2·x⁶+2·x⁵+3·x⁴+3·x³+3·x²+3·x+3
cyclic_reduction(a0+a1·x+a2·x²+a3·x³+a4·x⁴, 3, x)
  →  a2·x²+(a1+a4)·x+a0+a3
cyclic_reduction(a0+a1·x+a2·x²+a3·x³+a4·x⁴, 1, x)  →  a0+a1+a2+a3+a4
```

For convenience we will also write, if $f \in \mathbb{F}[X]$ and $a \in \mathbb{F}[x]_n$, fa as meaning the same as $\bar{f}a$.

Next we will consider the functions that allow us to deal with the product in $\mathbb{F}[x]_n$, and a couple of examples (**Cyclic product of polynomials**).

In terms of vectors of $\mathbb{F}^n$, the product in $\mathbb{F}[x]_n$ can be expressed directly as follows:

┌───┐
│ ▲│ Library │ Cyclic product of two polynomials with respect to a variable   ▲│
├───┤
│ ‖ cyclic_product(a, b, n:ℤ, x:Identifier):=cyclic_reduction(a·b, n, x)   │
│   Special call for univariate polynomials   │
│ ‖ cyclic_product(a, b, n:ℤ):=cyclic_reduction(a·b, n)   │
└───┘

```
cyclic_product(a+b·x+c·x², x², 3, x)  →  a·x²+c·x+b
cyclic_product(1+2 x²+3 x³+5 x⁵, 1+x+7 x⁷, 6)
  →  5·x⁵+24·x⁴+19·x³+2·x²+8·x+41
```

┌───┐
│ ▲│ Library │ Cyclic product of vectors   ▲│
├───┤
│ ‖ cyclic_product(a:Vector, b:Vector)   │
│   check length(a)=length(b)   │
│   := begin   │
│   local v, x, n=length(a)   │
│   a=polynomial(a,x)   │
│   b=polynomial(b,x)   │
│   v=vector(cyclic_reduction(a·b, n))   │
│   pad(v,n)   │
│   end   │
└───┘

```
cyclic_product([a, b, c],[d, e, f])
  →  [a·d+b·f+c·e,, a·e+b·d+c·f,, a·f+b·e+c·d]
cyclic_product([1, −2, 3, −4, 5],[2, 3, 5, 7, 11])  →  [−4,, 29,, −4,, 53,, 10]
```

3.1 Lemma. *If we let σ denote the cyclic permutation* [3.1], *then* $\sigma(a) = xa$ *for all* $a \in \mathbb{F}[x]_n$.

Proof: With the representation of a as the polynomial

$$a_1 + a_2 x + \ldots + a_n x^{n-1},$$

the product xa is

$$a_1 x + a_2 x^2 + \ldots + a_{n-1} x^{n-1} + a_n x^n.$$

Since $x^n = 1$, we have

$$xa = a_n + a_1 x + a_2 x^2 + \ldots + a_{n-1} x^{n-1},$$

which is the polynomial corresponding to the vector $\sigma(a)$.   □

3.2 Proposition. *A linear code C of length n is cyclic if and only if it is an ideal of* $\mathbb{F}[x]_n$.

Proof: If $C \subseteq \mathbb{F}[x]_n$ is cyclic, then Lemma 3.1 implies that $xC \subseteq C$. Inductively we also have that $x^j C \subseteq C$ for all non-negative integers j. Since C is a linear subspace, we also have $aC \subseteq C$ for all $a \in \mathbb{F}[x]_n$, which means that C is an ideal of $\mathbb{F}[x]_n$. Conversely, if $C \subseteq \mathbb{F}[x]_n$ is an ideal, then in particular it is a linear subspace such that $xC \subseteq C$. But again by Lemma 3.1 this means that C is cyclic. $\qquad\square$

Given any polynomial $f \in \mathbb{F}[X]$, let $C_f = (\bar{f}) \subseteq \mathbb{F}[x]_n$ denote the ideal generated by $\bar{f} = f(x)$, which can also be described as the image in $\mathbb{F}[x]_n$ of the ideal $(f) \subseteq \mathbb{F}[X]$ by the map $\pi : \mathbb{F}[X] \to \mathbb{F}[x]_n$.

E.3.2. If g and g' are monic divisors of $X^n - 1$, show that
1) $C_g \subseteq C_{g'}$ if and only if $g'|g$.
2) $C_g = C_{g'}$ if and only if $g = g'$.

3.3 Proposition. *Given a cyclic code C of length n, there is a unique monic divisor g of $X^n - 1$ such that $C = C_g$.*

Proof: Exercise E.3.2 tells us that we must only prove the existence of g.

Given C, let $g \in \mathbb{F}[X]$ be a nonzero polynomial of minimum degree among those that satisfy $\bar{g} \in C$ (note that $\pi(X^n - 1) = x^n - 1 = 0 \in C$ so that g exists and $deg(g) \leqslant n$). After dividing g by its leading coefficient, we may assume that g is monic. Since C is an ideal, we have $C_g = (\bar{g}) \subseteq C$ and so it will be enough to prove that g is a divisor of $X^n - 1$ and that $C \subseteq C_g$.

To see that g is a divisor of $X^n - 1$, let q and r be the quotient and remainder of the Euclidean division of $X^n - 1$ by g:

$$X^n - 1 = qg + r.$$

Hence

$$0 = x^n - 1 = \bar{q}\bar{g} + \bar{r}$$

and so $\bar{r} = -\bar{q}\bar{g} \in C_g \subseteq C$. But $deg(r) < deg(g)$, by definition of r. Thus $r = 0$, by definition of g, and this means that g divides $X^n - 1$.

Now let $a \in C$. We want to prove that $a \in C_g$. Let

$$a_X = a_0 + a_1 X + \cdots + a_{n-1} X^{n-1},$$

so that

$$a = a_0 + a_1 x + \cdots + a_{n-1} x^{n-1} = \bar{a}_X.$$

Let q_a and r_a be the quotient and remainder of the Euclidean division of a_X by g:

$$a_X = q_a g + r_a.$$

This implies that

$$a = \bar{a}_X = \bar{q}_a \bar{g} + \bar{r}_a$$

and therefore

$$\bar{r}_a = a - \bar{q}_a \bar{g} \in C.$$

This and the definition of r_a and g yield that $r_a = 0$ and hence $a = \bar{q}_a \bar{g} \in C_g$, as claimed. $\qquad\square$

The monic divisor g of $X^n - 1$ such that $C = C_g$ is said to be the *generator polynomial* of C_g.

E.3.3. For any $f \in \mathbb{F}[X]$, prove that the generator polynomial of C_f is $g = gcd\,(f, X^n - 1)$. $\qquad\qquad\qquad\qquad\qquad\qquad\qquad\qquad\square$

In the sequel we shall write, if g is a monic divisor of $X^n - 1$,

$$\widehat{g} = (X^n - 1)/g$$

and we will say that it is the *control polynomial*, or *check polynomial*, of C_g. Note that $\widehat{\widehat{g}} = g$.

Further notations and conventions

Let us modify the notations used so far in order to better adapt them to the polynomial representation of vectors. Instead of the index set $\{1, \ldots, n\}$, we will use $\{0, 1, \ldots, n - 1\}$. In this way an n-dimensional vector $a = (a_0, \ldots, a_{n-1})$ is identified with the polynomial

$$a(x) = a_0 + a_1 x + \ldots + a_{n-1} x^{n-1}.$$

The scalar product $\langle a|b \rangle$ of $a, b \in \mathbb{F}[x]_n$ can be determined in terms of the polynomial representation as follows: Given $a \in \mathbb{F}[X]_n$, let $\ell(a) = a_{n-1}$ (the leading term of a) and

$$\widetilde{a} = a_{n-1} + a_{n-2} x + \ldots + a_0 x^{n-1}. \qquad\qquad [3.2]$$

Then we have the relation

$$\langle a|b \rangle = \ell(\widetilde{a}b), \qquad\qquad [3.3]$$

as it is easy to check.

In the remainder of this section we will assume that $gcd\,(n, q) = 1$ and we let p denote the unique prime number such that $q = p^r$, for some positive integer r. This hypothesis implies that $n \neq 0$ in $\mathbb{F}$, therefore

$$D(X^n - 1) = nX^{n-1} \sim X^{n-1}$$

does not have non-constant common factors with $X^n - 1$. Consequently the irreducible factors of $X^n - 1$ are simple (have multiplicity 1). In particular we have, if $f_1, \ldots, f_r$ are the distinct irreducible monic factors of $X^n - 1$, that $X^n - 1 = f_1 \cdots f_r$ (for the case in which n and q are not relatively prime, see E.3.13 on page 158).

Thus the monic divisors of $X^n - 1$ have the form

$$g = f_{i_1} \cdots f_{i_s}, \quad \text{with } 1 \leqslant i_1 < \ldots < i_s \leqslant r.$$

So there are exactly 2^r cyclic codes of length n, although some of these codes may be equivalent (for a first example, see Example 3.6).

Since the ideals (f_i) are maximal ideals of $\mathbb{F}[x]_n = \mathbb{F}[X]/(X^n - 1)$, the codes $M_i = (f_i)$ are said to be *maximal cyclic codes*. In the same way, the codes $\widehat{M}_i = (\widehat{f}_i)$, $\widehat{f}_i = (X^n - 1)/f_i$, are called *minimal cyclic codes*, as they are the minimal nonzero ideals of $\mathbb{F}[X]_n$.

E.3.4. How many ternary cyclic codes of length 8 are there?

E.3.5. Let C be a binary cyclic code of odd length. Show that C contains an odd weight vector if and only if it contains $11 \cdots 1$.

E.3.6. Let C be a binary cyclic code that contains an odd weight vector. Prove that the even-weight subcode of C is cyclic.

Dimension of C_g

Given a monic divisor g of $X^n - 1$, the elements of C_g are, by definition, the polynomials that have the form $a\bar{g}$, where $a \in \mathbb{F}[x]_n$ is an arbitrary element. In other words, the map $\mu : \mathbb{F}[X] \to C_g$ such that $u \mapsto \bar{u}\bar{g}$ is surjective.

On the other hand, we claim that $ker\,(\mu) = (\widehat{g})$. Indeed, $\bar{u}\bar{g} = 0$ if and only if $(X^n - 1)|ug$. But $X^n - 1 = g\widehat{g}$ and $gcd\,(g, \widehat{g}) = 1$, so $(X^n - 1)|ug$ if and only if $\widehat{g}|u$, that is, if and only if $u \in (\widehat{g})$.

Hence we have that

$$C_g = im(\mu) \simeq \mathbb{F}[X]/ker\,(\mu) \simeq \mathbb{F}[X]/(\widehat{g}) \simeq \mathbb{F}[X]_k,$$

where $k = deg\,(\widehat{g})$. Thus we have proved:

3.4 Proposition. *The dimension of C_g coincides with the degree of the control polynomial $\widehat{g}$ of C_g:*

$$dim\,(C_g) = deg\,(\widehat{g}).$$

Since $deg\,(g) = n - deg\,(\widehat{g})$, $deg\,(g)$ is the codimension of C_g. $\qquad\square$

E.3.7. Prove that the linear map $\mathbb{F}[X]_k \to C_g$ such that $u \mapsto \bar{u}\bar{g}$ is an $\mathbb{F}$-linear isomorphism.

Generating matrices

With the same notations as in the preceeding subsection, the polynomials $u_i = \{x^i\bar{g}\}_{0 \leqslant i < k}$ form a basis of C_g. Therefore, if

$$g = g_0 + g_1 X + \cdots + g_{n-k} X^{n-k},$$

then the matrix

$$
G = \begin{pmatrix}
g_0 & g_1 & \cdots & g_{n-k} & 0 & 0 & \cdots & 0 \\
0 & g_0 & g_1 & \cdots & g_{n-k} & 0 & \cdots & 0 \\
\vdots & \ddots & \ddots & \ddots & \cdots & \ddots & \ddots & \vdots \\
0 & \cdots & 0 & g_0 & g_1 & \cdots & g_{n-k} & 0 \\
0 & \cdots & \cdots & 0 & g_0 & g_1 & \cdots & g_{n-k}
\end{pmatrix}
$$

(k rows and n columns) generates C. Note that $g_{n-k} = 1$, since g is monic, but we retain the symbolic notation g_{n-k} in order to display explicitly the degree of g.

▲| Library | Cyclic generating matrix of a cyclic code |▲

```
cyclic_matrix(g:Vector, n:Z)
check n⩾length(g)
:= begin
      local k=n−length(g), G
      if k>0  then
         g=g | constant_vector(k,0)
      end
      G=g
      for i in 1..k do
         g=[0]| take(g,n−1)
         G=(G,g)
      end
      [G]
   end
cyclic_matrix(g:Univariate_polynomial, n:Z):= cyclic_matrix(vector(g), n)
```

```
v=[g0, g1, g2, g3, g4];
f=polynomial(v,x)  →  g4·x⁴+g3·x³+g2·x²+g1·x+g0
cyclic_matrix(f,5)  →  (g0 g1 g2 g3 g4)
```

$$
\text{cyclic\_matrix}(v,6) \rightarrow \begin{pmatrix} g0 & g1 & g2 & g3 & g4 & 0 \\ 0 & g0 & g1 & g2 & g3 & g4 \end{pmatrix}
$$

$$
\text{cyclic\_matrix}(f,7) \rightarrow \begin{pmatrix} g0 & g1 & g2 & g3 & g4 & 0 & 0 \\ 0 & g0 & g1 & g2 & g3 & g4 & 0 \\ 0 & 0 & g0 & g1 & g2 & g3 & g4 \end{pmatrix}
$$

E.3.8. Show that the coding map $\mathbb{F}^k \rightarrow C_g$, $u \mapsto uG$, can be described in algebraic terms as the map $\mathbb{F}[x]_k \rightarrow C_g$ such that $u \mapsto u\bar{g}$.

Normalized generating and control matrices

If we want a normalized generating matrix (say in order to facilitate decoding), we can proceed as follows: For $0 \leqslant j < k$, let

$$
x^{n-k+j} = q_j g + r_j,
$$

with $deg\,(r_j) < n - k$. Then the k polynomials

$$v_j = x^{n-k+j} - r_j$$

form a basis of C_g and the corresponding coefficient matrix G' is normalized, in the sense that the matrix formed with the last k columns of G' is the identity matrix I_k (See the listing **Normalized generating and control matrices**).

```
▲ Library │ Cyclic normalized generating and control matrices        ▲
cyclic_normalized_matrix(g, n)
check n>degree(g) & monic?(g)
:= begin
     local r, x=variable(g), k=n−degree(g), G=null
     for j in 0..k−1 do
        r=remainder(xⁿ⁻ᵏ⁺ʲ,g)
        r=pad(vector(r),n−k)
        G=(G,−r)
     end
     [G]|Iₖ
   end
cyclic_normalized_control_matrix(g, n)
check n>degree(g) & monic?(g)
:= begin
     local r=degree(g), k=n−r,
           G=cyclic_normalized_matrix(g,n),
           R=G₁..ᵣ
     Iₙ₋ₖ|−Rᵀ
   end
```

```
g=1+2 X+X³; n=7;
```

$$G=\text{cyclic\_normalized\_matrix}(g,n) \;\longrightarrow\; \begin{pmatrix} 1 & 2 & 0 & 1 & 0 & 0 & 0 \\ 0 & 1 & 2 & 0 & 1 & 0 & 0 \\ -2 & -4 & 1 & 0 & 0 & 1 & 0 \\ -1 & -4 & -4 & 0 & 0 & 0 & 1 \end{pmatrix}$$

$$H=\text{cyclic\_normalized\_control\_matrix}(g,n) \;\longrightarrow\; \begin{pmatrix} 1 & 0 & 0 & -1 & 0 & 2 & 1 \\ 0 & 1 & 0 & -2 & -1 & 4 & 4 \\ 0 & 0 & 1 & 0 & -2 & -1 & 4 \end{pmatrix}$$

E.3.9. Let $u \in \mathbb{F}^k \simeq F[x]_k$. Show that the coding of u using the matrix G' can be obtained by replacing the monomials x^j of u by v_j ($0 \leqslant j < k$):

$$u_0 + u_1 x + \cdots + u_{k-1} x^{k-1} \mapsto u_0 v_0 + \ldots + u_{k-1} v_{k-1}.$$

E.3.10. Let H' be the control matrix of C_g associated to G'. Prove that the syndrome of $s \in \mathbb{F}^{n-k} \simeq F[x]_{n-k}$ of an element $a \in \mathbb{F}^n \simeq F[x]_n$ coincides with the remainder of the Euclidian division of a by g.

The dual code and the control matrix

Using the same notations, the control polynomial $\widehat{g}$ satisfies that $\widehat{g}a = 0$ is equivalent to $a \in C_g$ (for $a \in \mathbb{F}[x]_n$). Indeed, the relation $\widehat{g}a = 0$ is equivalent to say that $X^n - 1 = \widehat{g}g|\widehat{g}a_X$. Since $\mathbb{F}[X]$ is a domain, the latter is equivalent to $g|a_X$, and so we have $\bar{g}|a$, or $a \in C_g$.

Let us see that from this it follows that

$$C_g^{\perp} = \widetilde{C_{\widehat{g}}}, \qquad\qquad\qquad [3.4]$$

where $\widetilde{C_{\widehat{g}}}$ is the image of $C_{\widehat{g}}$ by the map $a \mapsto \tilde{a}$ (cf. [3.2]). Indeed, since both sides have dimension $n - k$, it is enough to see that

$$\widetilde{C_{\widehat{g}}} \subseteq C_g^{\perp}.$$

But if $a \in C_{\widehat{g}}$ and $b \in C_g$, then clearly $ab = 0$ and then $\langle \tilde{a}|b \rangle = \ell(ab) = 0$ (cf. [3.3]).

Since $\widehat{g}x^{n-k-1}, \ldots, \widehat{g}x, \widehat{g}$ form a basis of $C_{\widehat{g}}$, if we set

$$\widehat{g} = h_0 + h_1 X + \cdots + h_k X^k,$$

then

$$H = \begin{pmatrix} h_k & h_{k-1} & \cdots & h_0 & 0 & 0 & \cdots & 0 \\ 0 & h_k & h_{k-1} & \cdots & h_0 & 0 & \cdots & 0 \\ \vdots & \ddots & \ddots & \ddots & \cdots & \ddots & \ddots & \vdots \\ 0 & \cdots & 0 & h_k & h_{k-1} & \cdots & h_0 & 0 \\ 0 & \cdots & & 0 & h_k & h_{k-1} & \cdots & h_0 \end{pmatrix}.$$

is a control matrix of C_g (here $h_k = 1$, but we write h_k in order to display explicitly the indices of the coefficients; see **Cyclic control matrix**).

3.5 Remark. There are authors that call the ideal $(\widehat{g})$ the dual code of C_g. This is confusing, because the formula [3.4] shows that it does not go along well with the general definition of the dual code. On the other hand, note that the same formula shows that $C_{\widehat{g}}$ is *equivalent* to the dual of C_g.

In the next example we just need that the polynomial

$$g = X^5 - X^3 + X^2 - X - 1$$

is a factor of $X^{11} - 1$ over $\mathbb{Z}_3$. This can be checked directly, but note that in the next section we will prove more, namely, that g is irreducible and that the other irreducible factors of $X^{11} - 1$ are $X - 1$ and $X^5 + X^4 - X^3 + X^2 - 1$.

Library | Cyclic control matrix ▲ ▲

```
cyclic_control_matrix(h:Vector, n:Z) check n≥length(h)
:=  cyclic_matrix(reverse(h), n);
cyclic_control_matrix(h:monic?, n:Z) check n>degree(g)
:=  cyclic_control_matrix(vector(h), n);
```

---

h=[h0, h1, h2, h3]; n=7;

$$\text{cyclic\_control\_matrix}(h,n) \rightarrow \begin{pmatrix} h3 & h2 & h1 & h0 & 0 & 0 & 0 \\ 0 & h3 & h2 & h1 & h0 & 0 & 0 \\ 0 & 0 & h3 & h2 & h1 & h0 & 0 \\ 0 & 0 & 0 & h3 & h2 & h1 & h0 \end{pmatrix}$$

3.6 Example (The ternary Golay code). Let $q = 3$, $n = 11$ and $C = C_g$, where g is the factor of $X^{11} - 1$ described in the preceding paragraph. The type of the cyclic code C is $[11, 6]$.

Now we will prove that the minimum distance of this code is 5. If we let G denote the normalized generating matrix of C corresponding to g, then it is easy to check that $\overline{G}\,\overline{G}^T = 0$, where $\overline{G}$ is the parity completion of G (in order to preserve the submatrix I_6 at the end, we place the parity symbols on the left):

n=11; g=factor(x^n−1, Z_3)$_{2,1}$ → x^5+2·x^3+x^2+2·x+2

G=cyclic_normalized_matrix(g, n);

$$G=\text{left\_parity\_completion}(G) \rightarrow \begin{pmatrix} 1 & 2 & 2 & 1 & 2 & 0 & 1 & 0 & 0 & 0 & 0 & 0 \\ 1 & 0 & 2 & 2 & 1 & 2 & 0 & 1 & 0 & 0 & 0 & 0 \\ 2 & 2 & 2 & 0 & 1 & 1 & 0 & 0 & 1 & 0 & 0 & 0 \\ 1 & 1 & 0 & 1 & 1 & 1 & 0 & 0 & 0 & 1 & 0 & 0 \\ 0 & 1 & 2 & 2 & 2 & 1 & 0 & 0 & 0 & 0 & 1 & 0 \\ 2 & 1 & 2 & 1 & 0 & 2 & 0 & 0 & 0 & 0 & 0 & 1 \end{pmatrix}$$

$$G \cdot G^T \rightarrow \begin{pmatrix} 0 & 0 & 0 & 0 & 0 & 0 \\ 0 & 0 & 0 & 0 & 0 & 0 \\ 0 & 0 & 0 & 0 & 0 & 0 \\ 0 & 0 & 0 & 0 & 0 & 0 \\ 0 & 0 & 0 & 0 & 0 & 0 \\ 0 & 0 & 0 & 0 & 0 & 0 \end{pmatrix}$$

Thus the code $\overline{C} = \langle \overline{G} \rangle$ is selfdual and in particular the weight of any vector of $\overline{C}$ is a multiple of 3. Since the rows $\overline{G}$ have weight 6, the minimum distance of $\overline{C}$ is either 3 or 6. But each row of $\overline{G}$ has exactly one zero in the first 6 columns, and the position of this 0 is different for the different rows, so it is clear that a linear combination of two rows of $\overline{G}$ has weight at least $2 + 2$, hence at least 6. Since the weight of this combination is clearly not more than $12 - 4 = 8$, it has weight exactly 6. In particular, for each such combination there appear exactly two zeros in the first 6 positions. Now a linear combination of 3 rows will have weight at least 1+3, and so at least 6.

All other linear combinations of rows of $\overline{G}$ have weight at least 4, and so at least 6. So $\overline{C}$ has type $[12, 6, 6]$ and as a consequence C has type $[12, 6, 5]$. It is a perfect code. It is known (but rather involved) that all codes of type $(11, 3^6, 5)$ are equivalent to C (cf. [11]) and any such code will be said to be a *ternary Golay code*.   □

E.3.11. Find all binary cyclic codes of length 7. In each case, give the generating and control polynomials, a normalized generating matrix and identify the dual code.

Summary

- Cyclic ring structure of $\mathbb{F}^n$.

- Generating and control polynomials of a cyclic code.

- The dimension of a cyclic code is the degree of the control polynomial, hence the degree of the generating polynomial is the codimension.

- Cyclic and normalized generating matrices. Systematic encoding.

- The dual of the cyclic code C_g, g a divisor of $X^n - 1$, is the cyclic code $C_{\hat{g}}$, but written in reverse order. This allows to produce a control matrix of G from the generating matrix of $C_{\hat{g}}$.

- The ternary Golay code, which has type $[11, 6, 5]_3$. It is perfect and its parity completion, which has type $[12, 6, 6]_3$, is a selfdual code.

Problems

P.3.1 Consider two linear codes C_1 and C_2 of the same length and with control matrices H_1 and H_2.

1) Check that $C_1 \cap C_2$ is a linear code and find a control matrix for it in terms of H_1 and H_2.

2) Prove that if C_1 and C_2 are cyclic with generating polynomials g_1 and g_2, then $C_1 \cap C_2$ is cyclic. Find its generating polynomial in terms of g_1 and g_2.

P.3.2. Let $G = I_6|(\frac{1_5}{S_5})$, where S_5 is the Paley matrix of $\mathbb{Z}_5$. Prove that $\langle G \rangle$ is a ternary Golay code.

P.3.3. Use the MacWilliams identities to determine the weight enumerator of the parity completion of the ternary Golay code.

P.3.4. The order of $p = 2 \bmod n = 9$ is $k = 6$. Let $q = 2^k = 64$ and α a primitive n-th root of 1 in $K = \mathbb{F}_q$. Let $\tau : K \to \mathbb{Z}_2^n$ be the map defined by $\tau(\xi) = (Tr(\xi), Tr(\alpha\xi), \ldots, Tr(\alpha^{n-1}\xi))$.

1) Show that τ is linear and injective.

2) If C is the image of τ, prove that C is a cyclic code $[n, k]$.

3) Find the generator polynomial of C.

4) What is the minimum distance of C?

5) Generalize the construction of τ for any prime p and any positive integer n not divisible by p and show that 1 and 2 still hold true for this general τ.

> E. Prange [as early as 1962] seems to have been the first to study cyclic codes.
>
> F. MacWilliams and N.J.A. Sloane, [11], p. 214.

3.2   Effective factorization of $X^n - 1$

> Cyclic codes are the most studied of all codes.
> F. MacWilliams and N.J.A. Sloane, [11], p. 188.

Essential points

- The splitting field of $X^n - 1$ over a finite field.
- Ciclotomic classes modulo n and the factorization theorem.
- The factorization algorithm.
- Cyclotomic polynomials.

Introduction

In the previous section we learnt that the generator polynomials g of cyclic codes of length n over $\mathbb{F}_q$ are, assuming $gcd\,(n, q) = 1$, the monic divisors of $X^n - 1$ and that if $f_1, \ldots, f_r$ are the distinct monic irreducible factors of $X^n - 1$ over $\mathbb{F}_q$ then g has the form $g = f_{i_1} \cdots f_{i_s}$, with $0 \leqslant s \leqslant r$, $1 \leqslant i_1 < \cdots < i_s \leqslant r$.

The goal of this section is to explain an algorithm that supplies the factors $f_1, \ldots, f_r$ in an efficient way. It is basically the method behind the function factor(f,K). **Examples of factorization** illustrates how it works.

Examples of factorization

F2= $\mathbb{Z}_2$; F3=$\mathbb{Z}_3$; F4=F2[t]/(t^2+t+1);

F8=F2[u]/(u^3+u+1); F16=F2[s]/(s^4+s^3+s^2+s+1);

factor(x^5−1, F2) $\longrightarrow$ (x+1)·(x^4+x^3+x^2+x+1)

factor(x^5−1, F4) $\longrightarrow$ (x+1)·(x^2+t·x+1)·(x^2+(t+1)·x+1)

factor(x^5−1, F16) $\longrightarrow$ (x+1)·(x+s)·(x+s^2)·(x+s^3)·(x+(s^3+s^2+s+1))

factor(x^7−1, F2) $\longrightarrow$ (x+1)·(x^3+x+1)·(x^3+x^2+1)

factor(x^7−1, F3) $\longrightarrow$ (x+2)·(x^6+x^5+x^4+x^3+x^2+x+1)

factor(x^{23}−1, F2)

$\longrightarrow$ (x+1)·(x^{11}+x^9+x^7+x^6+x^5+x+1)·(x^{11}+x^{10}+x^6+x^5+x^4+x^2+1)

factor(x^{11}−1, F3) $\longrightarrow$ (x+2)·(x^5+2·x^3+x^2+2·x+2)·(x^5+x^4+2·x^3+x^2+2)

3.7 Example. We first illustrate the basic ideas by factoring $X^7 - 1$ over $\mathbb{Z}_2$. Since we always have $X^n - 1 = (X - 1)(X^{n-1} + \cdots + X + 1)$ it is enough to factor $f = X^6 + \cdots + X + 1$.

First let us consider a direct approach, using the special features of this case. Since f has 7 terms, it has no roots in $\mathbb{Z}_2$ and hence it does not have linear factors. Since the only irreducible polynomial of degree 2 is

$$h = X^2 + X + 1$$

and it does not divide f (the remainder of the Euclidean division of f by h is 1), f is either irreducible or factors as the product of two irreducible polynomials of degree 3. But there are only two irreducible polynomials of degree 3 over $\mathbb{Z}_2$ ($f_1 = X^3 + X + 1$ and $f_2 = X^3 + X^2 + 1$) and if f factors the only possibility is $f = f_1 f_2$ (since f does not have multiple factor, $f \neq f_1^2$ and $f \neq f_2^2$). Finally, this relation is easily checked and so we have that $X^7 - 1 = (X - 1)f_1 f_2$ is the factorization we were seeking.

Now let us consider an indirect approach. It consists in trying to find a finite field extension $\mathbb{F}'/\mathbb{F}$ (here $\mathbb{F} = \mathbb{F}_q = \mathbb{Z}_2$) that contains all roots of $X^7 - 1$. If $\mathbb{F}'$ exists, it contains, in particular, an element of order 7. If q' is the cardinal of $\mathbb{F}'$, we must have that $q' - 1$ is divisible by 7. But $q' = q^m$, for some positive integer m, and hence m must satisfy that $q^m - 1$ is divisible by 7. As $q = 2$, the first such m is 3, and so it is natural to try $\mathbb{F}' = \mathbb{F}_8$.

Now the group $\mathbb{F}_8^*$ is cyclic of order 7, and hence

$$X^7 - 1 = \prod_{\alpha \in \mathbb{F}_8^*} (X - \alpha).$$

The root 1 corresponds to the factor $X - 1$. Of the other 6 roots, 3 must be the roots of f_1 and the other 3 the roots of f_2. The actual partition depends, of course, on the explicit construction of $\mathbb{F}'$. Since $[\mathbb{F}' : \mathbb{F}] = 3$, we can define $\mathbb{F}'$ as the extension of $\mathbb{F}$ by an irreducible polynomial of degree 3, say f_1: $\mathbb{F}' = \mathbb{F}/(X^3 + X + 1)$. If we let $u = [X]$ (to use the notation in **Examples of factorization**), then u is a root of f_1. Being in characteristic 2, u^2 and $u^4 = u^2 + u$ are the other roots of f_1 and $f_1 = (X - u)(X - u^2)(X - u^4)$. So $u^3 = u + 1$, $(u+1)^2 = u^2 + 1$ and $(u^2 + 1)^2 = u^4 + 1 = u^2 + u + 1$ are the roots of f_2 and $f_2 = (X - (u+1))(X - (u^2+1))(X - (u^2+u+1))$. $\square$

The ideas involved in the last two paragraphs of the preceeding example can be played successfully to find the factorization of $X^n - 1$ over $\mathbb{F}_q$ for all n and q (provided $gcd\,(n, q) = 1$), as we will see in the next subsection.

The splitting field of $X^n - 1$

The condition $gcd\,(n, q) = 1$ says that $[q]_n \in \mathbb{Z}_n^*$ and hence we may consider the order m of $[q]_n$ in that group. By definition, m is the least positive integer such that $q^m \equiv 1 \pmod{n}$. In other words, m is the least positive integer such that $n\,|\,(q^m - 1)$ and we will write $e_n(q)$ to denote it. For example, $e_7(2) = 3$, because $2^3 = 8 \equiv 1 \pmod 7$. The value of $e_n(q)$ can be obtained with the function order(q,n).

```
| order(2,7)  → 3
| order(2,15) → 4
| order(2,23) → 11
| order(3,11) → 5
```

Let now $h \in \mathbb{F}[X]$ be any monic irreducible polynomial of degree $m = e_n(q)$ and define $\mathbb{F}' = \mathbb{F}_{q^m}$ as $\mathbb{F}[X]/(h)$. Let α be a primitive element of $\mathbb{F}'$ (if h were a primitive polynomial, we could choose $\alpha = [X]_h$). Then $ord(\alpha) = q^m - 1$ is divisible by n, by definition of m. Let $r = (q^m - 1)/n$ and $\omega = \alpha^r$.

3.8 Proposition. *Over $\mathbb{F}'$ we have*

$$X^n - 1 = \prod_{j=0}^{n-1} (X - \omega^j).$$

Proof: Since $ord(\omega) = (q^m - 1)/r = n$, the set $R = \{\omega^j \,|\, 0 \leqslant j \leqslant n - 1\}$ has cardinal n. Moreover, ω^j is a root of $X^n - 1$ for all j, because $(\omega^j)^n = (\omega^n)^j = 1$. So the set R supplies n distinct roots of $X^n - 1$ and hence $\prod_{j=0}^{n-1}(X - \omega^j)$ is monic polynomial of degree n that divides $X^n - 1$. Thus these two polynomials must be the same.   □

3.9 Proposition. $\mathbb{F}' = \mathbb{F}[\omega]$ *and so $\mathbb{F}'$ is the splitting field of $X^n - 1$ over $\mathbb{F}$.*

Proof: Indeed, if $|\mathbb{F}[\omega]| = q^s$, then $n = ord(\omega)$ divides $q^s - 1$ and by definition of m we must have $s = m$.   □

Cyclotomic classes

To proceed further we need the notion of cyclotomic classes. Given an integer j in the range $0..(n-1)$, the q-cyclotomic class of j modulo n is defined as the set

$$C_j = \{j, qj \ldots, q^{r-1}j\} \pmod{n},$$

where r is the least positive integer such that $q^r j \equiv j \pmod{n}$. For example, $C_0 = \{0\}$ always, and if $n = 7$ and $q = 2$, then $C_1 = \{1, 2, 4\}$ and $C_3 = \{3, 6, 5\}$. If $n = 11$ and $q = 3$, then $C_1 = \{1, 3, 9, 5, 4\}$ and $C_2 = \{2, 6, 7, 10, 8\}$.

E.3.12. Show that if $C_j \cap C_k \neq \emptyset$ then $C_j = C_k$. So the distinct cyclotomic classes form a partition of $\mathbb{Z}_n$. □

The function cyclotomic_class(j,n,q) yields the cyclotomic class C_j. In the case $q = 2$ we can use cyclotomic_class(j,n).

```
cyclotomic_class(1,7)    →  {1, 2, 4}
cyclotomic_class(3,7)    →  {3, 6, 5}
cyclotomic_class(1,11,3) →  {1, 3, 9, 5, 4}
cyclotomic_class(2,11,3) →  {2, 6, 7, 10, 8}
```

The function cyclotomic_classes(n,q) yields the list of q-cyclotomic classes modulo n. In the case $q = 2$ we can use cyclotomic_classes(n).

```
cyclotomic_classes(7)    →  {{0}, {1, 2, 4}, {3, 6, 5}}
cyclotomic_classes(11,3) →  {{0}, {1, 3, 9, 5, 4}, {2, 6, 7, 10, 8}}
```

The factorization theorem

If C is a q-cyclotomic class modulo n, define f_C as the polynomial with roots ω^j, for $j \in C$:

$$f_C = \prod_{j \in C} (X - \omega^j).$$

3.10 Lemma. *The polynomial f_C has coefficients in $\mathbb{F}$ for all C.*

Proof: Let us apply the Frobenius automorphism $x \mapsto x^q$ to all coefficients of f_C. The result coincides with

$$\prod_{j \in C} (X - \omega^{jq}).$$

But from the definition of q-cyclotomic class we have $\{jq \mid j \in C\} = C$ and so

$$\prod_{j \in C} (X - \omega^{jq}) = f_C.$$

This means that $a^q = a$ for all coefficients a of f_C and we know that this happens if and only if $a \in \mathbb{F}$. □

3.11 Theorem. *The correspondence $C \mapsto f_C$ is a bijection between the set of q-cyclotomic classes modulo n and the set of monic irreducible factors of $X^n - 1$ over $\mathbb{F}_q$.*

Proof: The factorization of $X^n - 1$ over $\mathbb{F}'$ established in lemma 3.8 and the fact that the q-cyclotomic classes modulo n form a partition of $\mathbb{Z}_n$ imply that

$$X^n - 1 = \prod_C f_C,$$

where the product runs over all q-cyclotomic classes modulo n. Thus it is enough to prove that $f_C \in \mathbb{F}[X]$ is irreducible. To see this, note that $\{\omega^j \mid j \in C\}$ is the conjugate set of any of its elements and so f_C is, by proposition 2.40, the minimum polynomial over $\mathbb{F}$ of ω^j for any $j \in C$. Therefore f_C is irreducible.   □

E.3.13. If n is divisible by p (the characteristic of $\mathbb{F}$), the factorization of $X^n - 1$ can be reduced easily to the case $gcd\,(n, q) = 1$. Indeed, if $n = n'p^s$, where p is prime and $gcd\,(n', p) = 1$, show that $X^n - 1 = (X^{n'} - 1)^{p^s}$. Consequently the number of distinct irreducible factors of $X^n - 1$ coincides with the number of q-cyclotomic classes modulo n' and each such factor has multiplicity p^s.

The factorization algorithm

Here are the steps we can follow to factor $X^n - 1$ over $\mathbb{F}_q$. If $n = n'p^s$ with $gcd\,(n', q) = 1$, factor $X^{n'} - 1$ by going through the steps below for the case $gcd\,(n, q) = 1$ and then count every factor with multiplicity p^s.

So we may assume that $gcd\,(n, q) = 1$ and in this case we can proceed as follows:

1) Find $m = e_n(q)$, the order of q in $\mathbb{Z}_n^*$. Define $r = (q^m - 1)/n$.
2) Find and irreducible polynomial $h \in \mathbb{F}[X]$ of degree m and let $\mathbb{F}' = \mathbb{F}[X]/(h)$.
3) Find a primitive element α of $\mathbb{F}'$ and let $\omega = \alpha^r$.
4) Form the list $\mathcal{C}$ of q-cyclotomic classes C modulo m, and for each such class C compute $f_C = \prod_{j \in C}(X - \omega^j)$.
5) The $\{f_C\}_{C \in \mathcal{C}}$ is the set of monic irreducible factors of $X^n - 1$ (hence $X^n - 1 = \prod_{C \in \mathcal{C}} f_C$).

It is instructive to compare this algorithm with the listing **Script for the factorization algorithm** and to look at the example attached there.

For another example, see **Factorization of $X^{17} - 1$ over $\mathbb{Z}_{13}$**.

```
┌──────────────────────────────────────────────────────────┐
│ ▲│ Library | Script  for the factorization algorithm of Xⁿ−1 over K │ ▲ │
├──────────────────────────────────────────────────────────┤
```

m=order(q,n)
h=irreducible_polynomial(K,m,x)
F=K[x]/(h); r=(q^m−1)/n
α=primitive_element(F); ω=α^r
C=cyclotomic_classes(n,q)

$$\left\{ \prod_{j\,in\,c} (X-\omega^j) \text{ with c in C} \right\}$$

Example

n=11; q=3; K=$\mathbb{Z}_q$;

m=order(q,n);

h=irreducible_polynomial(K, m, x);

F=K[x]/(h); r=(q^m−1)/n ⟶ 22

α=primitive_element(F); ω=α^r; order(ω) ⟶ 11

C=cyclotomic_classes(n,q);

$$\left\{ \prod_{j\,in\,c} (X-\omega^j) \text{ with c in C} \right\}$$

⟶ {X+2, X^5+X^4+2·X^3+X^2+2, X^5+2·X^3+X^2+2·X+2}

factor(X^{11}−1,$\mathbb{Z}_3$) ⟶ (X+2)·(X^5+2·X^3+X^2+2·X+2)·(X^5+X^4+2·X^3+X^2+2)

Factorization of X^{17}−1 over F=$\mathbb{Z}_{13}$

n=17; q=13; K=$\mathbb{Z}_q$; m=order(q,n) ⟶ 4

h=irreducible_polynomial(K,m,x) ⟶ x^4+11

F=K[x]/(h); r=(q^m−1)/n ⟶ 1680

α=primitive_element(F) ⟶ 6·x^3+3·x^2+2·x+1

ω=α^r ⟶ 10·x^3+2·x^2+8·x+1

order(ω) ⟶ 17

C=cyclotomic_classes(n,q)

⟶ {{0}, {1, 13, 16, 4}, {2, 9, 15, 8}, {3, 5, 14, 12}, {6, 10, 11, 7}}

$$f(c) := \prod_{j\,in\,c} (X-\omega^j);$$

f(C$_1$), f(C$_2$), f(C$_3$) ⟶ X+12, X^4+9·X^3+9·X+1, X^4+10·X^3+9·X^2+10·X+1

f(C$_4$), f(C$_5$) ⟶ X^4+2·X^3+5·X^2+2·X+1, X^4+6·X^3+6·X^2+6·X+1

Summary

- The splitting field of $X^n - 1$ over a finite field $\mathbb{F} = \mathbb{F}_q$ is, assuming $\gcd(n, q) = 1$ and setting $m = e_n(q)$, the field $\mathbb{F}' = \mathbb{F}_{q^m}$. If α is a primitive element of $\mathbb{F}'$, $r = (q^m - 1)/n$ and $\omega = \alpha^r$, then $X^n - 1 = \prod_{j=1}^n (X - \omega^j)$.

- Cyclotomic classes modulo n and the factorization theorem, which asserts (still with the condition $\gcd(n, q) = 1$) that the monic irreducible factors of $X^n - 1$ over $\mathbb{F}$ are the polynomials $f_C = \prod_{j \in C}(X - \omega^j)$, where C runs over the set of q-cyclotomic classes modulo n.

- The factorization algorithm. If p is the characteristic of $\mathbb{F}$ and $n = n'p^s$ with $\gcd(n', p) = 1$, then $X^n - 1 = (X^{n'} - 1)^{p^s}$ reduces the problem to the factorization of $X^{n'} - 1$. If n is not divisible by p (this is equivalent to saying that $\gcd(n, q) = 1$), then the factorization theorem gives an effective factorization procedure.

- Cyclotomic polynomials (they are introduced and studied in the problems P.3.7–P.3.8).

Problems

P.3.5. Show that a q-ary cyclic code C of length n is invariant under the permutation σ such that $\sigma(j) = qj \pmod{n}$.

P.3.6. We have seen that over $\mathbb{Z}_3$ we have $X^{11} - 1 = (X - 1)g_0 g_1$, where

$$g_0 = (X^5 - X^3 + X^2 - X - 1) \quad \text{and} \quad g_1 = (X^5 + X^4 - X^3 + X^2 - 1).$$

1) With the notations introduced in the example included in **Script for the factorization algorithm**, show that the roots of g_0 (g_1) are ω^j, where j runs over the nonzero quadratic residues (over the quadratic non-residues) of $\mathbb{Z}_{11}$.

2) If $k \in \mathbb{Z}_{11}^*$ is a quadratic non-residue (so $k \in \{2, 6, 7, 8, 10\}$) and π_k is the permutation $i \mapsto ki$ of $\mathbb{Z}_{11} = \{0, 1, ..., 10\}$, prove that $\pi_k(C_{g_0}) = C_{g_1}$ (here then are two distinct cyclic codes that are equivalent).

P.3.7. In the factorization of $X^n - 1 \in \mathbb{F}_q[X]$ given by lemma 3.8, let

$$Q_n = \prod_{\gcd(j,n)=1} (X - \omega^j)$$

(this polynomial has degree $\varphi(n)$ and is called the n-th *cyclotomic polynomial* over $\mathbb{F}_q$). Prove that:

1) $X^n - 1 = \prod_{d \mid n} Q_d$.

2) Deduce from 1 that $Q_n \in \mathbb{Z}_p[X]$ for all n, where p is the characteristic of $\mathbb{F}_q$. *Hint:* Use induction on n.

3) Prove that

$$Q_n = \prod_{d \mid n} (X^{n/d} - 1)^{\mu(d)}.$$

Note in particular that Q_n is independent of the field $\mathbb{F}_q$, and that it can be computed for any positive integer n, although it can be the n-th cyclotomic polynomial over $\mathbb{F}_q$ (or over $\mathbb{Z}_p$) if and only if n is not divisible by p. *Hint:* Use the multiplicative version of Möbius inversion formula.

4) Use the formula in 3 to prove that if n is prime then

$$Q_n = X^{n-1} + X^{n-2} + \cdots + X + 1$$

and

$$Q_{n^k}(X) = Q_n(X^{n^{k-1}}).$$

For example, $Q_5 = X^4 + X^3 + X^2 + X + 1$ and

$$Q_{25}(X) = Q_5(X^5) = X^{20} + X^{15} + X^{10} + X^5 + 1.$$

5) If m denotes the order of q modulo n (so we assume $\gcd(n, q) = 1$), prove that Q_n factors over $\mathbb{F}_q$ into $\varphi(n)/m$ distinct irreducible polynomials of degree m each. For example, $e_{25}(7) = 4$ and $\varphi(25) = 20$, and so Q_{25} splits over $\mathbb{Z}_7$ as the product of five irreducible polynomials of degree 4 each (see **Computation of cyclotomic polynomials**). *Hint:* the degree over $\mathbb{F}_q$ of any root of Q_n is equal to m.

P.3.8. If $P_q(n, X)$ is the product of all monic irreducible polynomials of degree n over $\mathbb{F}_q$, prove that

$$P_q(n, X) = \prod_m Q_m(X)$$

for all $n > 1$, where the product is extended over the set $M(q, n)$ of all divisors $m > 1$ of $q^n - 1$ for which n is the multiplicative order of q modulo m (see **Computation of the set M(q,n)**). *Hint:* First show that the roots of $P_q(n, X)$ are the elements of degree n in $\mathbb{F}_{q^n}$.

```
┌─┬─────────┬──────────────────────────────────────────────────┬──┐
│▲│ Library │ Computation of cyclotomic polynomials            │▲ │
├─┴─────────┴──────────────────────────────────────────────────┴──┤
│ Q(n:ℤ, X:Identifier) check n⩾0  :=                               │
│    begin                                                         │
│       local μ=mu_moebius                                         │
│       if n=1  then                                               │
│          X⁰                                                      │
│       else                                                       │
│              ∏        (Xⁿ/ᵈ−1)^μ(d)                              │
│          d in divisors(n)                                        │
│       end                                                        │
│    end                                                           │
└──────────────────────────────────────────────────────────────┘
```

$\{Q(n, X)$ with n in 1..10 such_that not prime?(n)$\}$
$\rightarrow \{1, X^2+1, X^2-X+1, X^4+1, X^6+X^3+1, X^4-X^3+X^2-X+1\}$

$Q(25, X) \rightarrow X^{20}+X^{15}+X^{10}+X^5+1$

$f=\text{factor}(X^{20}+X^{15}+X^{10}+X^5+1, \mathbb{Z}_7);$ $f=\{r_1$ with r in f$\};$

f_1, f_2, f_3
$\rightarrow X^4+2 \cdot X^3+4 \cdot X^2+2 \cdot X+1, X^4+4 \cdot X^3+4 \cdot X+1, X^4+4 \cdot X^3+3 \cdot X^2+4 \cdot X+1$

$f_4, f_5 \rightarrow X^4+5 \cdot X^3+5 \cdot X^2+5 \cdot X+1, X^4+6 \cdot X^3+5 \cdot X^2+6 \cdot X+1$

```
┌─┬─────────┬──────────────────────────────────────────────────┬──┐
│▲│ Library │ Computation of the set M(q,n)                    │▲ │
├─┴─────────┴──────────────────────────────────────────────────┴──┤
│ M(q,n):=                                                         │
│    begin                                                         │
│       local D=tail(divisors(qⁿ−1)), M={ }                        │
│       for m in D do                                              │
│          if gcd(q,m)=1 & order(q,m)=n  then                      │
│             M=M | {m}                                            │
│          end                                                     │
│       end                                                        │
│    end                                                           │
│ Special call for the binary case                                 │
│ M(n):=M(2,n)                                                     │
└──────────────────────────────────────────────────────────────┘
```

$\{M(n)$ with n in 2..7$\} \rightarrow \{\{3\}, \{7\}, \{5, 15\}, \{31\}, \{9, 21, 63\}, \{127\}\}$

$\{M(3,n)$ with n in 2..5$\}$
$\rightarrow \{\{4, 8\}, \{13, 26\}, \{16, 5, 10, 20, 40, 80\}, \{11, 22, 121, 242\}\}$

3.12 Remark (On computing cyclotomic polynomials). The procedure explained in P.3.7 to find the n-th cyclotomic polynomial involves cumbersome computations for large n. For example, if $n = 11!$ then $Q(n, X)$ is a polynomial of degree $\varphi(n) = 8294400$ with 540 terms (the number of divisors of n) and it took 45 seconds to evaluate in a Compaq Armada E500. Fortunately there is a better scheme based on P.3.9. It is presented in **Efficient computation of cyclotomic polynomials** and is the one implemented in the internal function cyclotomic_polynomial(n,T).

For the key mathematical properties of the cyclotomic polynomials on which this function is based, see P.3.9.

With the internal function, and $n = 11!$, $Q(n, X)$ was obtained in 0.23 seconds, while the same function coded externally (the last quoted listing) took 0.25 seconds. If we want to calculate a list of cyclotomic polynomials, the direct method may be much better, because it does not have to factor and find the divisors of many integers.

```
▲ Library │ Efficient computation of cyclotomic polynomials        ▲

cyclotomic_polynomial(n:ℤ, T:Identifier) check n>0 :=
   begin
                                                           p
      local P=factor(n), s=length(P), p=P₁,₁, m=p, i=1, f= ∑ Tᵖ⁻ʲ
                                                          j=1
      while i<s do
         i=i+1; p=Pᵢ,₁; m=m·p
               evaluate(f,Tᵖ)
         f=───────────────────
                    f
      end
      evaluate(f,Tⁿ/ᵐ)
   end
```

$\Vert$ cyclotomic_polynomial(6!,T) $\longrightarrow$ $T^{192}+T^{168}-T^{120}-T^{96}-T^{72}+T^{24}+1$

P.3.9. Let $Q_n(X)$ be the n-th cyclotomic polynomial ($n \geqslant 1$). Prove that:

1) $Q_{mn}(X)$ divides $Q_n(X^m)$ for any positive integers m and n.

2) $Q_{mn}(X) = Q_n(X^m)$ if every prime divisor of m also divides n.

3) For any n, $Q_n(X) = Q_r(X^{n/r})$, where r is the product of the prime divisors of n.

3.3 Roots of a cyclic code

An alternative specification of a cyclic code can be made in terms of the roots, possibly in an extension field, of the generator $g(X)$ of the ideal.

W.W. Peterson and E.J. Weldon, Jr., [16], p. 209.

Essential points

- Determination of a cyclic code by the set of roots of its generating polynomial.

- Bose–Chaudhuri–Hocquenghem (*BCH*) codes. The *BCH* lower bounds on the minimum distance and the dimension.

- The Reed–Solomon primitive codes are *BCH* codes.

Introduction

Let C be the cyclic code of length n and let g be its generating polynomial. By definition, the *roots* of $C = C_g$ are the roots of g in the splitting field $\mathbb{F}' = \mathbb{F}_{q^m}$ of $X^n - 1$ over $\mathbb{F}_q$. If $\omega \in \mathbb{F}_{q^m}$ is a primitive n-th root of unity and we let E_g denote the set of all $k \in \mathbb{Z}_n$ such that ω^k is a root of g, then we know that E_g is the union of the q-cyclotomic classes corresponding to the monic irreducible factors of g.

Let $E'_g \subseteq E_g$ be a subset formed with one element of each q-cyclotomic class contained in E_g. Then we will say that $M = \{\omega^k \mid k \in E'_g\}$ is a *minimal set of roots* of C_g.

3.13 Proposition. *If M is a minimal set of roots of a cyclic code C of length n, then*

$$C = \{a \in \mathbb{F}[x]_n \mid a(\xi) = 0 \text{ for all } \xi \in M\}.$$

Proof: If $a \in C$, then a is a multiple of $\bar{g}$, where g is the generating polynomial of C, and hence it is clear that $a(\xi) = 0$ for all $\xi \in M$. So we have the inclusion

$$C \subseteq \{a \in \mathbb{F}[x]_n \mid a(\xi) = 0 \text{ for all } \xi \in M\}.$$

Now let $a \in \mathbb{F}[x]_n$ and assume that $a_X(\xi) = a(\xi) = 0$ for all $\xi \in M$. By definition of M, there is $j \in E'_g$ such that $\xi = \omega^j$, and so ω^j is a root of a_X. Since the minimal polynomial of ω^j over $\mathbb{F}$ is f_{C_j}, we have that $f_{C_j}|a_X$ and hence a_X is divisible by $\prod_{j \in E'_g} f_{C_j} = g$, and this means that $a \in (\bar{g}) = C_g$. $\qquad\qquad\square$

E.3.14. If we define a *generating set of roots* of C_g as a set of roots of C_g that contains a minimal set of roots, check that Proposition 3.13 remains true if we replace 'minimal set' by 'generating set'. $\qquad\qquad\square$

The description of C_g given in Proposition 3.13 can be inverted. If

$$\xi = \xi_1, \ldots, \xi_r \in \mathbb{F}_{q^m}$$

are n-th roots of unity, then the polynomials $a \in \mathbb{F}[x]_n$ such that $a(\xi_i) = 0$, $i = 1, \ldots, r$, form an ideal C_ξ of $\mathbb{F}[x]_n$, which will be called the *cyclic code determined by ξ*. If g_i is the minimum polynomial of ξ_i over $\mathbb{F}_q$, then $C_\xi = C_g$, where g is the *lcm* of the polynomials $g_1, \ldots, g_r$. This observation will be used below for the construction of codes with predetermined properties.

Finding a control matrix from the roots

Let us find a control matrix of C_ξ. For each $i \in \{1, \ldots, r\}$, the condition $a(\xi_i) = 0$ can be understood as a linear relation, with coefficients $1, \xi_i, \ldots, \xi_i^{n-1} \in \mathbb{F}_q(\xi_i)$, that is to be satisfied by the coefficients (that we will regard as unknowns) $a_0, \ldots, a_{n-1}$ of a. If we express each ξ_i^j as a linear combination of a basis of $\mathbb{F}_{q^m}$ over $\mathbb{F}_q$, the relation in question can be understood as m linear relations that are to be satisfied by $a_0, \ldots, a_{n-1}$. Therefore the conditions that define C_ξ are equivalent to mr linear equations with coefficients in $\mathbb{F}_q$. A control matrix for C_ξ can be obtained by selecting a submatrix H of the matrix A of the coefficients of this linear system of equations such that the rows of H are linearly independent and with $rank(H) = rank(A)$ (for an implementation of this method, see the listing in Remark 4.2).

3.14 *Example* (The Hamming codes revisited). Let m be a positive integer such that $gcd\,(m, q-1) = 1$, and define $n = (q^m - 1)/(q-1)$. Let $\omega \in \mathbb{F}_{q^m}$ be a primitive n-th root of unity (if α is a primitive element of $\mathbb{F}_{q^m}$, take $\omega = \alpha^{q-1}$). Then C_ω is equivalent to the Hamming code of codimension m, $Ham_q(m)$.

Indeed, $n = (q-1)(q^{m-2} + 2q^{m-3} + \ldots + m - 1) + m$, as it can easily be checked, and hence $gcd\,(n, q-1) = 1$. It follows that ω^{q-1} is a primitive n-th root of unity and hence that $\omega^{i(q-1)} \neq 1$ for $i = 1, \ldots, n-1$. In particular, $\omega^i \notin \mathbb{F}_q$. Moreover, ω^i and ω^j are linearly independent over $\mathbb{F}_q$ if $i \neq j$. Since n is the maximum number of vectors of $\mathbb{F}_{q^m}$ that are linearly independent over $\mathbb{F}_q$ (cf. Remark1.39), the result follows from the above description of the control matrix of C_ω and the definition of the Hamming code $Ham_q(m)$ (p. 51).

BCH codes

> An important class of cyclic codes, still used a lot in practice, was discovered by R. C. Bose and D. K. Chaudhuri (1960) and independently by A. Hocquenghem (1959).
>
> J. H. van Lint, [25], p. 91.

Let $\omega \in \mathbb{F}_{q^m}$ be a primitive n-th root of unity. Let $\delta \geqslant 2$ and $\ell \geqslant 1$ be integers. Write $BCH_\omega(\delta, \ell)$ to denote the cyclic code of length n generated by the least common multiple g of the minimal polynomials $g_i = p_{\omega^{\ell+i}}$, $i \in \{0, \ldots, \delta - 2\}$. This code will be said to be a *BCH* code (for Bose–Chaudhuri–Hocquenghem) with *designed distance* δ and *offset* ℓ. In the case $\ell = 1$, we will write $BCH_\omega(\delta)$ instead of $BCH_\omega(\delta, 1)$ and we will say that these are *strict BCH* codes. A BCH code will be said to be *primitive* if $n = q^m - 1$ (note that this condition is equivalent to say that ω is a primitive element of $\mathbb{F}_{q^m}$).

3.15 Theorem (*BCH* bound). *If d is the minimum distance of $BCH_\omega(\delta, \ell)$, then $d \geqslant \delta$.*

Proof: First note that an element $a \in \mathbb{F}[x]_n$ belongs to $BCH_\omega(\delta, \ell)$ if and only if $a(\omega^{\ell+i}) = 0$ for all $i \in \{0, \ldots, \delta - 2\}$. But the relation $a(\omega^{\ell+i}) = 0$ is equivalent to

$$a_0 + a_1 \omega^{\ell+i} + \ldots + a_{n-1} \omega^{(n-1)(\ell+i)} = 0$$

and therefore

$$(1, \omega^{\ell+i}, \omega^{2(\ell+i)}, \ldots, \omega^{(n-1)(\ell+i)})$$

is a control vector of $BCH_\omega(\delta, \ell)$. But the matrix H whose rows are these vectors has the property, by the Vandermonde determinant, that any $\delta - 1$ of its columns are linearly independent. Indeed, the determinant of the columns $j_1, \ldots, j_{\delta-1}$ is equal to

$$\begin{vmatrix} \omega^{j_1 \ell} & \cdots & \omega^{j_{\delta-1} \ell} \\ \omega^{j_1(\ell+1)} & \cdots & \omega^{j_{\delta-1}(\ell+1)} \\ \vdots & & \vdots \\ \omega^{j_1(\ell+\delta-2)} & \cdots & \omega^{j_{\delta-1}(\ell+\delta-2)} \end{vmatrix} = \omega^{j_1 \ell} \cdots \omega^{j_{\delta-1} \ell} \begin{vmatrix} 1 & \cdots & 1 \\ \omega^{j_1} & \cdots & \omega^{j_{\delta-1}} \\ \vdots & & \vdots \\ \omega^{j_1(\delta-2)} & \cdots & \omega^{j_{\delta-1}(\delta-2)} \end{vmatrix}$$

which is nonzero if $j_1, \ldots, j_{\delta-1}$ are distinct integers.   □

3.16 *Example*. The minimum distance of a *BCH* code can be greater than the designed distance. Let $q = 2$ and $m = 4$. Let ω be a primitive element of $\mathbb{F}_{16}$. Since ω has order 15, we can apply the preceeding results in the case $q = 2$, $m = 4$ and $n = 15$. The 2-cyclotomic classes modulo n are $\{1, 2, 4, 8\}$, $\{3, 6, 12, 9\}$, $\{5, 10\}$, $\{7, 14, 13, 11\}$. This shows, if we write $C_\delta = BCH_\omega(\delta)$ and $d_\delta = d_{C_\delta}$ for simplicity, that $C_4 = C_5$, hence $d_4 = d_5 \geqslant 5$, and $C_6 = C_7$, hence $d_6 = d_7 \geqslant 7$. Note that the dimension of $C_4 = C_5$ is $15 - 2 \cdot 4 = 7$, and the dimension of $C_6 = C_7$ is $15 - 2 \cdot 4 - 2 = 5$.

3.17 *Example*. This example is similar to the previous one, but now $q = 2$ and $m = 5$. Let ω be a primitive element of $\mathbb{F}_{32}$. The 2-cyclotomic classes modulo 31 are

$$\{1, 2, 4, 8, 16\}, \{3, 6, 12, 24, 17\}, \{5, 10, 20, 9, 18\},$$
$$\{7, 14, 28, 25, 19\}, \{11, 22, 13, 26, 21\}, \{15, 30, 29, 27, 23\}$$

and so we see, with similar conventions as in the previous example, that $C_4 = C_5$, $C_6 = C_7$, $C_8 = C_9 = C_{10} = C_{11}$ and $C_{12} = C_{13} = C_{14} = C_{15}$. Therefore $d_4 = d_5 \geqslant 5$, $d_6 = d_7 \geqslant 7$, $d_8 = d_9 = d_{10} = d_{11} \geqslant 11$ and $d_{12} = d_{13} = d_{14} = d_{15} \geqslant 15$. If we write $k_\delta = dim\,(C_\delta)$, then we have $k_4 = 31 - 2 \cdot 5 = 21$, $k_6 = 31 - 3 \cdot 5 = 16$, $k_8 = 31 - 4 \cdot 5 = 11$ and $k_{12} = 31 - 5 \cdot 5 = 6$.

E.3.15. Let C be a cyclic code of length n and assume that ω^i is a root of C for all $i \in I$, where $I \subseteq \{0, 1, \ldots, n-1\}$. Suppose that $|I| = \delta - 1$ and that $(i + 1 \mod n) \in I$ for all $i \in I$. Show that the minimum distance of C is not less than δ.

E.3.16. If ω is a primitive element of $\mathbb{F}_{64}$, prove that the minimum distance of $BCH_\omega(16)$ is $\geqslant 21$ and that its dimension is 18. $\qquad\square$

In relation to the dimension of $BCH_\omega(\delta, \ell)$, we have the following bound:

3.18 Proposition. *If* $m = ord_n(q)$, *then*

$$dim\ BCH_\omega(\delta, \ell) \geqslant n - m(\delta - 1).$$

Proof: If g is the generating polynomial of $BCH_\omega(\delta, \ell)$, then

$$dim\ BCH_\omega(\delta, \ell) = n - deg\,(g).$$

Since g is the least common multiple of the minimum polynomials

$$p_i = p_{\omega^{\ell+i}}, \quad i = 1, \ldots, \ell - 1,$$

and $deg\,(p_{\omega^{\ell+i}}) \leqslant [\mathbb{F}_{q^m} : \mathbb{F}_q] = m$, it is clear that $deg\,(g) \leqslant m(\delta - 1)$, and this clearly implies the stated inequality. $\qquad\square$

Improving the dimension bound in the binary case

The bound in the previous proposition can be improved considerably for binary *BCH* codes in the strict sense. Let C_i be the 2-cyclotomic class of i modulo n. If we write p_i to denote the mimimum polynomial of ω^i, where ω is a primitive n-th root of unity, then $p_i = p_{2i}$, since $2i \bmod n \in C_i$. It turns out, if $t \geqslant 1$, that

$$lcm(p_1, p_2, \ldots, p_{2t}) = lcm(p_1, p_2, \ldots, p_{2t-1}) = lcm(p_1, p_3, \ldots, p_{2t-1}). \quad [3.5]$$

Now the first equality tells us that $BCH_\omega(2t + 1) = BCH_\omega(2t)$, so that it is enough to consider, among the binary *BCH* codes in the strict sense, those that have odd designed distance.

3.19 Proposition. *In the binary case, the dimension k of a binary code $BCH_\omega(2t + 1)$ of length n and designed distance $\delta = 2t + 1$ satisfy the following inequality:*

$$k \geqslant n - tm,$$

where $m = ord_n(2)$.

Proof: Let g be the polynomial [3.5]. Since the first expression of g in [3.5] is the generating polynomial of $BCH_\omega(2t + 1)$, we know that $k = n - deg(g)$. But looking at the third expression of g in [3.5] we see that $deg(g)$ is not higher than the sum of the degrees of $p_1, p_3, \ldots, p_{2t-1}$ and therefore $deg(g) \leqslant tm$ because the degree of each p_i is not higher than m. $\qquad\square$

Table 3.1: *Strict BCH binary codes of length 15 and 31, where $p_{i_1,\ldots,i_r} = p_{i_1} \cdots p_{i_r}$.*

δ	g	k	d
1	1	15	1
3	p_1	11	3
5	$p_{1,3}$	7	5
7	$p_{1,3,5}$	5	7
9–15	$X^{15} - 1$	0	–

δ	g	k	d
1	1	31	1
3	p_1	26	3
5	$p_{1,3}$	21	5
7	$p_{1,3,5}$	16	7
9,11	$p_{1,3,5,7}$	11	11
13,15	$p_{1,3,5,7,9}$	6	15
17–31	$X^{31} - 1$	0	–

3.20 Example. The lower-bound of the above proposition, when applied to the code $BCH_\omega(8) = BCH_\omega(9)$ of the example 3.17, gives us that

$$k \geqslant n - tm = 31 - 4 \cdot 5 = 11.$$

Since the dimension of this code is exactly 11, we see that the bound in question cannot be improved in general (cf. P.3.18). $\qquad\square$

E.3.17. Let $f = x^4 + x^3 + x^2 + x + 1 \in \mathbb{Z}_2[X]$, $F = \mathbb{Z}_2[X]/(f)$ and α a primitive element of F. Find the dimension and a control matrix of $BCH_\alpha(5)$.

> The Golay code is probably the most important of all codes, for both practical and theoretical reasons.
>
> F.J. MacWilliams, N.J.A. Sloane, [11], p. 64.

3.21 *Example* (The binary Golay code). Let $q = 2, n = 23$ and $m = ord_n(2)$ $= 11$. So the splitting field of $X^{23} - 1 \in \mathbb{Z}_2[X]$ is $L = \mathbb{F}_{2^{11}}$. The 2-cyclotomic classes modulo 23 are as follows:

$$C_0 = \{0\}$$
$$C_1 = \{1, 2, 4, 8, 16, 9, 18, 13, 3, 6, 12\}$$
$$C_5 = \{5, 10, 20, 17, 11, 22, 21, 19, 15, 7, 14\}.$$

If $\omega \in L$ is a primitive 23rd root of unity, the generating polynomial of $C = BCH_\omega(5)$ is

$$g = lcm(p_1, p_2, p_3, p_4) = p_1.$$

As $deg(p_1) = |C_1| = 11$, it turns out that $dim(C) = 23 - 11 = 12$. Moreover, C has miminum distance 7 (see E.3.18), and so C is a perfect binary code of type $[23, 12, 7]$. It is known that all codes of type $(23, 2^{12}, 7)$ are equivalent to C (cf. [25] for an easy proof) and any such code will be said to be a *binary Golay code*.

E.3.18. Show that the minimum distance of the binary code introduced in the example 3.21 is 7. *Hint:* Prove that the parity extension of C has minimum distance 8 arguing as in Example 3.6 to prove that the extended ternary Golay code has minimum distance 6 (for another proof, see P.3.11).

Primitive Reed–Solomon codes

> NASA uses Reed–Solomon codes extensively in their deep space programs, for instance on their *Galileo, Magellan* and *Ulysses* missions.
>
> Roman [1992], pàg. 152.

We have introduced *RS* codes in Example 1.24. It turns out that in the case $n = q - 1$ the *RS* codes are a special type of *BCH* primitive codes. Before proving this (see Proposition 3.22), it will help to have a closer look at such codes. Let $\mathbb{F}$ be a finite field and let $\omega \in \mathbb{F}$ be a primitive element.

The binary Golay code
n=23; g=factor$(x^n-1,\mathbb{Z}_2)_{2,1}$;
G=cyclic_normalized_matrix(g,n)

$$
\rightarrow
\begin{pmatrix}
1 & 1 & 0 & 0 & 0 & 1 & 1 & 1 & 0 & 1 & 0 & 1 & 0 & 0 & 0 & 0 & 0 & 0 & 0 & 0 & 0 & 0 & 0 \\
0 & 1 & 1 & 0 & 0 & 0 & 1 & 1 & 1 & 0 & 1 & 0 & 1 & 0 & 0 & 0 & 0 & 0 & 0 & 0 & 0 & 0 & 0 \\
1 & 1 & 1 & 1 & 0 & 1 & 1 & 0 & 1 & 0 & 0 & 0 & 0 & 1 & 0 & 0 & 0 & 0 & 0 & 0 & 0 & 0 & 0 \\
0 & 1 & 1 & 1 & 1 & 0 & 1 & 1 & 0 & 1 & 0 & 0 & 0 & 0 & 1 & 0 & 0 & 0 & 0 & 0 & 0 & 0 & 0 \\
0 & 0 & 1 & 1 & 1 & 1 & 0 & 1 & 1 & 0 & 1 & 0 & 0 & 0 & 0 & 1 & 0 & 0 & 0 & 0 & 0 & 0 & 0 \\
1 & 1 & 0 & 1 & 1 & 0 & 0 & 1 & 1 & 0 & 0 & 0 & 0 & 0 & 0 & 0 & 1 & 0 & 0 & 0 & 0 & 0 & 0 \\
0 & 1 & 1 & 0 & 1 & 1 & 0 & 0 & 1 & 1 & 0 & 0 & 0 & 0 & 0 & 0 & 0 & 1 & 0 & 0 & 0 & 0 & 0 \\
0 & 0 & 1 & 1 & 0 & 1 & 1 & 0 & 0 & 1 & 1 & 0 & 0 & 0 & 0 & 0 & 0 & 0 & 1 & 0 & 0 & 0 & 0 \\
1 & 1 & 0 & 1 & 1 & 1 & 0 & 0 & 0 & 1 & 1 & 0 & 0 & 0 & 0 & 0 & 0 & 0 & 0 & 1 & 0 & 0 & 0 \\
1 & 0 & 1 & 0 & 1 & 0 & 0 & 1 & 0 & 1 & 1 & 0 & 0 & 0 & 0 & 0 & 0 & 0 & 0 & 0 & 1 & 0 & 0 \\
1 & 0 & 0 & 1 & 0 & 0 & 1 & 1 & 1 & 1 & 0 & 0 & 0 & 0 & 0 & 0 & 0 & 0 & 0 & 0 & 0 & 1 & 0 \\
1 & 0 & 0 & 0 & 1 & 1 & 1 & 0 & 1 & 0 & 1 & 0 & 0 & 0 & 0 & 0 & 0 & 0 & 0 & 0 & 0 & 0 & 1
\end{pmatrix}
$$

Let δ be an integer such that $2 \leqslant \delta \leqslant n$. Consider the code $C = BCH_\omega(\delta)$. The length of this code is $n = q - 1$ and its generating polynomial is

$$
g = (X - \omega)(X - \omega^2) \cdots (X - \omega^{\delta-1}),
$$

as the minimum polynomial of ω^i over $\mathbb{F}$ is clearly $X - \omega^i$. In particular, $k = n - \delta + 1$, where k is the dimension of C. Thus $\delta = n - k + 1$, and since $d \geqslant \delta$, by the BCH bound, and $d \leqslant n - k + 1$, by the Singleton bound, we have that $d = \delta = n - k + 1$, and so C is an MDS code.

3.22 Proposition. *The code $BCH_\omega(\delta)$ coincides with the Reed–Solomon code $RS_{1,\omega,\ldots,\omega^{n-1}}(n - \delta + 1)$.*

Proof: According to Example 1.30, the Vandermonde matrix

$$
H = V_{1,\delta-1}(1, \omega, \ldots, \omega^{n-1})
$$

is a control matrix of $C = RS_{1,\omega,\ldots,\omega^{n-1}}(n - \delta + 1)$. The i-th row of H is $(1, \omega^i, \ldots, \omega^{i(n-1)})$ and so the vectors $a = (a_0, a_1, \ldots, a_{n-1})$ of C are those that satisfy $a_0 + a_1\omega^i + \cdots + a_{n-1}\omega^{i(n-1)} = 0$, $i = 1, \ldots, \delta - 1$. In terms of the polynomial a_X, this is equivalent to say that ω^i is a root of a_X for $i = 1, \ldots, \delta - 1$ and therefore C agrees with the cyclic code whose roots are $\omega, \ldots, \omega^{\delta-1}$. But this code is just $BCH_\omega(\delta)$. $\qquad\square$

Summary

- Determination of a cyclic code by the set of roots of its generating polynomial. Construction of a control matrix from a generating set of roots.

- The *BCH* codes and the lower bound on the minimum distance (the minimum distance is at least the designed distance). Lower bounds on the dimension (the general bound and the much improved bound for the binary case).

- The primitive *RS* codes are *BCH* codes.

Problems

P.3.10 (Weight enumerator of the binary Golay code). Let $a = \sum_{i=0}^{23} a_i z^i$ and $\bar{a} = \sum_{i=0}^{24} \bar{a}_i z^i$ be the weight enumerators of the binary Golay code C and its parity completion $\bar{C}$, respectively.

1) Prove that $a_i = a_{23-i}$ for $i = 0, \ldots, 23$ and $\bar{a}_i = \bar{a}_{24-i}$ for $i = 0, \ldots, 24$.

2) Since $\bar{C}$ only has even weight vectors and its minimum distance is 8, we have

$$\bar{a} = 1 + \bar{a}_8 z^8 + \bar{a}_{10} z^{10} + \bar{a}_{12} z^{12} + \bar{a}_{10} z^{14} + \bar{a}_8 z^{16} + z^{24}.$$

Use the MacWilliams identities for $\bar{C}$ to show that $\bar{a}_8 = 759$, $\bar{a}_{10} = 0$ and $\bar{a}_{12} = 2576$.

3) Show that $a_7 + a_8 = \bar{a}_8$, $a_9 = a_{10} = 0$ and $a_{11} = a_{12} = \bar{a}_{12}/2$.

4) Use the MacWilliams identities for C to prove that $a_7 = 253$ and $a_8 = 506$. Hence the weight enumerator of C is

$$1 + 253z^7 + 506z^8 + 1288z^{11} + 1288z^{12} + 506z^{15} + 253z^{16} + z^{24}.$$

P.3.11. The complete factorization over $\mathbb{Z}_2$ of $X^{23} - 1$ has the form

$$X^{23} - 1 = (X - 1)g_0 g_1.$$

where g_0 and g_1 have degree 11. The binary Golay code C has been defined as C_{g_0}.

1) With the notations of the example 3.21, show that the roots of g_0 (g_1) have the form ω^j, where j is a nonzero quadratic residue (a quadratic non-residue) of $\mathbb{Z}_{23}$.

2) Using that $k = -1$ is not a quadratic residue modulo 23, check that π_{-1} is the map $a \mapsto \tilde{a}$ that reverses the order of vectors. In particular we have $\tilde{g}_0 = g_1$ (this can be seen directly by inspection of the factorization of $X^{23} - 1$ over $\mathbb{Z}_2$).

P.3.12 A code C of length n is called *reversible* if $a = (a_0, a_1, \ldots, a_{n-1}) \in C$ implies that $\tilde{a} = (a_{n-1}, a_{n-2}, \ldots, a_0) \in C$.

1) Show that a cyclic code is reversible if and only if the inverse of each root of the generating polynomial g of C is again a root of g.

2) Describe the form that the divisor g of $X^n - 1$ must have in order that the cyclic code C_g is reversible.

P.3.13. For $n = 15$ and $n = 31$, justify the values of g, k and d on the table 3.1 corresponding to the different values of δ.

P.3.14 Find a control matrix for a binary *BCH* code of length 31 that corrects 2 errors.

P.3.15. Find a generating polynomial of a *BCH* code of length 11 and designed distance 5 over $\mathbb{Z}_2$.

P.3.16. Find the dimension of a *BCH* code over $\mathbb{F}_3$ of length 80 and with the capacity to correct 5 errors.

P.3.17. Calculate a generating polynomial of a ternary *BCH* code of length 26 and designed distance 5 and find its dimension.

P.3.18. Let $m \geqslant 3$ and $t \geqslant 1$ be integers, α a primitive element of $\mathbb{F}_{2^m}$ and $k(m, t)$ the dimension of the code $BCH_\alpha(2t + 1)$ (thus we assume that $2t < 2^m - 1$). Let $f = \lfloor m/2 \rfloor$ and $t_m = 2^{f-1}$. Prove that $k(m, t) = 2^m - 1 - mt$ if $t \leqslant t_m$.

3.4 The Meggitt decoder

The basic decoder of Section 8.9 was first described by Meggitt (1960).

W.W. Peterson and E.J. Weldon, Jr., [16], p. 265.

Essential points

- Syndrome of the received vector (polynomial).
- The Meggitt table.
- The Meggitt decoding algorithm.

Syndromes

Let $g \in \mathbb{F}[x]$ be the generating polynomial of a cyclic code C of length n over $\mathbb{F}$. We want to implement the Meggitt decoder for C (as presented, for example, in [15], Ch. XVII). In this decoder, a received vector $y = [y_0, \ldots, y_{n-1}]$ is seen as a polynomial $y_0 + y_1 x + \cdots + y_{n-1} x^{n-1} \in \mathbb{F}[x]_n$ and by definition the *syndrome* of y, $S(y)$, is the remainder of the Euclidean division of y by g (in computational terms, remainder(y,g). The vectors with zero syndrome are, again by definition, the vectors of C. Note that since g divides $X^n - 1$, $S(y)$ coincides with the n-cyclic reduction of remainder(y(X),g(X). In the sequel we will not distinguish between both interpretations.

3.23 Proposition. *We have the identity* $S(xy) = S(xS(y))$.

Proof: By definition of $S(y)$, there exists $q \in \mathbb{F}[x]_n$ such that $y = qg + S(y)$. Multiplying by x, and taking residue mod g, we get the result. $\qquad\square$

3.24 Corollary. *If we set* $S_0 = S(y)$ *and* $S_j = S(x^j y)$, $j = 1, \ldots, n-1$, *then* $S_j = S(xS_{j-1})$. $\qquad\square$

The Meggitt table

If we want to correct t errors, where t is not greater than the error-correcting capacity, then the Meggitt decoding scheme presupposes the computation of a table E of the syndromes of the error-patterns of the form $ax^{n-1} + e$, where $a \in \mathbb{F}^*$ and $e \in \mathbb{F}[x]$ has degree $n - 2$ (or less) and at most $t - 1$ non-vanishing coefficients.

3.25 Example (Meggitt table of the binary Golay code). The binary Golay code can be defined as the length $n = 23$ cyclic code generated by

$$g = x^{11} + x^9 + x^7 + x^6 + x^5 + x + 1 \in \mathbb{Z}_2[x]$$

and in this case, since the error-correcting capacity is 3 (see Example 3.21), the Meggitt table can be encoded in a Divisor as follows:

```
Meggitt table for the binary Golay code
n=23; R=0..(n−2) ;
g=x¹¹+x⁹+x⁷+x⁶+x⁵+x+1 : ℤ₂[x] ;
The table
E1= [remainder(xⁿ⁻¹, g) →xⁿ⁻¹];
E2= [remainder(xⁿ⁻¹+xⁱ, g) →xⁿ⁻¹+xⁱ with i in R];
E3= [remainder(xⁿ⁻¹+xⁱ+xʲ, g) →xⁿ⁻¹+xⁱ+xʲ with (i,j) in (R,R ) where j<i ];
E=E1+E2+E3 ;
Example
s=remainder(xⁿ⁻¹+x¹⁴+x³,g) →  x¹⁰+x⁹+x⁸+x⁷+x⁵+x²+x+1
E(s) → x²²+x¹⁴+x³
```

Thus we have that $E(s)$ is 0 for all syndromes s that do not coincide with the syndrome of x^{22}, or of $x^{22} + x^i$ for $i = 0, \ldots, 21$, or of $x^{22} + x^i + x^j$ for $i, j \in \{0, 1, \ldots, 21\}$ and $i > j$. Otherwise $E(s)$ selects, among those polynomials, the one that has syndrome s.

3.26 Example (Meggitt table of the ternary Golay code). The ternary Golay code can be defined as the length 11 cyclic code generated by

$$g = x^5 + x^4 + 2x^3 + x^2 + 2 \in \mathbb{Z}_3[x]$$

and in this case, since the error-correcting capacity is 2, the Meggitt table can be defined as in the listing **Meggitt table for the ternary Golay code**.

The Meggitt algorithm

If y is the received vector (polynomial), the Meggitt algorithm goes as follows:

1) Find the syndrome $s = s_0$ of y.

2) If $s = 0$, return y (we know y is a code vector).

```
Meggitt table for the ternary Golay code
n=11; R=0..(n−2) ;
U={1,−1};
g=x⁵+x⁴−x³+x²−1 : Z₃[x] ;
The table
E1=[remainder(a·xⁿ⁻¹, g)→xⁿ⁻¹ with a in U];
E2= [remainder(a·xⁿ⁻¹+b·xⁱ, g)→a·xⁿ⁻¹+b·xⁱ with (i,a,b) in (R,U,U)];
E=E1+E2;
Example
s=remainder(−xⁿ⁻¹+x⁵,g)  ⟶  x⁴+2·x+1
E(s) ⟶ −x¹⁰+x⁵
```

3) Otherwise compute, for $j = 1, 2, \ldots, n-1$, the syndromes s_j of $x^j y$, and stop for the first $j \geqslant 0$ such that $e = E(s) \neq 0$.

4) Return $y - e/x^j$.

In **The Meggitt decoder** we can see the definition of the function meggitt(y,g,n) which implements the Meggitt decoding algorithm. Its parameters are y, the polynomial to be decoded, g, the polynomial generating the code, and n, the length. That listing also contains an example of decoding the ternary Golay code. This example can be easily modified so as to produce a decoding example for the binary Golay code. Note that Meggitt decoding of the Golay codes is complete, as these codes are perfect.

Summary

- The syndrome $S(y)$ of the received polynomial y is defined as remainder(y,g). It is zero if and only if y is a code polynomial. The syndromes $S_j = S(x^j y)$ satisfy the relation $S_j = S(x S_{j-1})$ for $j > 0$.

- The Meggitt table. The entries in this table have the form

$$S(ax^{n-1} + e) \to ax^{n-1} + e,$$

where e is a polynomial of degree $< n - 1$ and with at most t nonzero terms (t is the correcting capacity) and a is a nonzero element of the field.

- The Meggitt decoding algorithm. If $S(y) \neq 0$, finds the smallest shift $x^j y$ such that $S_j = S(x^j y) \neq 0$. Then we know that the term of degree $n - 1 - j$ of y is an error and that the error coefficient is the leading term of $E(S_j)$.

```
┌─▲─Library─┬─The Meggit decoder (presupposes a Meggitt table E)──────────▲─┐
│ meggitt(y,g,n)
│ := begin
│      local x=variable(g), s=remainder(y,g), j=0, e
│      if s=0 then
│        say("Code vector " | y); return y
│      end
│      while E(s)=0 do
│        j=j+1; s=remainder(x·s,g)
│      end
│      e=E(s)/xʲ; say("Error pattern: "|e)
│      y=y−e
│    end
```

Example relative to the ternary Golay code
n=11; R=0..(n−2);
U={1,−1};
$g=x^5+x^4-x^3+x^2-1 : \mathbb{Z}_3[x]$;
S(y):=remainder(y,g);
Setting up the Meggit table
E1=[S(a·xⁿ⁻¹)→a·xⁿ⁻¹ with a in U];
E2=[S(a·xⁿ⁻¹+b·xⁱ) → a·xⁿ⁻¹+b·xⁱ with (i,a,b) in (R,U,U)];
E=E1+E2;
Received vector and decoding
$y=x^4+x^3+x+1 \longrightarrow x^4+x^3+x+1$
meggitt(y,g,n) $\longrightarrow x^9+x^5+x^4+x^3+x+1$

Problems

P.3.19. Consider the Hamming code $Ham_2(r)$, regarded as the cyclic code C_α, where α is a primitive root of $\mathbb{F}_{2^r}$. What is the Megitt table in this case? For $r = 4$, with $\alpha^4 = \alpha + 1$, decode the vector $1_5|0_{10}$ with the Meggitt algorithm.

P.3.20. Let α be a primitive element of $\mathbb{F}_{16}$ such that $\alpha^4 = \alpha + 1$ and let

$$g(x) = x^{10} + x^8 + x^5 + x^4 + x^2 + x + 1$$

be the generating polynomial of a binary *BCH* code of type $[15, 5]$. Assuming that we receive the vector

$$v = 000101100100011,$$

find the nearest code vector and the information vector that was sent.

P.3.21 Let $C = BCH_\alpha(5)$, where $\alpha \in \mathbb{F}_{32}$ is a root of the irreducible polynomial $X^5 + X^2 + 1 \in \mathbb{Z}_2[X]$. Thus C corrects two errors. What is the dimension of C? Assuming that the received vector has syndrome

1110011101, and that at most two errors have occurred, find the possible error polynomials.

P.3.22 Let $C = BCH_\alpha(7)$, where $\alpha \in \mathbb{F}_{32}$ is a root of $X^5 + X^2 + 1 \in \mathbb{F}_2[X]$.

1) Find the generating polynomial of C.

2) Decode the vector 000000011110101111011100010000.

4 Alternant Codes

Alternant codes are a [...] powerful family of codes. In fact, it turns out that they form a very large class indeed, and a great deal remains to be discovered about them

F. MacWilliams and N.J.A. Sloane, [11], p. 332

In this chapter we will introduce the class of alternant codes over $\mathbb{F}_q$ by means of a (generalized) control matrix. This matrix looks like the control matrix of a general Reed–Solomon code (Example 1.26), but it is defined over an extension $\mathbb{F}_{q^m}$ of $\mathbb{F}_q$ and the h_i need not be related to the α_i. Lower bounds for the minimum distance and the dimension follow easily from the definition. In fact these bounds generalize the corresponding lower bounds for *BCH* codes and the proofs are similar.

The main reasons to study alternant codes are that they include, in addition to general Reed–Solomon codes, the family of classical Goppa codes (a class that includes the strict *BCH* codes), that there are good decoding algorithms for them, and that they form a family of asymptotically good codes.

For decoding, the main concepts are the error-location and the error-evaluation polynomials, and the so-called key equation (a congruence) that they satisfy. Any method for solving the key equation amounts to a decoding algorithm. The most effective methods are due to Berlekamp and Sugiyama. Although Berlekamp's method is somewhat better for very large lengths, here we will present Sugiyama's method, because it is considerably easier to understand (it is a variation of the Euclidean algorithm for the *gcd*) and because it turns out that Berlekamp's method, which was discovered first (1967), can be seen as an optimized version of Sugiyama's method (1975). Since the decoder based on Berlekamp's method is usually called the Berlekamp–Massey decoder, here the decoder based on Sugiyama's method will be called Berlekamp–Massey–Sugiyama decoder.

We also present an alternative method to solve the key equation based on linear algebra, which is the basis of the Peterson–Gorenstein–Zierler decoder developed by Peterson (1960) and Gorenstein–Zierler (1961).

4.1   Definitions and examples

> The alternant codes were introduced by H. J. Helgert (1974).
>
> J. H. van Lint, [25], p. 147.

Essential points

- Alternant matrices. Alternant codes.

- Lower-bounds on the dimension and on the minimum distance (alternant bounds).

- General *RS* are alternant codes. Generalized Reed–Solomon codes (*GRS*) and its relation to alternant codes.

- *BCH* codes and classical Goppa codes are alternant codes.

- The classical Goppa codes include the *BCH* codes.

Let $K = \mathbb{F}_q$ and $\bar{K} = \mathbb{F}_{q^m}$. Let $\alpha_1, \ldots, \alpha_n$ and $h_1, \ldots, h_n$ be elements of $\bar{K}$ such that $h_i \neq 0$ for all i and $\alpha_i \neq \alpha_j$ for all $i \neq j$ and consider the matrix

$$H = V_r(\alpha_1, \ldots, \alpha_n) diag(h_1, \ldots, h_n) \in M_n^r(\bar{K}), \qquad [4.1]$$

that is,

$$H = \begin{pmatrix} h_1 & \cdots & h_n \\ h_1\alpha_1 & \cdots & h_n\alpha_n \\ \vdots & & \vdots \\ h_1\alpha_1^{r-1} & \cdots & h_n\alpha_n^{r-1} \end{pmatrix} \qquad [4.2]$$

We say that H is the *alternant control matrix* of order r associated with the vectors

$$h = [h_1, \ldots, h_n] \quad \text{and} \quad \alpha = [\alpha_1, \ldots, \alpha_n].$$

To make explicit that the entries of h and α (and hence of H) lie in $\bar{K}$, we will say that H is defined over $\bar{K}$.

In order to construct the matrix 4.2, observe that each of its r rows is, except the first, the componentwise product of the previous row by α. Hence we introduce the function componentwise_product(a, b), then use it to construct the Vandermonde matrix $V_r(\alpha)$, with vandermonde_matrix(α, r), and finally the formula [4.1] is used to define alternant_matrix(h, α, r), our main constructor of alternant matrices. See **Constructing alternant matrices**.

```
▲ Library │ Constructing alternant matrices                          ▲
  alternant_matrix(h:Vector, α:Vector, r:ℤ)
  check length(h)=length(α)& r>0
  :=vandermonde_matrix(α, r)·diagonal_matrix(h)

  alternant_matrix(α, r):=alternant_matrix(α, α, r)

  vandermonde_matrix(α:Vector, r:ℤ) check r>0
  :=begin
      local n=length(α), h=constant_vector(n,1), H=[h]
      for i in 1..r−1 do
        h=componentwise_product(h, α)
        H=H & h
      end
    end

  componentwise_product(a: Vector | List, b: Vector | List)
  check length(a)=length(b)
  :=[aᵢ·bᵢ with i in range(a)]
```

$$
\text{vandermonde\_matrix}([a, b, c, d, e],4) \;\rightarrow\;
\begin{pmatrix}
1 & 1 & 1 & 1 & 1 \\
a & b & c & d & e \\
a^2 & b^2 & c^2 & d^2 & e^2 \\
a^3 & b^3 & c^3 & d^3 & e^3
\end{pmatrix}
$$

```
p=19; K=ℤₚ;
a=[element(j,K)where prime?(j)with j in 2..p−1]  →  [2, 3, 5, 7, 11, 13, 17]
```

$$
\text{alternant\_matrix}(a,5) \;\rightarrow\;
\begin{pmatrix}
2 & 3 & 5 & 7 & 11 & 13 & 17 \\
4 & 9 & 6 & 11 & 7 & 17 & 4 \\
8 & 8 & 11 & 1 & 1 & 12 & 11 \\
16 & 5 & 17 & 7 & 11 & 4 & 16 \\
13 & 15 & 9 & 11 & 7 & 14 & 6
\end{pmatrix}
$$

We will also need (in the decoding processes studied in sections 2-4) the vector

$$
\beta = [\beta_1, \ldots, \beta_n], \quad \text{where } \beta_i = 1/\alpha_i \; (i = 1, \ldots, n),
$$

which of course is defined only if all the α_i are not zero. The computation of this vector is done with the function invert_entries, which yields, when applied to a vector v, the result of mapping the pure function $x \rightarrow 1/x$ to all components of v (see **The function invert_entries(v)**).

```
▲ Library │ The function invert_entries                              ▲
```

$$
\text{invert\_entries}(v):=\text{map}\!\left(x\mapsto\frac{1}{x}, v\right)
$$

$$
\text{invert\_entries}([1..7]) \;\rightarrow\; \left[1, \frac{1}{2}, \frac{1}{3}, \frac{1}{4}, \frac{1}{5}, \frac{1}{6}, \frac{1}{7}\right]
$$

Alternant codes

The K-code $A_K(h, \alpha, r)$ defined by the control matrix H is the subspace of K^n whose elements are the vectors x such that $xH^T = 0$. Such codes will be called *alternant codes*. If we define the H-*syndrome* of a vector $y \in \overline{K}^n$ as $s = yH^T \in \overline{K}^r$, then $A_K(h, \alpha, r)$ is just the subspace of K^n whose elements are the vectors with zero H-syndrome. If $h = \alpha$, we will write $A_K(\alpha, r)$ instead of $A_K(\alpha, \alpha, r)$. On the other hand, we will simply write $A(h, \alpha, r)$ or $A(\alpha, r)$ when $K = \mathbb{Z}_2$.

The main constructor of alternant codes is alternant_code(h, α, r). It returns a table C={h_=h, a_=α, r_=r, b_=β, H_=H}, where β=invert_entries(α) and H=alternant_matrix(h, α, r). The fields of this table, say H_, can be accessed as either C(H_) or, more conveniently, H_(C). As it will become clear later on, the table C stores the data constructed from h, α and r that are needed to decode alternant codes, and in particular by means of the Berlekamp–Massey–Sugiyama algorithm. If K is a finite field, the function call alternant_code(h, α, r, K) adds to the table the entry K_= K, which is meant to contain the base field of the code.

It should be remarked that in expressions such as C(H_), H_ is local to C, in the sense that there is no interference with another identifier H_ used outside C, even if this identifier is bound to a value. The alternative expression H_(C), however, only works properly if H_ has not been assigned a value outside C.

```
▲ Library │ Constructing alternant codes                                    ▲

alternant_code(h:Vector, α:Vector, r:ℤ)
check length(h)=length(α) & r>0 :=
    {h_=h, a_=α, r_=r, b_=invert_entries(α), H_=alternant_matrix(h, α, r)}
We can include a base−field as a fourth parameter
alternant_code(h:Vector, α:Vector, r:ℤ, K:Field)
:= alternant_code(h,α,r) | {K_=K}
h=a by default
alternant_code(α:Vector, r:ℤ):= alternant_code(α, α, r)
alternant_code(α:Vector, r:ℤ, K:Field):= alternant_code(α,r) | {K_=K}
```

```
C=alternant_code([2,3,4,5],3);
{h_(C), a_(C), r_(C), b_(C), H_(C)}
```
$$\rightarrow \left\{ [2, 3, 4, 5], [2, 3, 4, 5], 3, \left[\frac{1}{2}, \frac{1}{3}, \frac{1}{4}, \frac{1}{5}\right], \begin{pmatrix} 2 & 3 & 4 & 5 \\ 4 & 9 & 16 & 25 \\ 8 & 27 & 64 & 125 \end{pmatrix} \right\}$$
```
C(r_) → 3
D=alternant_code([1,−1, 1,−1], [2, 3, 4, 5], 3, ℤ₂);
K_(D) → ℤ₂
```

4.1 Proposition (Alternant bounds). *If* $C = A_K(h, \alpha, r)$, *then* $n - r \geqslant$ *dim* $C \geqslant n - rm$ *and* $d_C \geqslant r + 1$ (minimum distance alternant bound).

Proof: Let H' be the $rm \times n$ matrix over K obtained by replacing each element of H by the column of its components with respect to a basis of $\bar{K}$ over K. Then C is also the code associated to the control matrix H' and since the rank of H' over K is at most rm, it is clear that *dim* $(C) \geqslant n - rm$.

Now the $r \times r$ minor of H corresponding to the columns $i_1, \ldots, i_r$ is equal to

$$\begin{vmatrix} h_{i_1} & \cdots & h_{i_r} \\ h_{i_1}\alpha_{i_1} & \cdots & h_{i_r}\alpha_{i_r} \\ \vdots & & \vdots \\ h_{i_1}\alpha_{i_1}^{r-1} & \cdots & h_{i_r}\alpha_{i_r}^{r-1} \end{vmatrix} = h_{i_1} \cdots h_{i_r} \begin{vmatrix} 1 & \cdots & 1 \\ \alpha_{i_1} & \cdots & \alpha_{i_r} \\ \vdots & & \vdots \\ \alpha_{i_1}^{r-1} & \cdots & \alpha_{i_r}^{r-1} \end{vmatrix}$$

$$= h_{i_1} \cdots h_{i_r} D_r(\alpha_{i_1}, \ldots, \alpha_{i_r}),$$

which is non-zero (we write $D_r(\alpha_{i_1}, \ldots, \alpha_{i_r})$ to denote the determinant of the Vandermonde matrix $V_r(\alpha_{i_1}, \ldots, \alpha_{i_r})$). This means that any r columns of H are linearly independent over $\bar{K}$ and consequently the minimum distance of C is at least $r + 1$ (proposition 1.25).

Finally, *dim* $(C) \leqslant n + 1 - d_C$ by the Singleton bound, which together with $d_C \geqslant r + 1$ gives *dim* $(C) \leqslant n - r$. And with this the proof is complete. $\square$

4.2 Remark. The map $H \mapsto H'$ explained at the beginning of the previous proof (recall also the method for finding a control matrix for the cyclic code C_ξ explained on page 165) can be implemented as follows: Assume K is finite field and that $F = \bar{K}$ is obtained as F=extension(K,f(t)), where f is an irreducible polynomial of degree m over K. Let H be a matrix of type $r \times n$ over F. Then the matrix H' of type $rm \times n$ over K obtained by replacing each entry of H by the column of its components with respect to the basis $1, t, \ldots, t^{m-1}$ is provided by the function blow(H,F). Note that the call blow(h,F), where h is a vector with components in F, is the auxiliary routine that delivers the result of replacing each entry of h by the *row* of its components with respect to $1, t, \ldots, t^{m-1}$ (see **The function blow(H,F)**). We also include the function prune(H) that selects, given a matrix H of rank r, r rows of H that are linearly independent (see **The function prune(H)**).

4.3 *Example*. Let $\alpha \in \mathbb{F}_8$ and assume that $\alpha^3 = \alpha + 1$. Consider the matrix

$$H = \begin{pmatrix} 1 & 1 & 1 & 1 & 1 & 1 & 1 \\ 1 & \alpha & \alpha^2 & \alpha^3 & \alpha^4 & \alpha^5 & \alpha^6 \end{pmatrix}$$

```
▲ Library | The function calls blow(H, F) and blow(h, F)                    ▲
blow(H:Matrix, F:Field)
:= begin
      local X=null, H=Hᵀ
      for h in H do
          X=(X,blow(h,F))
      end
      [X]ᵀ
  end
blow(h:Vector, F:Field)
:= begin
      local h´ = [ ]
      for x in h do
          h´=h´ | components(x, F)
      end
  end
```

```
▲ Library | The function prune(H)                                          ▲
prune(H:Matrix)
:= begin
      local R, r, s=1
      if zero?(H)  then
          return [ ]
      end
      while zero?(H₁) do
          H=tail(H)
      end
      r=rank(H); R=H₁; H=tail(H)
      while H ≠ [ ] do
          if rank([R,H₁])>rank([R])  then
              R=(R,H₁); s=s+1
          end
          H=tail(H)
          if s=r  then
              break
          end
      end
      [R]
  end
```

and let C be the alternant binary code associated to H. Let us see that $C \equiv [7, 3, 4]$, so that $d = 4 > 3 = r + 1$.

First the minimum distance d of C is ≥ 4, as any three columns of H are linearly independent over $\mathbb{F}_2$. On the other hand, the first three columns and the column of α^5 are linearly dependent, for $\alpha^5 = \alpha^2 + \alpha + 1$, and so $d = 4$. Finally the dimension of C is 3, because it has a control matrix of rank 4 over

$\mathbb{F}_2$, as the listing **Computing the dimension of alternant codes** shows.

```
Computing the dimension of alternant codes
using the blow and prune functions
n=7; r=2;
K=Z₂; F=K[α]/(α³+α+1);
h=constant_vector(n,1); a=geometric_series(α, n);
H= [h, a];

                    ⎛1 1 1 1 1 1 1⎞
                    ⎜1 0 0 1 0 1 1⎟
prune(blow(H,F))  → ⎜0 1 0 1 1 1 0⎟
                    ⎝0 0 1 0 1 1 1⎠
Hence A_K(h,a,r) has dimension 3
```

Reed–Solomon codes

Given distinct elements $\alpha_1, \dots, \alpha_n \in K$, we know from Example 1.26 that the Reed–Solomon code $C = RS_{\alpha_1, \dots, \alpha_n}(k) \subseteq K^n$ (see Example 1.24) has a control matrix of the form

$$H = V_{n-k}(\alpha_1, \dots, \alpha_n)\,diag\,(h_1, \dots, h_n),$$

with

$$h_i = 1/(\prod_{j \neq i}(\alpha_j - \alpha_i)). \qquad [4.3]$$

Hence

$$RS_{\alpha_1, \dots, \alpha_n}(k) = A_K(h, \alpha, n - k),$$

where $h = (h_1, \dots, h_n)$ is given by [4.3]. Note that in this case $\overline{K} = K$, hence $m = 1$, and that the alternant bounds are sharp, because we know that the minimum distance of C is $n - k + 1 = r + 1$, where r is the number of rows of H, and $k = n - r$. The idea that Reed–Solomon codes can be defined as a special case of alternant codes is implemented in the function call RS(α,k) defined in **General RS codes**.

4.4 Remark (Generalized Reed–Solomon codes). The vector h in the definition of the code $RS_{\alpha_1, \dots, \alpha_n}(k)$ as an alternant code is obtained from α (formula [4.3]). If we allow that h can be chosen possibly unrelated to α, but still with components in K, the resulting codes $A_K(h, \alpha, n - k)$ are called *Generalized Reed–Solomon* codes, and we will write $GRS(h, \alpha, k)$ to denote them. See **GRS codes** for an implementation. It should be clear at this point that $GRS(h, \alpha, k)$ is scalarly equivalent to $RS_\alpha(k)$.

```
┌─────────────────────────────────────────────────────────────────────┐
│ ▲ Library │  General RS codes                                     ▲  │
├─────────────────────────────────────────────────────────────────────┤
```

RS(α:Vector, k:$\mathbb{Z}$)
check length(α)>k & k>0
:= begin
    local h, n=length(α)

$$h = \left[\prod_{j \text{ in } [1..i-1] \,|\, [i+1..n]} (\alpha_i - \alpha_j) \text{ with i in 1..n} \right]$$

    h=invert_entries(h)
    alternant_code(h,α,n−k) | {G_=vandermonde(α,k)} | {K_=field(α_1)}
  end

α=[element(j,$\mathbb{Z}_{11}$) with j in 1..9];

C=RS(α,5);

$$G\_(C), H\_(C) \rightarrow \begin{pmatrix} 1 & 1 & 1 & 1 & 1 & 1 & 1 & 1 & 1 \\ 1 & 2 & 3 & 4 & 5 & 6 & 7 & 8 & 9 \\ 1 & 4 & 9 & 5 & 3 & 3 & 5 & 9 & 4 \\ 1 & 8 & 5 & 9 & 4 & 7 & 2 & 6 & 3 \\ 1 & 5 & 4 & 3 & 9 & 9 & 3 & 4 & 5 \end{pmatrix}, \begin{pmatrix} 9 & 5 & 10 & 2 & 3 & 2 & 10 & 5 & 9 \\ 9 & 10 & 8 & 8 & 4 & 1 & 4 & 7 & 4 \\ 9 & 9 & 2 & 10 & 9 & 6 & 6 & 1 & 3 \\ 9 & 7 & 6 & 7 & 1 & 3 & 9 & 8 & 5 \end{pmatrix}$$

h_(C), a_(C) $\rightarrow$ [9, 5, 10, 2, 3, 2, 10, 5, 9],[1, 2, 3, 4, 5, 6, 7, 8, 9]

$$G\_(C) \cdot H\_(C)^T \rightarrow \begin{pmatrix} 0 & 0 & 0 & 0 \\ 0 & 0 & 0 & 0 \\ 0 & 0 & 0 & 0 \\ 0 & 0 & 0 & 0 \\ 0 & 0 & 0 & 0 \end{pmatrix}$$

Note that by definition of alternant codes, we have the following relation: if $\overline{K}$ is a finite field, r a positive integer and $h, \alpha \in \overline{K}^n$, the linear code over $\overline{K}$ defined by the alternating control matrix H of order r associated to h and α is the generalized Reed–Solomon code $GRS(h, \alpha, r)$ and

$$A_K(h, \alpha, r) = GRS(h, \alpha, r) \cap K^n.$$

```
┌─────────────────────────────────────────────────────────────────────┐
│ ▲ Library │ GRS codes                                             ▲  │
├─────────────────────────────────────────────────────────────────────┤
```

GRS(h:Vector, α:Vector, k:$\mathbb{Z}$)
check length(h)=length(α) & k$\geqslant$0 & k$\leqslant$length(h)
:= alternant_code(h, α, length(h)−k, field(h$_1$))

K=$\mathbb{Z}_7$; α=[element(j,K) with j in 1..6]; h=geometric_series(4:K ,6);

C=GRS(h,α,3);

H_(C), K_(C), h_(C), a_(C)

$$\rightarrow \begin{pmatrix} 1 & 4 & 2 & 1 & 4 & 2 \\ 1 & 1 & 6 & 4 & 6 & 5 \\ 1 & 2 & 4 & 2 & 2 & 2 \end{pmatrix}, \mathbb{Z}_7, [1, 4, 2, 1, 4, 2], [1, 2, 3, 4, 5, 6]$$

E.4.1. If $h' = (h_1, \ldots, h_n)$ is a nonzero vector in the kernel of

$$V_{n-1}(\alpha_1, \ldots, \alpha_n) diag\,(h_1, \ldots, h_n),$$

prove that the dual of $GRS(h, \alpha, r)$ is $GRS(h', \alpha, n - r)$. This result can be seen as a generalization of the formula for the control matrix of a Reed–Solomon code.

BCH codes

> BCH codes are of great practical importance for error corrections, paticularly if the expected number of errors is small compared with the length.
>
> F. MacWilliams and N. J. A. Sloane, [11], p. 257.

Let now α be an element of $\bar{K}$, d a positive integer and l a non-negative integer. Let n denote the order of α. Then we know that a control matrix of $C = BCH_\alpha(d, l)$ (the BCH code over K associated to α with *designed distance* d and *offset exponent* l) is

$$H = \begin{pmatrix} 1 & \alpha^l & \alpha^{2l} & \cdots & \alpha^{(n-1)l} \\ 1 & \alpha^{(l+1)} & \alpha^{2(l+1)} & \cdots & \alpha^{(n-1)(l+1)} \\ \vdots & \vdots & \vdots & & \vdots \\ 1 & \alpha^{(l+d-2)} & \alpha^{2(l+d-2)} & \cdots & \alpha^{(n-1)(l+d-2)} \end{pmatrix}$$

which is the alternant control matrix of order $d - 1$ associated to the vectors

$$h = (1, \alpha^l, \alpha^{2l}, \ldots, \alpha^{(n-1)l}) \quad \text{and} \quad \alpha = (1, \alpha, \alpha^2, \ldots, \alpha^{n-1}).$$

Thus we see that C can be defined as alternant_code(h, α, r) with h set to the vector geometric_series(α^l, n), α to the vector geometric_series(α, n), and r to $d - 1$ (see **Constructing BCH codes**). By default we take $l = 1$ (BCH codes in the strict sense). Note that the alternant bound coincides with the BCH bound.

E.4.2. Prove that the minimum distance of the BCH code defined in **Constructing BCH codes** is 7.

```
┌─────────────────────────────────────────────────────────────────────┐
│▲ Library │ Constructing BCH codes                                  ▲│
├─────────────────────────────────────────────────────────────────────┤
  BCH(α:Element(Field), d:ℤ, I:ℤ) check α≠0 & d>1
  := begin
        local n=order(α)
        alternant_code(geometric_series(αⁱ, n), geometric_series(α, n), d−1)
     end
  BCH(α:Element(Field), d:ℤ, I:ℤ, K:Field) check α≠0 & d>1
  :=   BCH(α, d, I) | {K_=K}
  I=1 by default (strict BCH)
  BCH(α:Element(Field), d:ℤ):= BCH(α, d, 1)
  BCH(α:Element(Field), d:ℤ, K:Field):= BCH(α, d, 1) | {K_=K}
```

```
  u=1:ℤ₁₇ ;
  a=7u; d=7; I=3;
  C=BCH(a, d, I); H=H_(C);
  rank(H), order(a)  →  6,16
  dimensions(H)  →  6,16
```

Primitive *RS* codes

> Beside serving as illuminating examples of BCH codes, they
> are of considerable practical and theoretical importance.
>
> F. MacWilliams and N. J. A. Sloane, [11], p. 294.

Given a finite field $K = \mathbb{F}_q$ and an integer k such that $0 \leqslant k \leqslant n$, $n = q - 1$, the Reed–Solomon code of dimension k associated to K is $RS_{1,\omega,\cdots,\omega^{n-1}}(k)$, where ω is a primitive element of K. But we know (Proposition 3.22) that this code coincides, if $r = n - k$, with $BCH_\omega(r + 1)$. This is the formula used in **Constructing primitive Reed–Solomon codes** to define the funtion RS(K,k). Since in this case we know for certain that K is the field over which the code is defined, we include the table-field K_$= K$ in the definition of the code.

```
┌─────────────────────────────────────────────────────────────────────┐
│▲ Library │ Constructing Reed−Solomon codes                        ▲│
├─────────────────────────────────────────────────────────────────────┤
  RS(K:Field, k:ℤ) check k⩾0 & k<cardinal(K)
  := BCH(primitive_element(K), n−k+1) | {K_=K}
```

$$
C=RS(\mathbb{Z}_{11},5);\ H_{\_}(C) \rightarrow
\begin{pmatrix}
1 & 2 & 4 & 8 & 5 & 10 & 9 & 7 & 3 & 6 \\
1 & 4 & 5 & 9 & 3 & 1 & 4 & 5 & 9 & 3 \\
1 & 8 & 9 & 6 & 4 & 10 & 3 & 2 & 5 & 7 \\
1 & 5 & 3 & 4 & 9 & 1 & 5 & 3 & 4 & 9 \\
1 & 10 & 1 & 10 & 1 & 10 & 1 & 10 & 1 & 10
\end{pmatrix}
$$

```
  K_(C)  →  Z11
```

E.4.3. Show that if α is a primitive element of a finite field F, then the matrix $V_n(\alpha, \ldots, \alpha^{n-k})^T$ is a control matrix for $RS(F, k)$.

E.4.4. Let ρ be a real number such that $0 < \rho < 1$ and t a positive integer. If K denotes an arbitrary finite field, show that the minimum $q = |K|$ such that the code $RS(K, k)$ has rate at least ρ and corrects t errors satisfies

$$q \geqslant 1 + \frac{2t}{1 - \rho}.$$

Conversely, given $\rho \in (0, 1)$ and a positive integer t, if q satisfies the condition and $K = \mathbb{F}_q$, then the code $RS(K, n - 2t)$, $n = q - 1$, has rate at least ρ and it corrects t errors. See **The function next_q(ρ, t)**. For example, if we want a rate of $3/4$ and that 7 errors are corrected, then we need $q = 59$. For correcting 10 errors we need $q = 81$.

▲| Library | The function next_q(ρ, t) |▲|

```
next_q (ρ, t):= next_q(1+ 2·t/1-ρ)

where for a given x ∈ℝ⁺
next_q (x)
:= begin
      local q=⌈x⌉
      while prime_power?(q)=false do
         q=q+1
      end
      q
   end
```

```
R=[0.50, 0.55, 0.60, 0.65, 0.70, 0.75, 0.80];
[[next_q(ρ,t) with t in 1..12] with ρ in R]
```

$$\longrightarrow \begin{pmatrix} 5 & 9 & 13 & 17 & 23 & 25 & 29 & 37 & 37 & 41 & 47 & 49 \\ 7 & 11 & 16 & 19 & 25 & 29 & 37 & 37 & 43 & 47 & 53 & 59 \\ 7 & 11 & 16 & 23 & 27 & 31 & 37 & 41 & 47 & 53 & 59 & 61 \\ 7 & 13 & 19 & 25 & 31 & 37 & 41 & 47 & 53 & 59 & 64 & 71 \\ 8 & 16 & 23 & 29 & 37 & 41 & 49 & 59 & 61 & 71 & 79 & 81 \\ 9 & 17 & 25 & 37 & 41 & 49 & 59 & 67 & 73 & 81 & 89 & 97 \\ 13 & 23 & 32 & 43 & 53 & 64 & 73 & 83 & 97 & 103 & 113 & 125 \end{pmatrix}$$

Generating matrix. Let ω be the primitive element of K used to construct the control matrix H of $C = RS(K, k)$. Since

$$C = RS_{1,\omega,\ldots,\omega^{n-1}}(k) \quad (n = q - 1, \ k = n - r),$$

a generating matrix of C is given by the Vandermonde matrix

$$V_k(1, \omega, \ldots, \omega^{n-1}).$$

This matrix is supplied by the function rs_G(ω, k).

```
▲ Library │ Generating matrix of RS₁,ω,...,ωⁿ⁻¹(k)                    ▲

rs_G(ω, k)
:= begin
      local n=order(ω)
      [geometric_series(ωʲ, n) with j in 0..k−1]
   end
```

Mattson–Solomon matrix. If α is an element of a finite field K, and $n = \operatorname{ord}(\alpha)$, the $n \times n$ matrix (α^{ij}), where $0 \leqslant i, j \leqslant n-1$, is called the *Mattson–Solomon matrix* of α. The Mattson–Solomon matrix of K is defined to be the Mattson–Solomon matrix of a primitive element of F.

```
▲ Library │ The Mattson−Solomon matrix                               ▲

mattson_solomon_matrix(α:Element(Field)):=
   begin
      local n=order(α)
      [[αⁱ·ʲ with j in 0..n−1] with i in 0..n−1]
   end
mattson_solomon_matrix(K:Field):=
   mattson_solomon_matrix(primitive_element(K))
```

$$\text{mattson\_solomon\_matrix}(2:\mathbb{Z}_7) \;\longrightarrow\; \begin{pmatrix} 1 & 1 & 1 \\ 1 & 2 & 4 \\ 1 & 4 & 2 \end{pmatrix}$$

$$\text{mattson\_solomon\_matrix}(\mathbb{Z}_5) \;\longrightarrow\; \begin{pmatrix} 1 & 1 & 1 & 1 \\ 1 & 2 & 4 & 3 \\ 1 & 4 & 1 & 4 \\ 1 & 3 & 4 & 2 \end{pmatrix}$$

E.4.5 (Mattson–Solomon transform). Let K be a finite field and $\alpha \in K$ an element of order n. Let $M = M_\alpha$ be the Mattson–Solomon matrix of α. Then the linear map $K^n \to K^n$ such that $x \mapsto xM$ is called the *Mattson–Solomon transform* of K^n relative to α. Show that if $y = xM$, then $yM = n\widehat{x}$, where $\widehat{x} = (x_0, x_{n-1}, \ldots, x_1)$.

Classical Goppa codes

> In order to solve the problem of coding and decoding powerful methods from modern algebra, geometry, combinatorics and number theory have been employed.
>
> V.D. Goppa, [8], p. 75.

Let $g \in \bar{K}[T]$ be a polynomial of degree $r > 0$ and let $\alpha = \alpha_1, \ldots, \alpha_n \in \bar{K}$ be distinct elements such that $g(\alpha_i) \neq 0$ for all i. Then the *classical*

Goppa code over K associated with g and α, which will be denoted $\Gamma(g, \alpha)$, can be defined (see [11], Chapter 12) as alternant_code(h, α, r), where h is the vector $[1/g(\alpha_1), \ldots, 1/g(\alpha_n)]$. In this way it is clear from Proposition 4.1 that the minimum distance of $\Gamma(g, \alpha)$ is $\geqslant r + 1$ and that its dimension k satisfies $n - rm \leqslant k \leqslant n - r$.

The code $\Gamma(g, \alpha)$ is provided by the function goppa(g,α). See **Constructing classical Goppa codes**.

▲ Library │ Constructing classical Goppa codes   ▲
goppa(g:Univariate_polynomial, a:Vector) := alternant_code(invert_entries(evaluate(g,a)), a, degree(g)) goppa(g:Univariate_polynomial, a:Vector, K:Field) := goppa(g, a) │ {K_=K}

```
Example of a Goppa code with d≥7
K=Z₅[x]/(x²−2); q=cardinal(K);
g=T⁶+T³+T+1;
a=[element(j, K)with j in 1..q−1];
a=[t in a such_that evaluate(g,t)≠0];
C=goppa(g,a);
The following computation shows that C has dimension 7
H=blow(H_(C), K);
length(a)−rank(H)  ⟶  7
```

E.4.6. Let $K = \mathbb{Z}_{11}$ and $\alpha = [4]_{11}$. Then α has order 5 and if $a = [1, \alpha, \ldots, \alpha^4]$, then $a = [1, 4, 5, 9, 3]$. Let $g = (X - 2)(X - 6)(X - 7) \in K[X]$ and $C = \Gamma(g, a)$. Show that $C \sim [5, 2, 4]_{11}$, hence MDS, and check that C is not cyclic.

The next proposition shows that the (strict) *BCH* codes are Goppa codes.

4.5 Proposition (Strict *BCH* codes are Goppa codes). *If ω is a primitive element of $\overline{K} = \mathbb{F}_{q^m}$ and δ is an integer such that $2 \leqslant \delta \leqslant n$, then the code $C = BCH_\omega(\delta)$ coincides with $C' = \Gamma(X^{\delta-1}, \alpha)$, where*

$$\alpha = (1, \omega^{-1}, \ldots, \omega^{-(n-1)}).$$

Proof: Since the h-vector of the control matrix H' of C' is

$$(1, \omega^{\delta-1}, \ldots, \omega^{(\delta-1)(n-1)}),$$

the i-th row of H' is equal to $(1, \omega^{\delta-i}, \ldots, \omega^{(\delta-i)(n-1)})$. Thus we see that H' is the control matrix H that defines C, but with the rows in reversed order (note that the number of rows of H' is $deg\,(X^{\delta-1}) = \delta - 1$). $\qquad\square$

4.6 Example (A non-strict *BCH* code that is not Goppa). Let C be the binary cyclic code of length 15 generated by $g = x^2 + x + 1$. Let $\alpha \in \mathbb{F}_{16}$ be such that $\alpha^4 = \alpha + 1$. Then the roots of g in $\mathbb{F}_{16}$ are $\beta = \alpha^5$ and $\beta^2 = \alpha^{10}$ and hence $C = C_\beta = BCH_\alpha(2,5)$ (the designed distance is 2 and the offset is 5). The (generalized) control matrix of this code is $[1, \beta, \beta^2, \ldots, 1, \beta, \beta^2]$. This cannot have the form $[1/g(\alpha_0), \ldots, 1/g(\alpha_{14})]$, with the α_i distinct and f a linear polynomial g with coefficients in $\mathbb{F}_{16}$. Hence C is not a Goppa code.

Summary

- Alternant matrices. Alternant codes. Reed–Solomon codes (not necessarily primitive) and generalized Reed–Solomon codes (*GRS*).

- Lower-bounds on the minimum distance ($d \geqslant r + 1$, where r is the order of the alternant control matrix used to define the code) and on the dimension ($k \geqslant n - mr$, where m is the degree of $\overline{K}$ over K). We also have the dimension upper-bound $k \leqslant n - r$.

- The classical Goppa codes are alternant codes. The *BCH* codes are classical Goppa codes (Proposition 4.5). Non-strict *BCH* codes, however, need not be alternant (4.6).

- In particular we will be able to decode Reed–Solomon codes and classical Goppa codes (hence also *BCH* codes) as soon as we know a general decoder for the alternant codes (fast decoders of such codes are constructed in the last two sections of this chapter).

Problems

P.4.1. Consider a Reed–Solomon code over $\mathbb{F}_q$ and assume its minimum distance is odd, $d = 2t + 1$.

1) If we fix q and need a rate of at least ρ, show that t is bounded above by $\lfloor (1 - \rho)(q - 1)/2 \rfloor$.

2) If we fix q and the error-correcting capacity t, show that the rate ρ is bounded above by $1 - 2t/(q - 1)$.

3) Fix $q = 256$ and regard the elements of $\mathbb{F}_{256}$ as *bytes* (8 bits) by taking components of the field elements with respect to a basis over $\mathbb{Z}_2$. Let C be a *RS* code over $\mathbb{F}_{256}$. What is the maximum number of bytes that can be corrected with C if we need a rate of at least $9/10$? What is the maximum rate of C if we need to correct 12 bytes? In each case, what is the length of C in bits?

P.4.2. Use $\bar{K} = F_{16}$ and $g = X^3 + X + 1$ to construct a binary Goppa code $C = \Gamma(g, \alpha)$ of maximal length and find its dimension and minimum distance.

P.4.3. Use $\bar{K} = F_{27}$ and $g = X^3 - X + 1$ to construct a ternary Goppa code $C = \Gamma(g, \alpha)$ of maximal length and find its dimension and minimum distance.

P.4.4. By the alternant bound, the minimum distance distance d of the Goppa code defined in the listing **Constructing Goppa codes** satisfies $d \geqslant 7$. Prove that $d = 8$.

P.4.5 (Examples of Goppa codes that are not cyclic). Let n be a positive divisor of $q - 1$ (q a prime power), $\alpha \in F_q$ a primitive n-th root of unity and $a = [1, \alpha, \ldots, \alpha^{n-1}]$. Let $C = \Gamma(g, a)$ be the Goppa code over F associated to a monic polynomial $g \in F[X]$ of degree r and the vector a (hence we assume $g(\alpha^i) \neq 0$ for $i = 0, \ldots, n - 1$).

1) Show that $G = (\alpha^{ij}g(\alpha^j))$, for $0 \leqslant i \leqslant n - r - 1$ and $0 \leqslant j \leqslant n - 1$, is a generating matrix of C.

2) Deduce that the elements of C have the form

$$v_f = [f(1)g(1), f(\alpha)g(\alpha), \ldots, f(\alpha^{n-1})g(\alpha^{n-1})],$$

with $f \in F[X]_{n-r}$.

3) Prove that if C is cyclic then $g = X^r$, and hence that C is a *BCH* code. *Hint:* If C is cyclic, the vector $[g(\alpha^{n-1}), g(1), \ldots, g(\alpha^{n-1})]$ has to coincide with v_f, for some f, but this can only happen if f is constant and from this one can deduce that $g = X^r$.

P.4.6 (Alternative definition of Goppa codes). Let $K = F_q$ and $\bar{K} = F_{q^m}$, m a positive integer. Let $g \in \bar{K}[T]$ be a polynomial of degree $r > 0$ and let $\alpha = \alpha_1, \ldots, \alpha_n \in \bar{K}$ be distinct elements such that $g(\alpha_i) \neq 0$ for all i. Originally, Goppa defined the code $\Gamma(g, \alpha)$ as

$$\{a \in K^n | \sum_{i=0}^{n} \frac{a_i}{x - \alpha_i} \equiv 0 \mod g\}. \qquad [4.4]$$

In this problem we will see that the code defined by this formula is indeed $\Gamma(g, \alpha)$.

1) Given $\alpha \in \bar{K}$ such that $g(\alpha) \neq 0$, show that $x - \alpha$ is invertible modulo g and that

$$\frac{1}{x - \alpha} = -\frac{1}{g(\alpha)} \frac{g(x) - g(\alpha)}{x - \alpha} \mod g$$

(note that $(g(x) - g(\alpha))/(x - \alpha)$ is a polynomial of degree $< r$ with coefficients in $\bar{K}$).

2) Show that the condition $\sum_{i=0}^{n} \dfrac{a_i}{x - \alpha_i} \equiv 0 \mod g$ in the definition [4.4] is equivalent to

$$\sum_{i=1}^{n} \frac{a_i}{g(\alpha_i)} \frac{g(x) - g(\alpha)}{x - \alpha} = 0.$$

3) Use the relation in 2 to prove that the code defined by [4.4] has a control matrix of the form

$$H^* = U \cdot H = U \cdot V_r(\alpha_1, \ldots, \alpha_n) \cdot \mathit{diag}\,(h_1, \ldots, h_n),$$

where $h_i = 1/g(\alpha_i)$ and, if $g = g_0 + g_1 X + \cdots + g_r X^r$, $U = (g_{r-i+j})$ $(1 \leqslant i, j \leqslant n)$, with the convention that $g_l = 0$ if $l > r$.

4) Since U is invertible, the code defined by [4.4] has also a control matrix of the form $H = V_r(\alpha_1, \ldots, \alpha_n)\mathit{diag}\,(h_1, \ldots, h_n)$, and hence that code coincides with $\Gamma(g, \alpha)$.

P.4.7 (Improved minimum distance bounds for binary Goppa codes). With the same notations as in the preceeding problem, let $\bar{g}$ be the *square closure* of g ($\bar{g}$ is the lowest degree perfect square that is divisible by g and can be obtained from g by replacing each odd exponent in the irreducible factorization of g by its even successor). We will see (cf. [11], Ch. 12,§3) that if $K = \mathbb{Z}_2$, then $\Gamma(g, \alpha) = \Gamma(\bar{g}, \alpha)$ and hence that $d \geqslant \bar{r} + 1$, where d is the minimum distance of $\Gamma(g, \alpha)$ and $\bar{r}$ the degree of $\bar{g}$. In particular we will have $d \geqslant 2r + 1$ if g has r distinct roots.

1) Let $a \in K^n$ satisfy $\sum_{i=0}^{n} \dfrac{a_i}{x - \alpha_i} \equiv 0 \mod g$ and set $S = S(a)$ to denote the support of a, so that $|S| = s$ is the weight of a. Let $f_a = \prod_{i \in S}(X - \alpha_i)$. Show that

$$f_a \sum_{i \in S} \frac{1}{X - \alpha_i} = f_a'$$

(the derivative of f_a).

2) Use the fact that f_a and f_a' have no common factors to show that

$$\sum_{i \in S} \frac{1}{X - \alpha_i} \equiv 0 \mod g \text{ if and only if } g \,|\, f_a'.$$

3) Show that f_a' is a square and hence that $g \,|\, f_a'$ if and only if $\bar{g} \,|\, f_a'$. Conclude from this that $\Gamma(g, \alpha) = \Gamma(\bar{g}, \alpha)$, as wanted.

P.4.8 (MacWilliams–Sloane, [11], p. 345). Let $C = \Gamma(g, \alpha)$ be a binary Goppa code and suppose $g = (X - \beta_1) \cdots (X - \beta_r)$, where $\beta_1, \ldots, \beta_r \in \bar{K} = \mathbb{F}_{2^m}$ are distinct elements. Show that the *Cauchy matrix* $(1/(\beta_i - \alpha_j))$, $1 \leqslant i \leqslant r, 1 \leqslant j \leqslant n$, is a control matrix for C.

4.2 Error location, error evaluation and the key equation

Helgert (1975) gave the key equation for decoding alternant codes.

F. MacWilliams and N.J.A. Sloane, [11], p. 367

Essential points

- Basic concepts: error locators, syndrome (vector and polynomial forms), error locator polynomial and error evaluator polynomial.
- Basic results: Forney's formula and the key equation.
- Main result: the solution of the key equation (Berlekamp–Massey–Sugiyama algorithm).

Basic concepts

Let C be the alternant code associated to the alternant matrix H of order r constructed with the vectors h and α (their components lie in $K' = \mathbb{F}_{q^m}$). Let $t = \lfloor r/2 \rfloor$, that is, the highest integer such that $2t \leqslant r$. Note that if we let $t' = \lceil r/2 \rceil$, then $t + t' = r$ (we will use the equivalent equality $r - t = t'$ in the proof of Lemma 4.12).

Let $x \in C$ (*sent vector*) and $e \in \mathbb{F}^n$ (*error vector*, or *error pattern*). Let $y = x + e$ (*received vector*). The goal of the decoders that we will present in this chapter is to obtain x from y and H when $s = |e| \leqslant t$.

Let $M = \{m_1, \ldots, m_s\}$ be the set of error positions, that is, $m \in M$ if and only if $e_m \neq 0$. Let us define the *error locators* η_i, $i = 1, \ldots, s$, by the relation $\eta_i = \alpha_{m_i}$. Since the α_j are different, the knowledge of the error locators is equivalent to the knowledge of the error positions (given the α's).

Define the *syndrome vector* $S = (S_0, \ldots, S_{r-1})$ by the formula

$$(S_0, \ldots, S_{r-1}) = yH^T.$$

Note that $S = eH^T$, as $xH^T = 0$. Consider also the *syndrome polynomial*

$$S(z) = S_0 + S_1 z + \cdots + S_{r-1} z^{r-1}.$$

Since $S = 0$ is equivalent to saying that y is a code vector, henceforth we will assume that $S \neq 0$.

E.4.7. Check that $S_j = \sum_{i=1}^s h_{m_i} e_{m_i} \eta_i^j$ $(0 \leqslant j \leqslant r-1)$.

E.4.8. Assuming $S \neq 0$, prove that the least j such that $S_j \neq 0$ satisfies $j < s$, hence also $j < t$. Since $\gcd(z^r, S(z)) = z^j$, $\gcd(z^r, S(z))$ has degree strictly less than s, hence also stricly less than t. Similarly, prove that $\deg(S(z)) \geqslant t$. *Hint:* If it were $j \geqslant s$ we would have $S_0 = \cdots = S_{s-1} = 0$ and the expressions for S_j in E.4.7 would imply a contradiction.   □

The *error locator polynomial* $\sigma(z)$ is defined by the formula

$$\sigma(z) = \prod_{i=1}^s (1 - \eta_i z).$$

Thus the roots of σ are precisely the inverses of the error locators.
We also define the *error evaluator polynomial* by the formula

$$\epsilon(z) = \sum_{i=1}^s h_{m_i} e_{m_i} \prod_{j=1, j \neq i}^s (1 - \eta_j z).$$

4.7 Remark. It is possible to define the error-locator polynomial as the monic polynomial whose roots are the error-locators. In that case the definition of the error-evaluator and syndrome polynomials can be suitably modified to insure that they satisfy the same key equation (Theorem 4.9), and that the errors can be corrected with an apropriate variation of Forney's formula (Proposition 4.8). For the details of this approach, see P.4.9

Basic results

4.8 Proposition (Forney's formula). *For $k = 1, \ldots, s$ we have*

$$e_{m_k} = -\frac{\eta_k \epsilon(\eta_k^{-1})}{h_{m_k} \sigma'(\eta_k^{-1})},$$

where σ' is the derivative of σ.

Proof: The derivative of σ is

$$\sigma'(z) = -\sum_i \eta_i \prod_{j \neq i} (1 - \eta_j z)$$

and from this expression we get that

$$\sigma'(\eta_k^{-1}) = -\eta_k \prod_{j \neq k} (1 - \eta_j / \eta_k).$$

On the other hand we have, from the definition of ϵ, that

$$\epsilon(\eta_k^{-1}) = h_{m_k} e_{m_k} \prod_{j \neq k} (1 - \eta_j/\eta_k).$$

Comparing the last two expressions we get the relation

$$\eta_k \epsilon(\eta_k^{-1}) = -h_{m_k} e_{m_k} \sigma'(\eta_k^{-1}),$$

which is equivalent to the stated formula. □

4.9 Theorem (key equation). *The polynomials $\epsilon(z)$ i $\sigma(z)$ satisfy the congruence*

$$\epsilon(z) \equiv \sigma(z) S(z) \bmod z^r.$$

Proof: From the definition of ϵ it is clear that we also have

$$\epsilon(z) = \sigma(z) \sum_{i=1}^{s} \frac{h_{m_i} e_{m_i}}{1 - \eta_i z}.$$

But

$$\sum_{i=1}^{s} \frac{h_{m_i} e_{m_i}}{1 - \eta_i z} = \sum_{i=1}^{s} h_{m_i} e_{m_i} \sum_{j \geq 0} (\eta_i z)^j$$

$$\equiv \sum_{i=1}^{s} h_{m_i} e_{m_i} \sum_{j=0}^{r-1} (\eta_i z)^j \bmod z^r$$

$$= \sum_{j=0}^{r-1} \left(\sum_{i=1}^{s} h_{m_i} e_{m_i} \eta_i^j \right) z^j$$

$$= \sum_{j=0}^{r-1} S_j z^j = S(z),$$

where we have used E.4.7 for the last step. □

4.10 Remark. The key equation implies that $deg\,(gcd\,(z^r, S)) < t$, because $gcd\,(z^r, S)$ divides ϵ (cf. the first part of E.4.8).

E.4.9. In the case of the code $BCH_\omega(\delta, l)$ over $\mathbb{F}_q$, prove that the syndromes $S_0, \ldots, S_{\delta-2}$ are the values of the received polynomial (or also of the error polynomial) on $\omega^l, \ldots, \omega^{l+\delta-2}$. Deduce from this that $S_{qj} = (S_j)^q$ if $j, qj \in \{l, \ldots, l + \delta - 2\}$.

Solving the key equation

> Nevertheless, decoding using the Euclidean algorithm is by far
> the simplest to understand, and is certainly at least comparable
> in speed with the other methods (for $n < 10^6$) and so it is the
> method we prefer.
>
> F. MacWilliams and N.J.A. Sloane, [11], p. 369.

The key equation shows that there exists a unique polynomial $\tau(z)$ such that

$$\epsilon(z) = \tau(z)z^r + \sigma(z)S(z).$$

This equation is equivalent to the key equation and one of the main steps in
the decoding of alternant codes is to find its solution (σ and ϵ) in terms of z^r
and $S(z)$.

Here we are going to consider the approach based on a modification of
the Euclidean algorithm to find the *gcd* of two polynomials. This modification is usually called *Sugiyama's algorithm* (for classical Goppa codes it was
published for the first time in [23]).

Sugiyama's variation on the Euclidean algorithm

The input of this algorithm is the pair of polynomials $r_0 = z^r$ and $r_1 = S(z)$
(remember that we have assumed $S \neq 0$) and it works as follows:

1) Let r_i, $i = 0, \ldots, j$, be the polynomials calculated by means of the
   Euclidean algorithm applied to r_0 and r_1, but with j the first index
   such that $deg\,(r_j) < t$. For $i = 2, \ldots, j$, let q_i be the quotient of the
   Euclidean division of r_{i-2} by r_{i-1}, so that $r_i = r_{i-2} - q_i r_{i-1}$.

   Note that since $deg\,(r_1) = deg\,(S) \geqslant t$ by E.4.8, we must have $j \geqslant 2$.
   On the other hand the integer j exists, for the greatest common divisor
   d of r_0 and r_1 has degree less than t (see E.4.8, or Remark 4.10) and
   we know that the iteration of the Euclidean algorithm yields d.

2) Define $v_0, v_1, \ldots, v_j$ so that $v_0 = 0$, $v_1 = 1$ and $v_i = v_{i-2} - q_i v_{i-1}$.

3) Output the pair $\{v_j, r_j\}$.

In order to prove that Sugiyama's algorithm yields the required solution
of the key equation (Lemma 4.12), it is convenient to compute, in parallel
with $v_0, v_1, \ldots, v_j$, the sequence $u_0, u_1, \ldots, u_j$ such that $u_0 = 1$, $u_1 = 0$,
and $u_i = u_{i-2} - q_i u_{i-1}$ for $i = 2, \ldots, j$.

E.4.10. Prove that for all $i = 0, \ldots, j$ we have $u_i r_0 + v_i r_1 = r_i$.

```
┌─┐
│▲│ Library │ The Sugiyama variation of the Euclidean division algorithm    │▲│
├─┴──────────────────────────────────────────────────────────────────────────┤
  sugiyama(r0,r1,t)
  :=begin
       local q, r,    v, v0=0, v1=1
       while t≤degree(r1) do
          {q,r}=quo_rem(r0,r1); v=v0−q·v1
          r0=r1; r1=r;  v0=v1; v1=v
       end
       {v1, r1}
  end
```

$$r=4; \; S=z+z^3; \; t=2;$$
$$\{\sigma, \varepsilon\}=\text{sugiyama}(z^r, S, t) \; \longrightarrow \; \{-z^2+1, z\}$$
$$\sigma\cdot S-\varepsilon \; \text{mod} \; z^r \longrightarrow 0$$

4.11 Remark (Extended Euclidean algorithm). If in the description of the Sugiyama algorithm we let j be the last integer such that $r_j \neq 0$, then $d = r_j$ is the $gcd\,(r_0, r_1)$ and E.4.10 shows that $a_0 = u_j$ and $a_1 = v_j$ yield a solution of a *Bezout identity* $a_0 r_0 + a_1 r_1 = d$. The vector $[d, a_0, a_1]$ is returned by the function bezout.

Using the function bezout(r0,r1)
$$r0=z^7; \; r1=z^3+z+1;$$
$$[d, a0, a1]=\text{bezout}(r0, r1)$$
$$\longrightarrow [1, -6\cdot z^2+4\cdot z-9, 6\cdot z^6-4\cdot z^5+3\cdot z^4-2\cdot z^3+z^2-z+1]$$
$$a0\cdot r0+a1\cdot r1 \longrightarrow 1$$
$$d \longrightarrow 1$$

Main results

We keep using the notations introduced in the description of Sugiyama's algorithm. Recall also that $t' = \lceil r/2 \rceil$ and that $r - t = t'$.

4.12 Lemma. Let $\bar{\varepsilon} = r_j$, $\bar{\tau} = u_j$, $\bar{\sigma} = v_j$. Then

$$\bar{\varepsilon}(z) = \bar{\tau}(z)z^r + \bar{\sigma}(z)S(z) \; and \; deg\,(\bar{\sigma}) \leqslant t', \quad deg\,(\bar{\varepsilon}) < t.$$

Proof: According to E.4.10, $u_i r_0 + v_i r_1 = r_i$ for $i = 0, \ldots, j$, which for $i = j$ is the claimed equality.

Now let us prove by induction on i that

$$deg\,(v_i) = r - deg\,(r_{i-1})$$

for $i = 1, \ldots, j$ (hence, in particular, that $deg\,(v_i)$ is strictly increasing with i). Indeed, the relation is certainly true for $i = 1$. We can thus assume that $i > 1$. Then $v_i = v_{i-2} - q_i v_{i-1}$ and

$$
\begin{aligned}
deg\,(v_i) &= deg\,(q_i) + deg\,(v_{i-1}) \quad (deg\,(v_{i-2}) < deg\,(v_{i-1})) \text{ by induction)}\\
&= deg\,(r_{i-2}) - deg\,(r_{i-1}) + r - deg\,(r_{i-2})\\
&= r - deg\,(r_{i-1})
\end{aligned}
$$

(in the second step we have used the definition of q_i and the induction hypothesis). In particular we have

$$
deg\,(\bar{\sigma}) = deg\,(v_j) = r - deg\,(r_{j-1}) \leqslant r - t = t',
$$

which yields the first inequality. The second inequality is obvious, by definition of j and $\bar{\epsilon}$. $\qquad\square$

E.4.11. With the same notations as in Lemma 4.12, use induction on i to show that

$$
u_i v_{i-1} - v_i u_{i-1} = (-1)^i,
$$

$i = 1, \ldots, j$, and deduce from it that $gcd\,(u_i, v_i) = 1$. In particular we obtain that $gcd\,(\bar{\tau}, \bar{\sigma}) = 1$.

4.13 Theorem. *With the notations as in* Lemma 4.12, *there exists* $\rho \in \mathbb{F}_{q^m}^*$ *such that* $\sigma = \rho\bar{\sigma}$ *and* $\epsilon = \rho\bar{\epsilon}$.

Proof: Multiplying the key equation (Theorem 4.9) by $\bar{\sigma}$, the equality in Lemma 4.12 by σ, and subtracting the results, we obtain the identity

$$
\bar{\sigma}\epsilon - \bar{\epsilon}\sigma = (\bar{\sigma}\tau - \bar{\tau}\sigma)z^r.
$$

Now the degree of the polynomial on the left is $< r$, because

$$
deg\,(\bar{\sigma}\epsilon) = deg\,(\bar{\sigma}) + deg\,(\epsilon) \leqslant t' + t - 1 = r - 1
$$

and

$$
deg\,(\sigma\bar{\epsilon}) = deg\,(\sigma) + deg\,(\bar{\epsilon}) \leqslant t + t - 1 \leqslant r - 1.
$$

Since the right hand side of the identity contains the factor z^r, we infer that

$$
\bar{\sigma}\epsilon = \bar{\epsilon}\sigma, \quad \bar{\sigma}\tau = \bar{\tau}\sigma.
$$

Therefore

$$
\sigma|\bar{\sigma}\epsilon, \quad \bar{\sigma}|\bar{\tau}\sigma.
$$

Since $gcd\,(\sigma, \epsilon) = 1$, for no root of σ is a root of ϵ, and $gcd\,(\bar{\tau}, \bar{\sigma}) = 1$, by E.4.11, we get that $\sigma|\bar{\sigma}$ and $\bar{\sigma}|\sigma$, and hence $\sigma = \rho\bar{\sigma}$ and $\epsilon = \rho\bar{\epsilon}$ for some $\rho \in \mathbb{F}_{q^m}^*$. $\qquad\square$

4.14 Remark. Theorem 4.13 shows that $\bar{\sigma}$ and σ have the same roots, so we can use $\bar{\sigma}$ instead of σ in order to find the error locators. In addition, Forney's formula proves that we can use $\bar{\sigma}$ and $\bar{\epsilon}$ instead of σ and ϵ to find the error values, because it is clear that

$$\frac{\eta_k \bar{\epsilon}(\eta_k^{-1})}{h_{m_k} \bar{\sigma}'(\eta_k^{-1})} = \frac{\eta_k \epsilon(\eta_k^{-1})}{h_{m_k} \sigma'(\eta_k^{-1})}.$$

Summary

- Basic concepts: error locators (η_i), syndrome (vector S and polynomial $S(z)$), error locator polynomial $(\sigma(z))$ and error evaluator polynomial $(\epsilon(z))$.

- Basic results: Forney's formula (Proposition 4.8) and the key equation (Theorem 4.9).

- Main result: Sugiyama's algorithm (which in turn is a variation of the Euclidean *gcd* algorithm) yields a solution of the key equation (Theorem 4.13) that can be used for error location and error correction (Remark 4.14).

- The error-locator polynomial $\sigma(z)$ can also be defined so that its roots are the *error-locators* (rather than the inverses of the error locators). If we do so, and adopt suitable definitions of $S(z)$ and $\epsilon(z)$, we still get a Forney's formula for error-correction and a key equation $\epsilon(z) \equiv \sigma(z)S(z) \bmod z^r$ that can be solved with the Sugiyama algorithm (P.4.9).

Problems

P.4.9 (Alternantive syndrome, Forney's formula and key equation). Suppose we define the syndrome polynomial $S(z)$, the error locator $\sigma(z)$ and the error evaluator as follows:

$$S(z) = S_0 z^{r-1} + \cdots + S_{r-1},$$

$$\sigma(z) = \prod_{i=1}^{s} (z - \eta_i),$$

$$\epsilon(z) = -\sum_{i=1}^{s} h_{m_i} e_{m_i} \eta_i^r \prod_{j \neq i} (z - \eta_i)$$

(now σ is the polynomial whose roots ara $\eta_1, \ldots, \eta_s$).

1) Show that

$$\sigma'(\eta_k) = \prod_{j\neq k}(\eta_k - \eta_j) \text{ and } \epsilon(\eta_k) = -h_{m_k}e_{m_k}\eta_k^r\prod_{j\neq k}(\eta_k - \eta_j)$$

and hence that we have the following alternative Forney's formula:

$$e_{m_k} = -\frac{\epsilon(\eta_k)}{h_{m_k}\eta_k^r\sigma'(\eta_k)}.$$

2) Prove that we still have the same key equation

$$\epsilon(z) \equiv \sigma(z)S(z) \bmod z^r.$$

Note that if we let $\{\bar{\epsilon}, \bar{\sigma}\}$ be the pair returned by sugiyama(z^r,S,t), then the zeros of $\bar{\sigma}$ give the error locators, and the alternative Forney's formula with $\bar{\epsilon}$ and $\bar{\sigma}$ finds the error values.

P.4.10. With the notations introduced in P.4.6, prove that the syndrome S_y^* with respect to the control matrix $H^* = U \cdot H = U \cdot V_r(\alpha) \cdot diag\,(h)$ of the received vector y corresponds to the polynomial

$$S_y^*(z) = \sum_{i=0}^{n-1} \frac{y_i}{g(\alpha_i)} \frac{g(z) - g(\alpha_i)}{z - \alpha_i},$$

which is the unique polynomial of degree $< r$ such that

$$S_y^*(z) \equiv -\sum_{i=0}^{n-1} \frac{y_i}{z - \alpha_i} \bmod g(z).$$

Show also that $S_y^*(z) = S_e^*(z)$, where e is the error vector.

P.4.11. Continuing with the previous problem, define an error-locator polynomial $\sigma^*(z) = \prod_{i=1}^{s}(z - \eta_i)$ (as in P.4.9) and an error-evaluator polynomial $\epsilon^*(z) = \sum_{i=1}^{s} e_{m_i} \prod_{j\neq i}(z - \eta_i)$. Then $deg\,(\sigma^*) = s$, $deg\,(\varepsilon^*) < s$ and $gcd\,(\sigma^*, \varepsilon^*) = 1$. Prove that $e_{m_i} = \epsilon^*(\eta_i)/\sigma^{*\prime}(\eta_i)$ (the form taken by Forney's formula in this case) and the following key equation:

$$\sigma^*(z)S^*(z) \equiv \epsilon^*(z) \bmod g(z).$$

P.4.12. Modify Sugiyama's algorithm to solve the key equation in the preceeding problem.

4.3 The Berlekamp–Massey–Sugiyama algorithm

> *Decoding.* The main source is Berlekamp [*Algebraic Coding Theory*, [3] in our references].
>
> F. MacWilliams and N.J.A. Sloane, [11], p. 292.

Essential points

- Algorithm BMS (Berlekamp–Massey–Sugiyama)

- Main result: BMS corrects at least $t = \lfloor r/2 \rfloor$ errors.

- Alternant decoder (AD): the function alternant_decoder(y) that implements BMS.

- Auxilary functions for experimenting with AD.

Introduction

> It is fairly clear that the deepest and most impressive result in coding theory is the algebraic decoding of BCH-Goppa codes.
>
> R.J. McEleice, [12], p. 264.

As indicated at the beginning of this chapter, one of the nice features of alternant codes is that there are fast decoding algorithms for them.

Our function alternant_decoder (AD) is a straightforward implementation of the algorithm, obtained by putting together the results in the last section, with some extras on error handling. This is a fundamental algorithm which we will call, for lack of a better choice, *Berlekamp–Massey–Sugiyama algorithm*, BMS (cf. [29], §1.6, p. 12-15). There are other fast decoders for alternant codes. However, they can be regarded, with the benefit of hindsight, as variations of the BMS algorithm, or even of the original Berlekamp algorithm. The next section is devoted to one of the earliest schemes, namely the Peterson–Gorenstein–Zierler algorithm.

Let H be the alternant control matrix of order r associated to the vectors $h, \alpha \in \bar{K}^n$, and let $C = C_H \subseteq K^n$ be the corresponding code. As explained at the beginning of Section 4.1 (page 181) we will set $\beta = [\beta_1, \ldots, \beta_n]$ to denote the vector formed with the inverses of the elements α_i, that is, $\beta_i = \alpha_i^{-1}$. We also set $t = \lfloor r/2 \rfloor$.

Next we will explain in mathematical terms the BMS decoding algorithm to decode C_H. It takes as input a vector $y \in K^n$ and it outputs, if y turns out to be decodable, a code vector x. Morevover, x coincides with the sent vector if not more than t errors occurred.

The BMS algorithm

In the description below, *Error* means "a suitable decoder-error message".

1) Find the syndrome $s = yH^T$, say $s = [s_0, \ldots, s_{r-1}]$.

2) Transform s into a polynomial $S = s(z)$ in an indeterminate z (S is called *polynomial syndrome*):

$$S = s_0 + s_1 z + \cdots + s_{r-1} z^{r-1}.$$

3) Let $\{\sigma, \epsilon\}$ be the pair returned by sugiyama(z^r,S,t).

4) Make a list $M = \{m_1, \ldots, m_s\}$ of the indices $j \in \{1, \ldots, n\}$ such that $\sigma(\beta_j) = 0$. These indices are called *error locations*. If s is less than the degree of σ, return *Error*.

5) Let x be the result of replacing the value y_j by $y_j + e_j$, for each $j \in L$, where

$$e_j = \frac{\alpha_j \cdot \epsilon(\beta_j)}{h_j \sigma'(\beta_j)}$$

and σ' is the derivative of σ, provided e_j lies in K. If for some j we find that $e_j \notin K$, return *Error*. Otherwise return x if x is a code-vector, or else *Error*.

4.15 Theorem. *The Euclidean decoding algorithm corrects at least $t = \lfloor r/2 \rfloor$ errors.*

Proof: Clear from the discussions in the previous sections. $\square$

Implementing BMS

Given an alternant code C, we wish to be able to call alternant_decoder(C) and have it return the pure function $y \mapsto x'$, where x' is the result of applying the BMS algorithm for the code C to y. If this were the case, we could set g=alternant_decoder(C) and then apply g to the various vectors y that we need to decode.

Auxiliary functions

To work out this plan, we first explain two auxiliary functions. The first, zero_positions(f,a), returns, given a univariate polynomial f and a vector a, the list of indices j for which a_j is a root of f. It goes through all the a_j and retains the j for which the value of f is 0 (this is often referred to as the *Chien search*).

```
▲ Library │ The function zero_positions(f, a)                      ▲
 zero_positions (f:Univariate_polynomial, a:Vector) :=
    {j in range(a) such_that evaluate(f, aⱼ)=0}
```
```
 f=(x−1)·(x−2)·(x−3); a=[0,1,7,2,9,3,11,1,2];
 zero_positions(f , a)  →  {2, 4, 6, 8, 9}
```

We will also need flip(v,L). This function replaces, for all j in the list L, the value v_j of the vector v by $1 - v_j$. In fact, v can also be a list, in which case the value of flip is a list, and in either case L can be a vector.

```
▲ Library │ The function flip(v, L)                                ▲
 flip(v:List | Vector, L:List | Vector | Range):= for j in L do vⱼ = 1−vⱼ  end
 By default, L=[range v]
 flip(v:Vector | List) :=flip(v,range(v))
```
```
 flip([1,1,0,0,1,0])  →  [0, 0, 1, 1, 0, 1]
 flip([1,1,0,0,1,0], {3,4})  →  [1, 1, 1, 1, 1, 0]
```

The function alternant_decoder

As we said before, this function (AD in shorthand) implements the BMS decoding algorithm and can be found in the listing **The function alternant_decoder** at the end of this section. Following an observation by S. Kaplan, the present version includes the computation of Forney values, and checking whether they lie in the base field, even for the base field $\mathbb{Z}_2$. Knowing the error positions in the binary case would be enough to correct the received vector *if we knew that the error values given by Forney's formula are* 1, but the point is that these values may lie in some extension of $\mathbb{Z}_2$ if the number of errors exceeds the correcting capacity.

If C is an alternant code of length n over the field K and $y \in K^n$, then alternant_decoder(y,C) yields the result of decoding y according to the BMS algorithm relative to C (or the value 0 and some suitable error message if some expected conditions are not satisfied during decoding). The call alternant_decoder(C) yields the function $y \mapsto$ alternant_decoder(y,C).

The listing **A BCH code that corrects 3 errors** is adapted from [20] (Examples 8.1.6 and 8.3.4). The reader may wish to work it by hand in order to follow the operation of all the functions. Note that in polynomial notations the received vector is $1 + x^6 + x^7 + x^8 + x^{12}$ and the error vector is $x^4 + x^{12}$.

```
A BCH code that corrects 3 errors
F=Z₂[x]/(x⁴+x+1);
C=BCH(x,5,Z₂);
x=alternant_decoder([1, 0, 0, 0, 0, 0, 1, 1, 1, 0, 0, 0, 1, 0, 0],C)
  ↠  [1, 0, 0, 0, 1, 0, 1, 1, 1, 0, 0, 0, 0, 0, 0]
 −corrected positions : {5,13}
alternant_decoder(x,C)  ↠  [1, 0, 0, 0, 1, 0, 1, 1, 1, 0, 0, 0, 0, 0, 0]
 − x is a code vector
```

A few more examples can be found in **Goppa examples of the alternant_decoder**.

```
A Goppa code that corrects 3 errors
F=Z₂[t]/(t⁴+t+1); n=cardinal(F)−1;
g=X⁶; a=[element(j,F)with j in 1..n]; C=goppa(g,a,Z₂);
 − C corrects 3 errors
y=[0,1,1,0,0,0,0,0,0,0,0,0,0,0,1] : Vector(Z₂);
alternant_decoder(y,C)  ↠  [0,0,0,0,0,0,0,0,0,0,0,0,0,0,0]
 −corrected positions : {2,3,15}
y=[0,1,1,0,0,1,0,0,0,0,0,0,0,0,1] : Vector(Z₂);
x=alternant_decoder(y,C)  ↠  [1,1,1,0,0,1,1,1,0,0,0,0,0,0,1]
 −corrected positions : {1,7,8}
alternant_decoder(x,C)  ↠  [1,1,1,0,0,1,1,1,0,0,0,0,0,0,1]
 − x is a code vector
y=x+epsilon_vector(n,7)  ↠  [1,1,1,0,0,1,0,1,0,0,0,0,0,0,1]
alternant_decoder(y,C)  ↠  [1,1,1,0,0,1,1,1,0,0,0,0,0,0,1]
 −corrected positions : {7}
y=y+epsilon_vector(n,10)  ↠  [1,1,1,0,0,1,0,1,0,1,0,0,0,0,1]
alternant_decoder(y,C)  ↠  [1,1,1,0,0,1,1,1,0,0,0,0,0,0,1]
 −corrected positions : {7,10}
y=y+epsilon_vector(n,12)  ↠  [1,1,1,0,0,1,0,1,0,1,0,1,0,0,1]
alternant_decoder(y,C)  ↠  [1,1,1,0,0,1,1,1,0,0,0,0,0,0,1]
 −corrected positions : {7,10,12}
[0,1,1,0,0,1,0,0,0,1,0,0,0,0,1];
x=alternant_decoder(y,C)  ↠  [1,1,1,0,0,1,1,1,0,0,0,0,0,0,1]
 −nondecodable vector : too few roots
alternant_decoder(x,C)  ↠  [1,1,1,0,0,1,1,1,0,0,0,0,0,0,1]
 −0 is not a vector
```

Some tools for experimenting with the alternant decoder

To experiment with alternant_decoder, it is convenient to introduce a few auxiliary functions, which incidentally may be useful in other contexts. We will use the function call random(1,k) (k a positive integer), which produces a (pseudo)random integer in the range 1..k.

Random choice of m elements of a given set. If X is a non empty set, the function rd_choice(X) chooses an element of X at random, and rd_choice(X,m) (for a positive integer m less then the cardinal of X) extracts a random subset of m elements of X. Finally, if n is a positive integer, then rd_choice(n,m) (for a positive integer m less than n) chooses m elements at random of the set $\{1, \ldots, n\}$.

```
▲ Library │ Choosing elements of a set at random (rd_choice)          ▲

‖ rd_choice(X:List) := X_random(1, length(X))
‖ rd_choice(X:List, m:ℤ) check m>0 & m<length(X)
‖ := begin
        local x=rd_choice(X), C
        if m=1  then
           C={x}
        else
           X=X/{x};  C={x} | rd_choice(X, m−1)
        end
        set(C)
     end
‖ rd_choice(n:ℤ, m:ℤ) check m>0& m<n :=rd_choice(m, list(1..n))

‖ X={p in 2..30 such_that prime?(p)};
‖ rd_choice(X)  →  19
‖ rd_choice(X,3)  →  {5, 7, 17}
```

Random list of given length of elements of a finite field. For a finite field (or ring) K of cardinal q, the function element(j,K) returns the j-th element of K, where $0 \leqslant j \leqslant q - 1$, with a built-in natural order (for example, in the range $j = 0, \ldots, p - 1$, p the characteristic of K, element(j,K) is the residue class of $j \bmod p$ in $\mathbb{Z}_p \subseteq K$). Hence we can generate a random element of K by picking j at random in the range $0 \leqslant j \leqslant q - 1$. Similarly, we can choose a nonzero element of K by choosing j at random in the range $1 \leqslant j \leqslant q - 1$.

Using rd (or rd_nonzero) s times (s a positive integer) we can produce a list of s elements (or s nonzero elements) of K (see the function calls rd(s,K) and rd_nonzero(s,K)) in **Choosing a list of elements of a finite field at random**.

```
▲ Library │ Choosing an element of a finite field at random                    ▲
│ rd(K:Field):=element(random(0,cardinal(K)-1), K)
│ rd_nonzero(K:Field):=element(random(1,cardinal(K)-1), K)
```

```
│ rd(Z₁₇) → 5
```

```
▲ Library │  Choosing a random list of elements of a field                     ▲
│ rd(s:Z, K:Field) check s>0:={rd(K) with i in 1..s}
│ rd_nonzero(s:Z, K:Field) check s>0:={rd_nonzero(K) with i in 1..s}
```

```
│ rd(7,Z₃) → {0, 2, 1, 0, 2, 2, 1}
│ rd_nonzero(7,Z₃) → {1, 2, 2, 2, 1, 1, 1}
```

Random error patterns.   Given the length of the code, n, a weight s with
$1 \leqslant s \leqslant n$, and the field (or even a ring) K, a random error pattern of
length n, weight s and entries in K can be produced by picking randomly s
positions in the range $1..n$ and s nonzero elements of K and using them to
form a length n vector whose nonzero entries are precisely those s elements
of K placed at those s positions.

```
▲ Library │  Generating random error-patterns                                  ▲
│ rd_error_vector(n:Z, s:Z, K:Ring) check s>0 & s≤n
│   :=begin
│         local k, E=rd_nonzero(s,K), L=rd_choice(n, s), e=constant_vector(n,0)
│         for i in range(E) do
│            k=Lᵢ
│            eₖ=eₖ+Eᵢ
│         end
│      end
│   By default, n=card(K)-1
│   rd_error_vector(s:Z, K:Ring):=rd_error_vector(cardinal(K)-1, s, K)
│   By default, K=Z₂
│   rd_error_vector(n:Z, s:Z):=rd_error_vector(n, s, Z₂)
```

```
│ rd_error_vector(7, 3, Z₅) → [0, 0, 3, 1, 0, 2, 0]
│ rd_error_vector(3, Z₅) → [1, 4, 0, 1]
│ rd_error_vector(5,2) → [1, 0, 0, 0, 1]
```

Finding code vectors.   If G is the generating matrix of a linear code C over
the finite field K, the function rd_linear_combination(G,K) returns a random
linear combination of the rows of G, that is, a random vector of C.

Decoding trials.   To simulate the coding + transmission (with noise) + decoding situation for an alternant code we can set up a function that generates

```
┌─┬───────┬──────────────────────────────────────────────────┬─┐
│▲│Library│ Generating random linear combinations            │▲│
├─┴───────┴──────────────────────────────────────────────────┴─┤
│ rd_linear_combination(G:Matrix, K:Field) :=                   │
│   begin                                                       │
│     local k=length(G), u=vector(rd(k, K))                     │
│     u·G                                                        │
│   end                                                         │
└───────────────────────────────────────────────────────────────┘
```

```
│ a=[2,3,5,7]; G=vandermonde_matrix(a,3); K=Z_5;
│ rd_linear_combination(G,K)  →  [3, 4, 0, 3]
```

a random code vector x, that adds to it a random error pattern e of a prescribed number s of errors, and that calls alternant_decoder on $y = x + e$. If the result is a vector x', to make the results easier to interprete we return the matrix $[x, e, x + e]$ in case $x' = x$ and the matrix $[x, e, x + e, x']$ in case the simulation has resulted in an undetectable error. In case x' is not a vector (hence $x' = 0$), there has been a decoder error and then we simply return the matrix $[x, e]$.

```
┌─┬───────┬──────────────────────────────────────────────────┬─┐
│▲│Library│ A checker for the alternant_decoder              │▲│
├─┴───────┴──────────────────────────────────────────────────┴─┤
│ decoder_trial(C:Code, s:Z) check s>0 := decoder_trial(C, s, K_(C)) │
│ decoder_trial(C:Code, s:Z , K:Field) check s>0                │
│ := begin                                                      │
│     local x', G=kernel(H_(C))^T, n=n_columns(G),              │
│             x=rd_linear_combination(G, K),                    │
│             e=rd_error_vector(n, s, K)                        │
│     x'=alternant_decoder(x+e, C)                              │
│     if is?(x',Vector)  then                                   │
│       if x'=x  then                                           │
│         say( "Decoder trial was succesful" )                  │
│         [x, e, x+e]                                           │
│       else                                                    │
│         say( "Decoder trial produced an undetectable error" ) │
│         [x, e, x+e, x']                                       │
│       end                                                     │
│     else                                                      │
│       say( "Decoder trial produced a decoder error")          │
│       [x, e]                                                  │
│     end                                                       │
│   end                                                         │
└───────────────────────────────────────────────────────────────┘
```

Examples

In this subsection we will look at a few examples of *RS* codes that show how the machinery developed in this section works.

First we will consider *RS* codes of the form $C = RS(K, r)$, where K is a finite field and r is a positive integer less than $n = q - 1$, $q = |F|$ (see

E.4.3). Recall that C has type $[n, n - r, r + 1]$ and that it corrects $\lfloor r/2 \rfloor$ errors. We will also consider an example of the form $RS(a, k)$, where k is the dimension and $a \in K^n$.

Example: RS[12,6,7]

Suppose we want an RS code with rate at least $1/2$ which can correct 3 errors. The table in the listing after E.4.4 tells us that the least possible q is 13. So let us take $q = 13$, hence $n = 12$. Since $t = 3$, the least possible codimension is $r = 6$, thus $k = 6$.

We can construct this code with the function call $RS(\mathbb{Z}_{13}, 6)$, and we can observe its working by means of calls to decoder_trial for diverse s, as illustrated in the next example.

$$
\begin{aligned}
&\text{C=RS}(\mathbb{Z}_{13}, 6); \\[4pt]
&\text{decoder\_trial}(C,1) \;\rightarrow\;
\begin{pmatrix}
1 & 5 & 1 & 3 & 9 & 9 & 5 & 8 & 8 & 11 & 10 & 11 \\
0 & 11 & 0 & 0 & 0 & 0 & 0 & 0 & 0 & 0 & 0 & 0 \\
1 & 3 & 1 & 3 & 9 & 9 & 5 & 8 & 8 & 11 & 10 & 11
\end{pmatrix} \\[4pt]
&\text{decoder\_trial}(C,2) \;\rightarrow\;
\begin{pmatrix}
2 & 6 & 5 & 3 & 7 & 8 & 0 & 11 & 11 & 8 & 7 & 9 \\
1 & 5 & 0 & 0 & 0 & 0 & 0 & 0 & 0 & 0 & 0 & 0 \\
3 & 11 & 5 & 3 & 7 & 8 & 0 & 11 & 11 & 8 & 7 & 9
\end{pmatrix} \\[4pt]
&\text{decoder\_trial}(C,3) \;\rightarrow\;
\begin{pmatrix}
5 & 7 & 8 & 8 & 2 & 3 & 9 & 1 & 10 & 9 & 5 & 11 \\
0 & 0 & 10 & 0 & 0 & 0 & 0 & 1 & 0 & 0 & 0 & 5 \\
5 & 7 & 5 & 8 & 2 & 3 & 9 & 2 & 10 & 9 & 5 & 3
\end{pmatrix} \\[4pt]
&\text{decoder\_trial}(C,4) \;\rightarrow\;
\begin{pmatrix}
6 & 5 & 4 & 0 & 7 & 7 & 10 & 5 & 0 & 8 & 10 & 12 \\
10 & 0 & 0 & 0 & 0 & 6 & 0 & 0 & 0 & 9 & 11 & 0 \\
3 & 5 & 4 & 0 & 7 & 0 & 10 & 5 & 0 & 4 & 8 & 12 \\
3 & 8 & 4 & 9 & 7 & 0 & 3 & 5 & 0 & 4 & 8 & 12
\end{pmatrix} \\[4pt]
&\text{decoder\_trial}(C,4) \;\rightarrow\;
\begin{pmatrix}
5 & 12 & 11 & 12 & 9 & 5 & 9 & 8 & 2 & 10 & 4 & 6 \\
0 & 0 & 0 & 0 & 4 & 0 & 0 & 3 & 0 & 0 & 4 & 10
\end{pmatrix}
\end{aligned}
$$

We see that for an error-pattern of weight 4, which is beyond the correcting capacity of the code, in one case we got an undetectable error and in the other a decoder error. Error-correction in the case of 1, 2 or 3 errors works as expected.

Example: RS[26,16,11]

Suppose now that we want an RS code with rate at least $3/5$ which can correct at least 5 errors. This time the table in the listing after E.4.4 tells us that the least possible q is 27. So let us set $q = 27$, thus $n = 26$. Since $t = 5$, the least possible codimension is $r = 10$ and so $k = 16$.

The construction of this code, and also the observation of its working by means of calls to decoder_trial(s), is illustrated in the next example for various s. Note that we use the shorthand notation given by the ind_t table.

Note that up to 5 errors it behaves as expected, but that in the case of 6 errors it lead to a decoder error.

```
RS[26,16,11]. Corrects up to 5 errors
K=Z₃[t]/(t³+2·t+1); E=ind_table(t);
C=RS(K, 10);
shorthand(decoder_trial(C,3), E)
```

$$\rightarrow \begin{pmatrix} 12 & 25 & 3 & 23 & 20 & 15 & 9 & 19 & 8 & 5 & 2 & 17 & 2 & 13 & 6 & 2 & 9 & 10 & 0 & 15 & 4 & 13 & 7 & 11 & 2 & 22 \\ \_ & \_ & \_ & \_ & \_ & \_ & \_ & \_ & \_ & 14 & \_ & \_ & \_ & \_ & 7 & \_ & \_ & \_ & \_ & 6 & \_ & \_ & \_ & \_ & \_ & \_ \\ 12 & 25 & 3 & 23 & 20 & 15 & 9 & 19 & 8 & 8 & 2 & 17 & 2 & 13 & 15 & 2 & 9 & 10 & 0 & 15 & 25 & 13 & 7 & 11 & 2 & 22 \end{pmatrix}$$

```
shorthand(decoder_trial(C,5), E)
```

$$\rightarrow \begin{pmatrix} 19 & 1 & 15 & 8 & 7 & 2 & 2 & 25 & 12 & 5 & \_ & 15 & \_ & 10 & 6 & 4 & 6 & 1 & \_ & 9 & 6 & 13 & 20 & 22 & 19 & 23 \\ \_ & 21 & \_ & \_ & \_ & \_ & 14 & \_ & \_ & 15 & \_ & \_ & 4 & 23 & \_ & \_ & \_ & \_ & \_ & \_ & \_ & \_ & \_ & \_ & \_ & \_ \\ 19 & 6 & 15 & 8 & 7 & 2 & 4 & 25 & 12 & 11 & \_ & 15 & 4 & \_ & 6 & 4 & 6 & 1 & \_ & 9 & 6 & 13 & 20 & 22 & 19 & 23 \end{pmatrix}$$

```
shorthand(decoder_trial(C,6), E)
```

$$\rightarrow \begin{pmatrix} 4 & 18 & 8 & 25 & 23 & 8 & 10 & 12 & 0 & 24 & 9 & 11 & 19 & 4 & 11 & 5 & 8 & 1 & 25 & 8 & 24 & 10 & 7 & 22 & 13 & 18 \\ \_ & 23 & \_ & \_ & \_ & \_ & 9 & 9 & \_ & 0 & \_ & 20 & \_ & \_ & \_ & \_ & 3 & \_ & \_ & \_ & \_ & \_ & \_ & \_ & \_ & \_ \end{pmatrix}$$

Example RS[80,60,21]

Suppose we want a RS code with rate $3/4$ and correcting 10 errors. Then $d = 21$, hence $k = n - 20$. On the other hand $k = 3n/4$ and so $n = 80$. We could take $K = \mathbb{F}_{81}$, but since $\mathbb{Z}_{83}$ has simpler arithmetic, we can choose this latter field as K and use only 80 of its nonzero elements. We can let $n = 80$ and $a = (a_1, \ldots, a_n)$, where $a_j = [j]_n$ and use the code $RS(a, 60)$. Happily, this code is alternant (and in general such codes are neither Goppa nor BCH) and hence that can be decoded with the AD decoder. In the following example we have indicated a decoder trial of 10 errors, but we have not displayed the result because of its length.

```
RS[80,60,21]. Corrects 10 errors
K=Z₈₃; a=[element(j,K) with j in 1..80];
C=RS(a, 60);
decoder_trial(C,10)
```

Detractors of coding research in the early 1960s had used the lack of an efficient decoding algorithm to claim that Reed–Solomon codes were nothing more than a mathematical curiosity. Fortunately the detractors' assessment proved to be completely wrong.

Wicker–Bhargava, [29], p. 13.

Example: A Goppa code $[24, 9, \geqslant 11]$

In this example we construct the code $C = \Gamma(T^{10}, a)$, where a contains all the nonzero elements of $\mathbb{F}_{25}$. If ω is a primitive root of $\mathbb{F}_{25}$, C coincides with $BCH_\omega(11)$, which could be constructed as BCH(ω,11).

```
Example of a Goppa code with d⩾7
K=Z₅[x]/(x²−2); q=cardinal(K);
g=T⁶+T³+T+1;
a=[element(j, K)with j in 1..q−1];
a=[t in a such_that evaluate(g,t)≠0];
C=goppa(g,a);
The following computation shows that C has dimension 7
H=blow(H_(C), K);
length(a)−rank(H)  ⟶  7
```

Summary

- The Berlekamp–Massey–Sugiyama algorithm (BMS) for alternant codes, which corrects up to $t = \lfloor r/2 \rfloor$ errors. Here r is the order of the alternant matrix used to define the code. In the special case of (general) RS codes, r is the codimension. In the case of Goppa codes, r is the degree of the polynomial used to define the code. For BCH codes, possibly not strict, r is $\delta - 1$, where δ is the designed distance.

- Implementation of the BMS algorithm (the function alternant_decoder and several auxiliary functions that allow to experiment with the algorithm).

- Examples of decoding RS codes, both general and primitive, Goppa codes and BCH codes.

Problems

P.4.13. Find a RS code that has rate $3/5$ and which corrects 25 errors. What are its parameters? Which is the field with smallest cardinal that can be used? Estimate the number of field operations needed to decode a vector with the AD decoder.

P.4.14. Consider the code $C = \Gamma(T^{10} + T^2 + T + 3, a)$ over $\mathbb{Z}_5$, where a is the vector formed with the nonzero elements of $\mathbb{F}_{25}$. Show that $dim\,(C) = 4$. Note that this coincides with the lower bound $k \geqslant n - mr$ for the dimension of alternant codes, as $n = 24$, $r = 10$ and $m = 2$. Note also that with the polynomial T^{10} instead of $T^{10} + T^2 + T + 3$ the corresponding Goppa code has dimension 9, as seen in the last example above.

P.4.15. Write a function alternative_decoder to decode alternant codes using the alternative definitions of $S(z)$, $\sigma(z)$ and $\epsilon(z)$, and the alternative Forney formula and key equation that were introduced in P.4.9.

P.4.16. Write a function goppa_decoder to decode Goppa codes based on the polynomials $S^*(z)$, $\sigma^*(z)$ and $\epsilon^*(z)$ introduced in P.4.10 and P.4.11, on Forney's formula and the key equation established in P.4.11, and on the solution to this key equation given in P.4.11.

```
▲│ Library │ Implementing the BMS algoritm                          ▲

  alternant_decoder(C:Code):= y↦alternant_decoder(y,C)
  alternant_decoder(y, C:Code)
  := begin
      local H, h, r, a, b, K,  s, z, S, σ, ε,  M, E,  x, y0
      if y ∉ Vector  then
          say("AD> Argument is not a vector");  return(0)
      end
      H=H_(C); h=h_(C); r=r_(C) ; a=a_(C); b=b_(C); K=K_(C)
      if length(y)≠n_columns(H)  then
          say("AD> Vector argument has the wrong length");  return(0)
      end
      s=y·Hᵀ
      if zero?(s)  then
          say( "AD> Code vector");  return(y)
      end
      S=polynomial(s,z)
      if degree(S)<⌊r/2⌋ | trailing_degree(S)⩾⌊r/2⌋ then
          say("AD> Nondecodable vector: extraneous syndrome");  return(0)
      end
      {σ, ε}=sugiyama(zʳ, S, ⌊r/2⌋)
      M=zero_positions(σ,b)
      if length(M)<degree(σ)  then
          say( "AD> Nondecodable vector: too few roots"); return(0)
      end
      σ=σ'; E=null; y0=y
      for  m in M  do
                aₘ·evaluate(ε,bₘ)
          x=─────────────────────
                hₘ·evaluate(σ,bₘ)
          if  not subfield?(field(x),K)   then
              say("AD> Nondecodable vector: Forney value not in base field")
              return 0
          end
          E=(E,−x);  yₘ=yₘ+x
      end
      E={E}
      if zero?(y·Hᵀ) then
          say("AD> Successful decoding, with error table " | {M,E});  return(y)
      else
          say("AD> Decoder error: x' has nonzero syndrome");  return(0)
      end
  end
```

4.4 The Peterson–Gorenstein–Zierler algorithm

> Furthermore, a simply implemented decoding procedure has
> been devised for these codes [BCH codes].
> W. W. Peterson and E. J. Weldon, Jr., [16], p. 269.

Essential points

- Finding the number of errors by means of the matrices A_ℓ (formula [4.6]).
- Finding the error-locator polynomial by solving a system of linear equations.
- The algorithm of Peterson–Gorenstein–Zierler (PGZ).

We will present another algorithm for decoding alternant codes. Let us use the notations introduced at the beginning Section 4.2, but define the error-locator polynomial σ as:

$$\sigma(z) = \prod_{i=1}^{s}(z - \eta_i), \qquad [4.5]$$

so that the roots of σ are now the error locators.

We will assume henceforth that $s \leqslant t$.

Finding the number of errors

For any integer ℓ such that $s \leqslant \ell \leqslant t$, define

$$A_\ell = \begin{pmatrix} S_0 & S_1 & \cdots & S_{\ell-1} \\ S_1 & S_2 & \cdots & S_\ell \\ \vdots & \vdots & \ddots & \vdots \\ S_{\ell-1} & S_\ell & \cdots & S_{2\ell-2} \end{pmatrix} \qquad [4.6]$$

This matrix is usually called the *Hankel matrix* associated with the vector $(S_0, S_1, \ldots, S_{2\ell-2})$.

4.16 Lemma. *We have that* $\det(A_\ell) = 0$ *for* $s < \ell \leqslant t$ *and* $\det(A_s) \neq 0$. *In other words, s is the highest integer* (among those satisfying $s \leqslant t$) *such that* $\det(A_s) \neq 0$.

Proof: Let $M' = \{m'_1, ..., m'_\ell\} \subseteq \{0, ..., n-1\}$ be any subset such that $M \subseteq M'$. For $i = 1, ..., \ell$, set $\eta_i = \alpha_{m'_i}$. As we have seen, for $j = 0, ..., r-1$ we have

$$S_j = \sum_{k=1}^{s} h_{m_k} e_{m_k} \alpha_{m_k}^j = \sum_{k=1}^{\ell} h_{m'_k} e_{m'_k} \eta_k^j$$

Let

$$D = diag(h_{m'_1} e_{m'_1}, ..., h_{m'_\ell} e_{m'_\ell}),$$

so that it is clear that $\det(D) \neq 0$ if $\ell = s$ and that $\det(D) = 0$ if $\ell > s$. Let us also write

$$W = V_\ell(\eta_1, ..., \eta_\ell),$$

where $V_\ell(\eta_1, ..., \eta_\ell)$ is the Vandermonde matrix of ℓ rows associated with the elements η_i. Note that in particular we have $\det(W) \neq 0$. We also have

$$WDW^T = A_\ell,$$

since the i-th row of W ($i = 0, ..., \ell-1$) is $(\eta_1^i, ..., \eta_\ell^i)$, the j-th row of DW^T ($j = 0, ..., \ell-1$) is $(h_{m'_1} e_{m'_1} \eta_1^j, ..., h_{m'_\ell} e_{m'_\ell} \eta_\ell^j)^T$, and their product is $\sum_{k=1}^{\ell} h_{m'_k} e_{m'_k} \eta_k^{i+j} = S_{i+j}$.

Thus we have $\det(A_\ell) = \det(D) \det(W)^2$, which vanishes if $\ell > s$ (in this case $\det(D) = 0$), and is nonzero if $\ell = s$ (in this case $\det(D) \neq 0$ and $\det(W) \neq 0$).   □

Finding the error locator polynomial

Once the number of errors, s, is known (always assuming that it is not greater than t), we can see how the coefficients of the error locator polynomial [4.5] are obtained. Note that

$$\sigma(z) = \prod_{i=1}^{s}(z - \eta_i) = z^s + a_1 z^{s-1} + a_2 z^{s-2} + ... + a_s, \qquad [4.7]$$

where $a_j = (-1)^j \sigma_j = \sigma_j(\eta_1, ..., \eta_s)$ is the j-th elementary symmetric polynomial in the η_i ($0 \leqslant j \leqslant s$).

E.4.12. Show that $1 + a_1 z + \cdots + a_s z^s$ is the standard error locator polynomial $\prod_{i=0}^{s}(1 - \eta_i z)$.

4.17 Proposition. *If $a = (a_s, ..., a_1)$ is the vector of coefficients of σ and $b = (S_s, ..., S_{2s-1})$, then a satisfies the relation*

$$A_s a = -b$$

and is uniquely determined by it.

Proof: Since $\det A_s \neq 0$, we only have to show that the relation is satisfied. Substituting z by η_i in the identity

$$\prod_{i=1}^{s}(z - \eta_i) = z^s + a_1 z^{s-1} + \ldots + a_s$$

we obtain the relations

$$\eta_i^s + a_1 \eta_i^{s-1} + \cdots + a_s = 0,$$

where $i = 1, \ldots, s$. Multiplying by $h_{m_i} e_{m_i} \eta_i^j$ and adding with respect to i, we obtain the relations

$$S_{j+s} + a_1 S_{j+s-1} + \cdots + a_s S_j = 0,$$

where $j = 0, \ldots, s - 1$, and these relations are equivalent to the stated matrix relation. $\qquad\square$

Algorithm PGZ

Putting together Lemma 4.16 and Proposition 4.17 we obtain an algorithm to decode alternant codes. In essence this algorithm is due to Peterson, Gorenstein and Zierler (see [16]). In detail, and with the same conventions as in the BMS algorithm concerning the meaning of *Error*, we have:

1) Calculate the syndrome vector, $yH^T = (S_0, \ldots, S_{r-1})$. If $S = 0$, return y.

2) Thus we can assume that $S \neq 0$. Starting with $s = t$, and while $\det(A_s)$ is zero, set $s = s - 1$. At the end of this loop we still have $s > 0$ (otherwise S would be 0) and we assume that s is the number of errors.

3) Solve for a in the matrix equation $A_s a = -b$ (see Proposition 4.17). After this we have $a_1, \ldots, a_s$, hence also the error-locator polynomial σ.

4) Find the elements α_j that are roots of the polynomial σ. If the number of these roots is $< s$, return *Error*. Otherwise let $\eta_1, \ldots, \eta_s$ be the error-locators corresponding to the roots and set $M = \{m_1, \ldots, m_s\}$, where $\eta_i = \alpha_{m_i}$.

5) Find the error evaluator $\epsilon(z)$ by reducing the product

$$(1 + a_1 z + \cdots + a_s z^s)S(z)$$

mod z^r (see E.4.12).

6) Find the errors e_{m_i} using Forney's formula with the error-locator polynomial $1 + a_1 z + \cdots + a_s z^s$ and the error-evaluator $\varepsilon(z)$. If any of the error-values is not in K, return *Error*. Otherwise return $y - e$.

4.18 Proposition. *The algorithm PGZ corrects up to t errors.*

Proof: It is an immediate consequence of what we have seen so far.   □

4.19 Remark. In step 5 of the algorithm we could use the alternative polynomial evalurator $S(z) = S_0 z^r + S_1 z^{r-1} + \cdots + S_{r-1}$, find the alternative error evaluator as the remainder of the division of $\sigma(z)S(z)$ by z^r and then, in step 6, we would use the alternative Forney formula (see Remark 4.7).

4.20 Remark. In step 5 of the algorithm we could determine the error coefficients $e_{m_1}, ..., e_{m_s}$ by solving the system of linear equations

$$h_{m_1} e_{m_1} \eta_1{}^j + h_{m_2} e_{m_2} \eta_2{}^j + ... + h_{m_s} e_{m_s} \eta_s{}^j = S_j,$$

which is equivalent to the matrix equation

$$\begin{pmatrix} h_{m_1} & h_{m_2} & \cdots & h_{m_s} \\ h_{m_1}\eta_1 & h_{m_2}\eta_2 & \cdots & h_{m_s}\eta_s \\ h_{m_1}\eta_1^2 & h_{m_2}\eta_2^2 & \cdots & h_{m_s}\eta_s^2 \\ \vdots & \vdots & \ddots & \vdots \\ h_{m_1}\eta_1{}^{s-1} & h_{m_2}\eta_2{}^{s-1} & \cdots & h_{m_s}\eta_s{}^{s-1} \end{pmatrix} \begin{pmatrix} e_{m_1} \\ e_{m_2} \\ e_{m_3} \\ \vdots \\ e_{m_s} \end{pmatrix} = \begin{pmatrix} S_0 \\ S_1 \\ S_2 \\ \vdots \\ S_{s-1} \end{pmatrix}$$

and then return $y - e$ (or an error if one or more of the components of e is not in K).

Summary

- Assuming that the received vector has a nonzero syndrome, the number of errors is determined as the first ℓ in the sequence $t, t-1, ...$ such that $\det(A_\ell) \neq 0$, where A_ℓ is the matrix [4.6].

- The error-locator polynomial [4.7] is found by solving a system of linear equations $A_s a = -b$ defined in Proposition 4.17.

- The error values are obtained by finding the error evaluator polynomial via the key equation, and then applying Forney's formula, or by solving the system of linear equations in Remark 4.20.

- The three steps above summarize the PGZ algorithm.

```
▲ Library │ The Peterson–Gorenstein–Zierler decoder                    ▲

pgz(y, C:Code):=
  begin
    local H, h, a, K, r=r_(C), t=⌊r/2⌋, s=t+1,
          S, A, D, b, σ, z, M, ε, x, E, S'=0
    if y ∉ Vector  then
      say("pgz> " | y | " is not a vector")
      return 0
    end
    H=H_(C)
    if length(y)≠n_columns(H)  then
      say("pgz> Vector " | y | " has the wrong length")
      return 0
    end
    h=h_(C); a=a_(C); K=K_(C)
    S=y·Hᵀ
    if zero?(S)  then
      say("pgz> Code vector")
      return y
    end
    repeat
      s=s−1;  A=[[Sᵢ₊ⱼ₋₁with j in 1..s]with i in 1..s]; D=|A|
    until D≠0
    b=[Sⱼ with j in s+1..2 s]
    σ=solve(A,−b);  σ=1+z·polynomial(reverse(σ), z)
    b=invert_entries(a);  M=zero_positions(σ,b)
    if length(M)<degree(σ)  then
      say("pgz> Nondecodable vector: too few roots")
      return 0
    end
    ε=remainder(σ·polynomial(S,z), zʳ)
    E={ }
    σ=σ'
    for  m in M  do
             aₘ·evaluate(ε,bₘ)
        x=─────────────────────
             hₘ·evaluate(σ,bₘ)
      if  not subfield?(field(x),K)  then
        say("pgz> nondecodable vector: Forney's value not in base field")
        return 0
      end
      E=E|{−x}
      yₘ=yₘ+x
    end
    say("pgz> error pattern: ",{M, E})
    return y
  end
```

Problems

P.4.17. We have explained the PGZ algorithm under the assumption that $s \leqslant t$. If the received vector is such that $s > t$, analyze what can go wrong at each step of the algorithm and improve the algorithm so that such errors are suitably handled.

P.4.18. Let α be a primitive root of $\mathbb{F}_{16}$ satisfying $\alpha^4 = \alpha + 1$ and consider the code $C = BCH_\alpha(5)$. Decode the vector 100100110000100.

1) with the Euclidean algorithm;

2) with the PGZ algorithm.

Appendix: The WIRIS/*cc* system

The *syntax* rules of the language state how to form expressions, statements, and programs that *look* right.

C. Ghezzi and M. Jazayeri, [7], p. 25.

A.1 (User interface). The WIRIS user interface has several *palettes* and each palette has several *icons*. Palette icons are useful to enter expressions that are typeset on the screen according to standard mathematical conventions. In the **Operations** palette, for example, we have icons for fractions, powers, subindices, roots, and symbolic sums and products. Here is a sample expression composed with that palette:

$$\left\lVert \frac{\sqrt{a}+\sqrt[5]{\sum_{j=0}^{k} b_j + \frac{c}{d+1}}}{|x| \cdot \left(\sqrt{a+2} - \sqrt{\frac{b}{4}+3}\right)} \cdot \prod_{x \text{ in } L} x \right.$$

In addition to the above style (WIRIS or *palette style*) the WIRIS interface also supports a *keyboard style*, which only requires the keyboard to compose expressions. In some cases the keyboard style is indistinguishable from the palette style, as for example the $=$ operator, but generally it is quite different, as $a.i$ for the subindex expression a_i, or Zn(n) for $\mathbb{Z}_n$ (the ring of integers modulo n).

A.2 (Blocks and statements). At any given moment, a WIRIS session has one or more *blocks*. A block is delimited by a variable height '['. Each block contains one or more *statements*. A statement is delimited by a variable length |. A *statement* is either an empty statement or an expression or a succession of two or more expressions separated by semicolons. The semicolon is not required if next expression begins on a new line. An *expression* is either a formula or an assignment. A *formula* is a syntactic construct intended to compute a new value by combining operations and values in some explicit way. An *assignment* is an expression that binds a value to a variable (see A.5 for more details).

The *active block* (*active statement*) displays the cursor. Initially there is just one empty block (that is, a block that only contains the null statement). Blocks can be added with the [| icon in the **Edit** palette (the effect is a new empty block following the active block, or before the first if the cursor was at the beginning of the first block). Pressing the *Enter* key adds a null statement just before or just after the current statement according to whether the cursor is or is not at the beginning of that statement.

> The semantic rules of the languge tell us how to build *meaning-*
> *ful* expressions, statements, and programs.
>
> C. Ghezzi and M. Jazayeri, [7], p. 25.

A.3 (Evaluation). The *evaluation* of all the statements in the active block is requested with either *Ctrl+Enter* or by clicking on the red arrow. The value of each statement in the block is computed and displayed next to it with the red arrow in between (the value of a statement is the value of its last expression). After evaluation, the next block, if there is one, becomes the active block; otherwise an empty block is appended.

In the following we will typeset input expressions in a sans serif font, like in a+b, and the output values in the usual fonts for mathematics, like in $a + b$, if this distinction may possibly be helpful to the reader.

A.4. Integers, like 314, and operations with integers, like $(2 + 3) \cdot (11 - 7)$ or 2^{64}, are represented in the usual way:

```
Usual representation of integers and their opperations
| (2+3)·(11−7)  →  20
| 2^64  →  18446744073709551616
  WIRIS ignores white space between an operant and an operator
| ( 2  +  3 )  ·  ( 11      −  7 )  →  20
| 2      64  →  18446744073709551616
```

In the listing above there are two text lines. These lines were entered in the usual way and converted to text with the **T** icon in the **Edit** menu. In general, this icon converts the active statement into text and vice versa. Text is not evaluated and retains all its typographical features. In particular, all the services for composing expressions are available while editing it. Finally note that text is not delimited by a variable height | in the way statements are.

> Why should a book on *The Principles of Mathematics* contain a
> chapter on 'Proper Names, Adjectives and Verbs'?
>
> A. Wood, [30], p. 191.

A.5 (Naming objects). Let x be an *identifier* (a letter, or a letter followed by characters that can be a letter, a digit, an underscore or a question mark; capital and lower case letters are distinct). Let e be a formula expression. Then the construct x=e binds e (the value of e) to x. On evaluating, the value of x is e. On the other hand the construct x:=e binds e (not e) to x. When we request the value of x, we get the result of evaluating e at that momement, not the result of having evaluated e once for all at the moment of forming the construct (as it happens with x=e).

If instead of x=e we use let x=e, then all works as with x=e, but in addition x is used to denote the value e.

An identifier that has not been bound to a value is said to be a *free identifier* (or *free variable*). Otherwise it is said to be a *bound identifier* (or a *bound variable*).

Syntactic constructs that have the form x=e, x:=e or let x=e are called *assignments*, or *assignment expressions*.

```
a=2 → 2
let b=3 → b
3 → b
x := a → a
x → 2
a=5 → 5
x → 5
a=b → b
x → b
```

A.6 (Modular arithmetic and finite prime fields). The construction of the ring $\mathbb{Z}_p$ of integers modulo the positive integer p is performed by the function Zn(p), in keyboard style, and with the subindex operation to the symbol $\mathbb{Z}$ in WIRIS style. If we put A=Zn(p) and n is an integer, then the class of n mod p is the value of the expression n:A (for more details on this construct, see A.15). Operations in the ring A are denoted in the usual way:

```
A=ℤ₃₅;
x=7:A; y=22:A;
x¹²⁰⁰·y³⁵⁰+x¹⁰⁰⁰·y³⁰⁰+500 → 25
1/y → 8
1/x
```

Note that the inverse of 22 mod 35 is 8, and that the inverse of 7 mod 35 does not exist (this is why 1/x is left unevaluated and an error message is reported). Note also that WIRIS does not display the value of a statement which ends with a semicolon. The reason for this is that in this case the last expression of the statement actually is null (cf. A.2).

For $j = 0, \ldots, n - 1$, the value of element($j, \mathbb{Z}_n$) is the class $[j]_n$.

The ring $\mathbb{Z}_p$ is a field if and only if p is prime. Note, in particular, that Zn(2) constructs the field of binary integers.

$$[\,1, 1/2, 1/3, 1/4, 1/5, 1/6\,] : \text{Vector}(\mathbb{Z}_7) \;\rightarrow\; [1, 4, 5, 2, 3, 6]$$

A.7 (Construction of field and ring extensions). Suppose K is a ring and that f∈ K[x] is a monic polynomial of degree r. Then the function extension(K,f), which can be also called with the expression K[x]/(f), constructs the ring $K' = K[x]/(f)$. The variable x is assigned the class of x mod f and K is a subring of K'. The name of K[x]/(f) is still K[x], but here x is no longer a variable, but an element of K' that satisfies the relation $f(x) = 0$. When K is finite with q elements, then K' is finite with q^r elements. If K is a field and f is irreducible over K, then K' is a field. If we want to use a different name for the variable x of the polynomials and the variable α to which its class modulo f is bound, we can use extension(K,α,f).

If R is a finite ring, cardinal(R) yields the number of elements of R and characteristic(R) yields the minimum positive number n such that n:R is zero. If n is the cardinal of R, the elements of R can be generated with element(j,R), for j in the range 0..(n-1).

To test whether a ring K is a finite field, we can use the Boolean function Finite_field(K), or its synonym GF(K), which is defined as the expression is?(K,Field) & (cardinal(K)<infinity)?

```
f=x⁵-x+1;
let F=extension(ℤ₅,α,f) → F
x¹⁰ → x¹⁰
α¹⁰ → α²+3·α+1
f=β⁵-β+1;
let L=ℤ₅[β]/(f) → L
β¹⁰ → β²+3·β+1
```

A.8 (Sequences). Sequences are generated with the comma operator ',' . The only sequence with 0 terms is null. If s and t are sequences with m and n terms, respectively, then s,t is the sequence with $n + m$ terms obtained by appending t to s. In the latter construction, any object which is not a sequence acts as a sequence of one term. The number of terms of a sequence s is given by number_of_items(s) (or its synonym n_items(s)) and its j-th element by item(j,s).

A.9 (Ranges). If a, b and d are real numbers, $d \neq 0$, then the *range r* defined by a, b, d is the value of the expression a..b..d. It represents the sequence of real numbers x in the interval $[a, b]$ that have the form $x = a + jd$ for some nonnegative integer j. Note that the sequence is increasing if $d > 0$ and decreasing if $d < 0$. It is non-empty if and only if either $a = b$ or $b - a$ and d have equal sign. The expression a..b denotes a sequence of values in the interval $[a, b]$ that depends on the function using a..b as an argument. However, it is usually interpreted as a..b..1, as we will continue to do unless stated otherwise.

A.10 (Aggregates and pairings). In the sequel, we will say that an object is an *aggregate* if it is a range or a sequence of objects enclosed in braces or a sequence of objects enclosed in brackets.

The sequence of objects that we enclose in braces or brackets is unrestricted with the exception of a special kind that here will be called *pairings* and which may belong to four flavours: relations, tables, substitutions and divisors. The table summarizes how these flavours are constructed.

Pairing flavours

	a →b	x=a	a ⇒b
{ }	relation	table	substitution
[ ]	divisor	table	

Here is a more detailed description. If one term of an aggregate delimited with braces has the form $a \rightarrow b$ (a–>b in keyboard style), where a and b ara arbitrary objects, then all other terms must have the same form (otherwise we get an error on evaluation) and in this case the resulting object is called a *relation* (it implements a list of ordered pairs). *Divisors* are like relations, but using brakets instead of braces. *Tables* are like divisors (hence delimited by brakets), but instead of terms of the form $a \rightarrow b$ they are formed with terms of the form $x = a$, where x is an identifier and a any object (tables can also be delimited with braces). Substitutions are like relations (hence delimited with braces), but its terms have the form $x \Rightarrow a$ (x=>a in keyboard style), with x an identifier and a any object. In this case, x occurs more than once, only the last term in which it occurs is retained.

Given a pairing P, the set of values a for which there exists an object b such that a→b is a term in P is called the *domain* of P and it is the value of domain(P). In the case of divisors, we get the same value with support(P).

Pairings have a functional charater. To see this, let us consider each flavor separately. If R is a relation (respectively a table) and a is an object (an identifier), the value of the expression R(a) is the sequence (possibly the null sequence) formed with the values b such that a→b (a=b) is a term in R. The value of R is the relation (table) R whose terms have the form $a \rightarrow R(a)$ $(a = R(a))$ where a runs over domain(R).

```
R={1→1, b→2, 1→1}  →  {1→(1,1), b→2}
R(1)  →  1,1
R(c)  →  null
R(c)=5  →  {1→(1,1), b→2, c→5}
R  →  {1→(1,1), b→2, c→5}
```

A nice additional feature of tables is that if x is any identifier, then T(x) can also be obtained as x(T). This often makes clearer the meaning of expressions.

```
T={a=1, b=2, a=1}  →  {a=(1,1), b=2}
T(a)  →  1,1
T(c)  →  null
T(c)=5  →  {a=(1,1), b=2, c=5}
c(T)  →  5
```

Now let us consider a divisor D. If a is an object, the value of the expression D(a) is the sum (possibly 0) of the values b such that a→b is a term in D. The value of D is the divisor D whose terms have the form $a \rightarrow D(a)$, where a runs over domain(D). Divisors can be added. The sum D+E of the divisors D and E is characteristized by the rule (D+E)(a)=D(a)+E(a) for all a.

```
D=[ 1→1, b→2, 1→1]  →  [1→2, b→2]
D(1)  →  2
D(c)  →  0
D(c)=5  →  [1→2, b→2, c→5]
E=[c→−3]  →  [c→−3]
D+E  →  [1→2, b→2, c→2]
```

Finally if S is a substitution and b is any object, S(b) returns the result of replacing any occurrence of x in b by the value a corresponding to x in S, for all x in the domanin of S.

In all cases, the ordering of the value P of a pairing P is done according to the internal ordering of domain(P).

If x is an aggregate, the number of elements in x, say n, is the value of length(x). We remark that the number of terms in a sequence s is not given

$$
\begin{array}{l}
\mathsf{S}=\{a{\Rightarrow}1,\, b{\Rightarrow}2\} \;\rightarrow\; \{a{\Rightarrow}1, b{\Rightarrow}2\} \\
\mathsf{S}(a^4 \cdot b^3 \cdot c^2) \;\rightarrow\; 8 \cdot c^2 \\
\mathsf{S}(c) \;\rightarrow\; c \\
\mathsf{S}(c)=5 \;\rightarrow\; \{a{\Rightarrow}1, b{\Rightarrow}2, c{\Rightarrow}5\} \\
\mathsf{S}(a^4 \cdot b^3 \cdot c^2) \;\rightarrow\; 200
\end{array}
$$

by length(s), but by number_of_items(s) (with synonym n_items(s)). The function range(x) returns the range $1..n$. Moreover, if i is in the range $1..n$, then x_i (or x.i in keyboard style) yields the i-th element of r; otherwise it results in an error.

Let x still denote an aggregate of length n, and k a positive integer in the range $1..n$. Then take(x,k) returns an aggregate of the same kind (except, as we explain below, when x is a range) formed with the first k terms of x. To return the last k terms of x, write -k instead of k. The function tail(x) is defined to be equivalent to take(x,-(n-1)). When x is a range of length n, take(x,k) is the *list* of the first k terms of x (and take(x,-k) is the list of the last k terms of x; in particular, tail(x) is the list of the last $n-1$ terms of x).

If x and y are aggregates of the same kind, but not ranges, then x|y (or join(x,y)) yields the concatenation of x and y.

$$
\begin{array}{l}
\mathsf{R}=\{1{\rightarrow}1, b{\rightarrow}2, 1{\rightarrow}1, c{\rightarrow}3\} \;\rightarrow\; \{1{\rightarrow}(1,1), b{\rightarrow}2, c{\rightarrow}3\} \\
\text{length(R), range(R),}\; R_3 \;\rightarrow\; 3,\, 1..3,\, c{\rightarrow}3
\end{array}
$$

$$
\begin{array}{l}
\mathsf{D}=[\,1{\rightarrow}1, b{\rightarrow}2, 1{\rightarrow}1\,] \;\rightarrow\; [1{\rightarrow}2, b{\rightarrow}2] \\
D_1 \;\rightarrow\; 1{\rightarrow}2 \\
\text{domain(D), support(D)} \;\rightarrow\; \{1, b\},\, \{1, b\}
\end{array}
$$

$$
\begin{array}{l}
\mathsf{T}=\{a{=}1, b{=}2, a{=}1, c{=}5\} \;\rightarrow\; \{a{=}(1,1), b{=}2, c{=}5\} \\
\text{domain(T)} \;\rightarrow\; \{a, b, c\} \\
\text{take(T,}{-}2) \;\rightarrow\; \{b{=}2, c{=}5\}
\end{array}
$$

$$
\begin{array}{l}
\mathsf{S}=\{a{\Rightarrow}1, b{\Rightarrow}2\} \;\rightarrow\; \{a{\Rightarrow}1, b{\Rightarrow}2\} \\
\mathsf{S}{=}\mathsf{S} \mid \{b{\Rightarrow}3, c{\Rightarrow}4, d{\Rightarrow}5\} \;\rightarrow\; \{a{\Rightarrow}1, b{\Rightarrow}3, c{\Rightarrow}4, d{\Rightarrow}5\} \\
r{=}\text{range(S)} \;\rightarrow\; 1..4 \\
\text{take(r,3), tail(r)} \;\rightarrow\; \{1, 2, 3\},\, \{2, 3, 4\}
\end{array}
$$

A.11 (Lists and vectors). Aggregates delimited by braces (brakets) and which are not pairings are called *lists* (*vectors*).

Let x denote a range, a list or a vector. Then the call index(a,x) gives 0 if a does not appear in x and otherwise yields the position index of the first occurrence of a in x. Here we again remark that sequences behave differently, for the i-th element of a sequence s is given by item(i,s). To reverse the order of x we have the function reverse(x). In the case of a range, reverse(a..b..d) yields b..a..-d.[1]

```
r=1..17..3  →  1..17..3
length(r), reverse(r), index(7,r)  →  6, 17..1..−3, 3
l={a, 1, b, 2, c, 3}  →  {a, 1, b, 2, c, 3}
u=[a, 1, b, 2, c, 3]  →  [a, 1, b, 2, c, 3]
index(3,u)  →  6
reverse(l)  →  {3, c, 2, b, 1, a}
```

For a list or a range x, vector(x) returns the vector formed with the elements of x (this is the same as [x] if x is a range; when x is a list, however, [x] is a vector with x as unique component). On the other hand, list(x) yields the list with the same elements as x when x is a vector, and just a list whose only element is x when x is a range. More generally, if x is a sequence of ranges, then [x] is the vector formed with the elements of all the ranges in x, while {x} is just the list of the ranges in the sequence.

To append an object a to a list or vector x, we have the function append(x,a). It can also be obtained with x|{a} if x is a list or x|[a] if x is a vector.

```
r=1..9..2  →  1..9..2
vector(r), [r]  →  [1, 3, 5, 7, 9], [1, 3, 5, 7, 9]
list(r), {r}  →  {1, 3, 5, 7, 9}, {1..9..2}
l=list(r); vector(l), [l]  →  [1, 3, 5, 7, 9], [{1, 3, 5, 7, 9}]
append(l,11)  →  {1, 3, 5, 7, 9, 11}
[r] | [11]  →  [1, 3, 5, 7, 9, 11]
```

The function constant_vector(n,a) yields, if n is a positive integer and a any object, a vector of length n whose entries are all equal to a (it is denoted a_n). The function constant_list(n,a) behaves similarly, but produces a list instead of a vector.

If j lies in the range 1..n, then the expression epsilon_vector(n,j) yields the vector of length n whose entries are all 0, except the j-th, which is 1. The same expression for j not in the range 1..n yields the zero-vector of length n, which can also be constructed with the expression constant_vector(n,0), or simply zero_vector(n).

The sum of the elements of a range, list or vector x is given by sum(x). Similarly, min(x) yields the minimum element of x.

---

[1]The function index(a,x) also works for pairings, but the returned values are usually different from what we would expect from the order in which they were defined, as the terms in pairings are internally ordered according to the internal order given to the domain of the pairing. This is especially manifest with the function reverse(x), which returns x when x is a pairing.

To form the set associated to a list L we have set(L). This function discards repetions in L and returns the ordered list formed with the remaining (distinct) values.

Finally, the call zero?(x) returns true if the vector x is zero and false otherwise.

```
constant_vector(4,t), constant_list(4,t)  →  [t, t, t, t], {t, t, t, t}
epsilon_vector(5,3)  →  [0, 0, 1, 0, 0]
sum({1,2,3,4,5})  →  15
min([1000..500..−23])  →  517
zero?([0,0,1]), zero?([0,0,1−1])  →  false, true
```

A.12 (Hamming distance, weight and scalar product). The Hamming distance between vectors x and y of the same length is given by the function hamming_distance(x,y) (or by its synonym hd(x,y)). Similarly, weight(x) (or its synonym wt(x)) yields the weight of vector x. The scalar product of two vectors x and y is delivered by $\langle x,y \rangle$ (or x·y in keyboard style).

```
Hamming distance and weight / Examples
x=[1,0,1,0,1,0,1]  →  [1, 0, 1, 0, 1, 0, 1]
y=[1,1,1,1,1,1,1]  →  [1, 1, 1, 1, 1, 1, 1]
wt(x)  →  4
hd(x,y)  →  3
⟨x,y⟩, x·y  →  4, 4
```

A.13 (Matrices). Structurally, matrices are vectors whose entries are vectors of the same length. So if we set A=[[1,2],[2,-3]], then A is a 2×2 matrix whose rows are the vectors [1,2] and [2,-3]. Writing matrices and working with them can be done quickly using the **Matrices** palette (see **Matrix expressions**).

The expression dimensions(A), A a matrix, gives a sequence m, n such that m and n are the number of rows and columns of A, respectively. The integers m and n can also be obtained separately as the values of the functions n_rows(A) and n_columns(A), which are abbreviations of number_of_rows(A) and number_of_columns(A). The transpose of A is A^T (or transpose(A)).

If A is a matrix, x is a vector, and the number of columns (rows) of A is equal to the length of x, then the values Ax^T (respectively xA) are obtained with the expressions A·x (respectively x·A). In the expression A·x the vector x is automatically interpreted as a column vector. Similarly, if A and B are matrices, and the number of columns of A is equal to the number of rows of B, then A·B computes the product AB.

The element in row i column j of A can be obtained with the expression $A_{i,j}$ (or A.i.j in keyboard form). On the other hand, if I and J are lists, A_I forms

Matrix expressions

$$A = \begin{pmatrix} 1 & 2 & 3 & -2 \\ 2 & 4 & -6 & -1 \\ 5 & 4 & -3 & -5 \end{pmatrix};$$

dimensions(A) $\rightarrow$ 3,4

n_rows(A), n_columns(A) $\rightarrow$ 3,4

$[2,-3,5] \cdot A \rightarrow [21, 12, 9, -26]$

$A \cdot [1..4] \rightarrow [6, -12, -16]$

$$B = A \cdot A^T \rightarrow \begin{pmatrix} 18 & -6 & 14 \\ -6 & 57 & 49 \\ 14 & 49 & 75 \end{pmatrix}$$

$B_2 \rightarrow [-6, 57, 49]$

$B_{2,3} \rightarrow 49$

$$B_{\{2,3\}} | I_2 \rightarrow \begin{pmatrix} -6 & 57 & 49 & 1 & 0 \\ 14 & 49 & 75 & 0 & 1 \end{pmatrix}$$

$$B_{\{2,3\},\{1,3\}} \& I_2 \rightarrow \begin{pmatrix} -6 & 49 \\ 14 & 75 \\ 1 & 0 \\ 0 & 1 \end{pmatrix}$$

$|B|$, determinant(B) $\rightarrow$ 11628,11628

trace(B) $\rightarrow$ 150

$$B^{-1} \rightarrow \begin{pmatrix} \dfrac{937}{5814} & \dfrac{284}{2907} & -\dfrac{91}{969} \\ \dfrac{284}{2907} & \dfrac{577}{5814} & -\dfrac{161}{1938} \\ -\dfrac{91}{969} & -\dfrac{161}{1938} & \dfrac{55}{646} \end{pmatrix}$$

a matrix with the intersections of the rows of A specified by I (if these rows exist) and $A_{I,J}$ forms a matrix with the intersections of the rows and columns of A specified by I and J, respectively (provided these rows and columns exist).

If A and B are matrices with the same number of rows (columns), then the matrix $A|B$ (respectively $\dfrac{A}{B}$) is the value of the expression A|B (respectively A&B).

The expression identity_matrix(r) (or also I_r in WIRIS) yields I_r, the identity matrix of order r. Another basic constructor is constant_matrix(m,n,a), which yields an $m \times n$ matrix with all entries equal to a. The expression constant_matrix(n,a) is equivalent to constant_matrix(n,n,a).

In the case of square matrices A, its determinant is given by $|A|$ (or determinant(A)). Similarly, trace(A) returns the trace of A. If $|A| \neq 0$, then A^{-1} (or inverse(A)) gives the inverse of A.

A.14 (Boolean expressions). The basic Boolean values are true and false. In general, a *Boolean expression* (a *Boolean function* in particular) is defined as any expression whose value is true or false.

If a and b are Boolean expressions, then not a (or not(a)) is the *negation* of a, and a&b, a|b are the *conjunction* and *disjunction* of a and b, respectively. The negation of a is true if and only if a is false. Similarly, the conjunction (disjunction) of a and b is true (false) if and only if both expressions are true (false).

Another way of producing Boolean expressions is with the question mark (?) placed after a binary relation, as for example 3>2 ?, whose value is true, or 3==2?, whose value is false. Since ? is considered a literal character, and hence it can be used to form identifiers, the space before ? is required if, appended to the preceeding token, it would form a valid identifier. If in doubt, use parenthesis to surround the binary relation. On the other hand, ? may be omitted when the binary relation is used in a context that expects a Boolean expression, as for example in the if field of a conditional expression.

There are a few Boolean functions whose identifiers end with a question mark. For example:

even?(n)
Yields true if the integer n is even and false otherwise.

invertible?(x)
For a ring element x, it yields true when x is invertible in its ring, and false otherwise.

irreducible?(f,K)
For a field K and a polynomial f with coefficients in K, tests whether f is irreducible over K or not.

is?(x,T)
Tests whether x has type T (see page 232)

monic?(f)
Decides whether a polynomial f is monic or not.

odd?(n)
Yields true if the integer n is odd and false otherwise.

prime?(n)
Decides whether the integer n is prime or not.

zero?(x)
For a vector x, it returns true if x is the zero vector and false otherwise.

> Most of the deep differences among the various programming
> languages stem from their different semantic underpinnings.
>
> C. Ghezzi, M. Jazayeri, [7], p. 25

A.15 (Functions). A WIRIS function f (where f stands for any identifier) with argument sequence X is defined by means of the syntax f(X) := F, where F (the *body* of the function) is the expression of what the function is supposed to do with the parameters X. The same name f can be used for functions that have a different number of arguments.

Often the value of a function is best obtained through a series S of consecutive steps. Formally we take this to mean that S is a statement (see A.2, page 221). But a statement S with more than one expression is not considered to be an expression and hence cannot by itself form the body of a function. To make an expression out of S we have the construct begin S end.

As any statement, S may optionally be preceeded by a declaration of the form local V, where V is a sequence of indentifiers. These identifiers can be used as variables inside the statement S. This means that these variables will not interfere with variables outside S that might happen to have the same name. The identifiers v in V may be optionally replaced by assignments v=e, e any expression.

A.16 (Functional programming constructs). Each object is assigned a *type* (or *domain*), so that all objects are first class objects. Some of the basic predefined types are: Boolean, Integer, Float, Rational, Real, Complex, Field, Ring, Polynomial, Sequence, Range, List, Vector, Matrix, Relation, Table, Divisor, Function, Substitution, Rule. All predefined types begin with a capital letter.

The main use of types is in the definition of functions (A.15). It is important to realize that any object (in particular a function) can be returned as a value by functions (or passed as an argument to functions).

To test whether an object x is of a type T, we have the Boolean funtion is?(x,T), or x$\in$T in palette style. For a given T, this is a Boolean function of x.

$\Big[\!\!\Big[$ is?(3/14, Rational), is?(3.14, $\mathbb{Q}$), is?(3.14, Rational) $\rightarrow$ true, false, false

$\Big[\!\!\Big[$ is?(3.14, Float), is?(3.14, $\mathbb{R}$) $\rightarrow$ true, true

An argument x of a function can be checked to belong to a given type T with the syntax x:T. In fact T can be any Boolean expression formed with types and Boolean functions. This checking mechanism, which we call a *filter*, allows us to use the same name for functions that have the same number of arguments but which differ in the type of at least one of them.

A further checking mechanism for the arguments X of a function is that between X and the operator := it is allowed to include a directive of the form check b(X), where b(X) stands for a Boolean expression formed with the parameters of the function. Several of the features explained above are illustrated in the listing **A sample function** .

```
▲ Library │ A sample function                                       ▲
  Given a finite field K of q elements and a positive integer r that divides
    q−1, it returns an element of K of order r
  get_element_of_order(r:ℤ, K:GF)
  check r>0 & cardinal(K)<+∞ & remainder(cardinal(K)−1, r) =0
  := begin
       local ω=primitive_element(K), n=cardinal(K)−1
       ωⁿ/ʳ
     end
```

$\|$ \caretget_element_of_order(7, $\mathbb{Z}_{29}$) $\longrightarrow$ 16

A.17 (Structured programming constructs). Let us have a look at the basic constructs that allow a structured control of computations. In the descriptions below we use the following conventions: e is any expression, b is a Boolean expression, S and T are statements and X is an aggregate.

if b then S end
Evaluate S in case the value of b is true.

if b then S else T end
Evaluate of S or T according to whether the value of the Boolean expression b is true or false.

while b do S end
Keep evaluating S as long as b evaluates to true.

repeat S until b
Evaluate S once and keep doing it so long as the value of b is false.

for x in X do S end
Evaluate S for any x running over the items of X.

for I; b; U do S end
Start evaluating I, which must be a sequence of expressions. Then evaluate S and U (U must also be a sequence of expressions) as long as b evaluates to true.

continue
The presence of this directive somewhere in the S of a while, repeat or for iterators skips the evaluation process of the remaining expressions of S and continues from there on.

break
The presence of this directive somewhere in the S of a while, repeat or for

iterators exits the iteration and continues as if it had been completed.

with, such_that

The expression {e with x in X such_that b } groups with braces the expressions e that satisfy b, where the Boolean expressionn b depends on x, which in turn runs over the terms of X. The expression {x in X such_that b } groups with braces the terms x of X that satisfy b. There are similar constructions using brakets instead of braces. The directives such_that and where are synomyms.

return R

Exit a function with the value of the expression R.

A.18 (Symbolic sums and related expressions). Let e denote an expression, i a variable and r a list, vector or range. For each value i of r, let e(i) denote the result of substituting any occurrence of i in e by i. Then the expression sigma e with i in r (keyboard style) or $\sum\limits_{i\,in\,r} e$ (palette style) yields the sum $\sum_{i \in r} e(i)$. The expression product e with i in r works similarly, but taking products instead of sums.

If b is a Boolean expression, then sigma e with i in r such_that b yield a sum like $\sum_{i \in r} e(i)$, but restricted to the $i \in r$ such that b(i) is true. In palette style, $\prod\limits_{\substack{i\,in\,r \\ b}} e$ (to get a new line for the condition b, press *Shift-Enter*). The case of products works similarly.

$$n=10 \;\rightarrow\; 10$$

$$\sum_{i=1}^{n} i^3 \;\rightarrow\; 3025$$

$$\sum_{i\,in\,1..n} i^3 \;\rightarrow\; 3025$$

$$\sum_{i\,in\,1..n\ where\ prime?(i)} i^3 \;\rightarrow\; 503$$

$$\frac{1}{2} \cdot \prod_{\substack{p\,in\,3..20..2 \\ prime?(p)}} \frac{p-1}{p} \;\rightarrow\; \frac{55296}{323323}$$

Bibliography

[1] J. Adámek. *Foundations of Doding: Theory and Applications of Error-Correcting Codes with an Introduction to Cryptography and Information Theory*. John Wiley & Sons, 1991.

[2] I. Anderson. *Combinatorial Designs and Tournaments*. Clarendon Press, 1997.

[3] E.R. Berlekamp. *Algebraic Coding Theory*. Aegean Park Press, 1984.

[4] R.E. Blahut. *Algebraic Methods for Signal Processing and Communications Coding*. Springer-Verlag, 1992.

[5] M. Bossert. *Channel Coding for Telecommunications*. John Wiley & Sons, 1999.

[6] J.H. Conway and N.J.A. Sloane. *Sphere Packings, Lattices and Groups*. Number 290 in Grundlehren der mathematischen Wissenschaften. Springer-Verlag, 1988.

[7] C. Ghezzi and M. Jazayeri. *Programming Language Concepts*. John Wiley & Sons, 3rd edition edition, 1998.

[8] V.D. Goppa. *Geometry and Codes*. Mathematics and Its Applications. Kluwer Academic Publishers, 1988.

[9] R. Hill. *A first course in coding theory*. Oxford applied mathematics and computing science series. Clarendon Press at Oxford, 1988.

[10] R. Lidl and H. Niederreiter. *Introduction to finite fields and their applications*. Cambridge University Press, 1986.

[11] F.J. MacWilliams and N.J.A. Sloane. *The Theory of Error-correcting Codes*. Number 6 in Mathematical Library. North-Holland, 1977.

[12] R.J. McEliece. *The Theory of Information and Coding: A Mathematical Framework for Communication*. Number 3 in Encyclopedia of Mathematics and its Applications. Addison-Wesley, 1977.

[13] R.J. McEliece. *Finite Fields for Computer Scientists and Engineers*. Kluwer Academic Publishers, 1987.

[14] A.J. Menezes, editor. *Applications of finite fields*. Kluwer Academic Publishers, 1993.

[15] O. Papini and J. Wolfmann. *Algèbre discrète et codes correcteurs*. Springer-Verlag, 1995.

[16] W.W. Peterson and E.J. Weldon. *Error-Correcting codes*. MIT Press, 1972. 2nd edition.

[17] A. Poli. *Exercices sur les codes correcteurs*. Masson, 1995.

[18] A. Poli and LL. Huguet. *Codes correcteurs, théorie et applications*. Number 1 in LMI. Masson, 1989.

[19] J.G. Proakis and M. Salehi. *Communications Systems Engineering*. Prentice Hall, 2nd edition edition, 2002.

[20] S. Roman. *Coding and Information Theory*. Number 134 in Graduate Texts in Mathematics. Springer-Verlag, 1992.

[21] C.E. Shannon. A mathematical theory of communication. *Bell System Tech. J.*, 27:379–423,623–656, 1948.

[22] M. Sudan. *Algorithmic Introduction to Coding Theory*. MIT/Web, 2002.

[23] Y. Sugiyama, M. Kasahara, S. Hirasawa, and T. Namekawa. A method for solving the key equation for decoding goppa codes. *Information and Control*, 27:87–99, 1975.

[24] M. Tsfasman and S. Vladut. *Algebraic-geometric codes*. Kluwer, 1991.

[25] J.H. van Lint. *Introduction to Coding Theory*. Number 82 in GTM. Springer-Verlag, 1999. 3rd edition.

[26] J.H. van Lint and R.M. Wilson. *A course in combinatorics*. Cambridge University Press, 1992.

[27] S.A. Vanstone and P.C. van Oorschot. *An Introduction to Error Correcting Codes with Applications*. The Kluwer Intern. Ser. in Engineering and Computer Science. Kluwer Academic Publishers, 1989.

[28] S.B. Wicker. Deep space applications. In V.S. Pless and W.C. Huffman, editors, *Handbook of Coding Theory*, volume II, pages 2119–2169. Elsevier Science B.V., 1998.

[29] S.B. Wicker and V.K. Bhargava, editors. *Reed-Solomon codes and their applications*. The Institute of Electrical and Electronics Engineers, 1994.

[30] A. Wood. *Preliminary notes to* Russell's Philosophy: a Study of its Development. 1959. Appended to Bertrand Russell's *My Philosophical Developement*, Unwin Books.

[31] S. Xambó-Descamps. OMEGA: a system for effective construction, coding and decoding of block error-correcting codes. *Contributions to Science*, 1(2):199–224, 2001. An earlier version of this paper was published in the proceedings of the EACA-99 (Fifth "Encuentro de Algebra Computacional y Applications" (Meeting on Computational Algebra and Applications), Santa Cruz de Tenerife, September 1999).

Index of Symbols

$\binom{n}{k}$	$n!/k!(n-k)!$ (binomial number)
$\lfloor x \rfloor$	Greatest integer that is less than or equal x
$\lceil x \rceil$	Smallest integer that is greater than or equal x
t_n	Vector $[t, ..., t]$ of length n. Examples used often: 0_n and 1_n
$x\|y$	Concatenation of x and y (x and y may be vectors or lists)
$\langle x\|y \rangle$	$x_1 y_1 + \ldots + x_n y_n$ (scalar product of vectors x and y)
$\|X\|$	Number of elements of a set X
A^*	Group of invertible elements of the ring A
$A[X]$	Ring of polynomials in X with coefficients in A
$\mathbb{Z}_n$	Ring of integers modulo n
$\mathbb{F}_q$, $GF(q)$	Field of q elements (q must be a prime power)
$\mathbb{F}_q^*$	Group of nonzero elements of $\mathbb{F}_q$
$[L : K]$	Degree of a field L over a subfield K
I_r	Identity matrix of order r
$M_n^k(A)$	Matrices of type $k \times n$ with entries in A

Other than the general mathematical notations explained above, we list symbols that are used regularly in the book, followed by a short description and, enclosed in parenthesis, the key page number(s) where they appear. We do not list symbols that are used only in a limited portion of the text.

p	Probability of a symbol error in a symmetric channel (2)
S	Source alphabet (14)
T	Channel alphabet (14, 16)
C	A general code (15)
q	Cardinal of the channel alphabet T (16)
k, k_C	Dimension of a code C (16)
R, R_C	Rate of a code C (16)
$hd(x, y)$	Hamming distance between vectors x and y (17)
d, d_C	Minimum distance of a code C (17)
$(n, M, d)_q$	Type of a code, with short form $(n, M)_q$ (17)
$[n, k, d]_q$	Type of a code, with short form $[n, k]_q$ (17)
$\|x\|$	Weight of vector x (17, 36)
$x \cdot y$	Componentwise product of vectors x and y (17)
δ, δ_C	Relative distance of a code C (17)
t	Correcting capacity of a code (20)
$B(x, r)$	Hamming ball of center x and radius r (20)
D_C	Set of minimum distance decodable vectors (20)
$P_e(n, t, p)$	Upper bound for the code error probability (23)

$\rho(n, t, p)$	Upper bound for the error-reduction factor (24)
$R_q(n, d)$	Maximum rate R among codes $(n, M, d)_q$ (26)
$k_q(n, d)$	Maximum dimension k among codes $[n, k, d]_q$ (26)
$A_q(n, d)$	Maximum cardinal M among codes $(n, M, d)_q$ (26)
$B_q(n, d)$	Maximum q^k among linear codes $[n, k, d]_q$ (81)
$\overline{C}$	Parity completion of a binary code (27)
$vol_q(n, r)$	Cardinal of $B(x, r)$ (27)
$\langle G \rangle$	Linear code generated by the rows of matrix G (37)
$RS_\alpha(k)$	Reed–Solomon code of dimension k on α (39)
$V_k(\alpha)$	Vandermonde matrix of k rows on the vector α (39)
$D(\alpha)$	Vandermonde determinant of $\alpha = \alpha_1, \ldots, \alpha_n$ (41)
$Ham_q(r)$	The q-ary Hamming code of codimension r (51)
$Ham_q^\vee(r)$	The q-ary dual Hamming code (51)
$RM_q^x(m)$	First order Reed–Muller code of dimension $m + 1$ (62)
$\rho_z(C)$	Residue of C with respect to $z \in C$ (79)
$K_j(x, n, q)$	Krawtchouk polynomials (90)
$\varphi(n)$	Euler's function (104)
$\mu(n)$	Möbius' function (119)
$I_q(n)$	Number of irreducible polynomials of degree n over $\mathbb{F}_q$ (120)
$ord(\alpha)$	Order of α (124)
$ind_\alpha(x)$	Index of x with respect to the primitive element α (129)
$Tr_{L/K}(\alpha)$	Trace of $\alpha \in L$ over K (137)
$Nm_{L/K}(\alpha)$	Norm of $\alpha \in L$ over K (137)
C_f	Cyclic code defined by a polynomial f (145)
$BCH_\omega(\delta, \ell)$	BCH code of designed distance δ and offset ℓ, ω a primitive n-th root of unity (166)
$A_K(h, \alpha, r)$	Alternant code defined over K corresponding to the alternant control matrix of order r associated to the vectors $h, \alpha \in \bar{K}$ (182)
$\Gamma(g, \alpha)$	Classical Goppa code (191)
η_i	Error locators (195)
$S, S(z)$	Syndrome vector and polynomial for alternant codes (195)
$\sigma(z)$	Error-locator polynomial (196)
$\epsilon(z)$	Error-evaluator polynomial (196)

The table below includes the basic WIRIS/cc expressions explained in the Appendix which might be non-obvious. The first field contains the expression in a keyboard style, optionally followed by the same expression in palette style (A.1, page 221). The second field is a brief description of the expression. The WIRIS/cc functions used in the text are included in the alphabetical index.

=	Assignment operator
:=, $\leftarrow$	Delayed assignment operator
==, $=$	Binary operator to create equations
->, $\rightarrow$	Operator to create pairs in a relation or divisor
!=, $\neq$	Not equal operator
>=, $\geqslant$	Greater than or equal operator
<=, $\leqslant$	Less than or equal operator
=>, $\Rightarrow$	Operator to create pairs in a substitution or rule
\$\$	Operator to create compound identifiers (92)
|	Disjunction (and concatenation) operator (231)
&	Boolean conjunction operator (231)
Zn(p), $\mathbb{Z}_p$	Ring of integers mod p, $\mathbb{F}_p$ if p is prime (223)
n:A	Natural image of n in A (223)
x=e	Bounds the value of e to x (223)
a..b..d	$\{x \in [a,b] \mid x = a + jd, j \in \mathbb{Z}, j \geqslant 0\}$ (225)
a..b	Usually a..b..1 (225)
$[x_1, \cdots, x_n]$	Vector whose components are $x_1, \cdots, x_n$ (227)
$\{x_1, \cdots, x_n\}$	List of the objects $x_1, \cdots, x_n$ (227)
A^{-1}	The inverse of an invertible matrix A (230)
A^{T}	The transpose of a matrix A (229)
$\|A\|$	The determinant of a square matrix A (230)
A|B	If A and B are matrices with the same number of rows, the matrix obtained by concatenating each row of A with the corresponding row of B (230)
A&B	Matrix A stacked on matrix B (A.13, page 230)
x·y, $\langle x,y \rangle$	The scalar product of vectors x and y (A.12, page 229)

Alphabetic Index, Glossary and Notes

The entries of the index below are ordered alphabetically, with the convention that we ignore blanks, underscores, and the arguments in parenthesis of functions. Each entry is followed by information arranged in two fields. The first field, which is optional, consists of the page number or numbers, enclosed in parenthesis, indicating the more relevant occurrences of the entry. The second field may be a pointer to another index entry or a brief summary of the concept in question.

There are two kinds of of entries: those that refer to mathematical concepts, which are typeset like *alphabet*, and those that refer to WIRIS/*cc* constructs, which are typeset like **aggregate**. Those entries that do not refer directly to a concept introduced in the text (see, for example, *covering radius* or *Fermat's little theorem*) constitute a small set of supplementary notes.

In the summaries, terms or phrases written in *italics* are mathematical entries. Acronyms are written in italic boldface, like **BSC**. Terms an phrases typeset as aggregate refer to WIRIS/*cc* entries.

A

active block (222) In a **WIRIS** session, the block that displays the blinking cursor (a thick vertical bar).

active statement (222) In a **WIRIS** session, the statement that displays the blinking cursor (a thick vertical bar).

AD(y,C) (205) Equivalent to alternant_decoder(y,C).

additive representation (128) [of an element of a finite field with respect to a subfield] The vector of components of that element relative to a given linear basis of the field with respect to the sufield. Via this representation, addition in the field is reduced to vector addition in the subfield.

aggregate (225) A range, list, vector or pairing.

alphabet A finite set. See *source alphabet* and *channel alphabet*. See also *binary alphabet*.

alternant bounds (182) For an alternant code C, $d_C \geqslant r + 1$ (alternant bound for the minimum distance) and $n - r \geqslant k_C \geqslant n - rm$ (alternant bounds for the dimension), where r is the order on the alternant control matrix $H \in M_n^r(\bar{K})$ that defines C and $m = [\bar{K} : K]$.

alternant codes (182) An alternant code over a finite field K is a linear code of the form $\{x \in K^n \mid xH^T = 0\}$, where $H \in M_n^r(\bar{K})$ is an alternant control matrix defined over a finite field $\bar{K}$ that contains K. General *RS* codes and *BCH* codes are alternant codes (see page 185 for *RS* codes and page 187 for *BCH* codes).

alternant_code(h,a,r) (182) Creates the *alternant code* associated to the matrix control_matrix(h,a,r).

alternant control matrix (180) A matrix H of the form $H = V_r(\alpha)diag(h) \in M_n^r(\bar{K})$, where r is a positive integer (called the order of the control matrix) and $h, \alpha \in \bar{K}^n$ (we say that H is defined over $\bar{K}$).

alternant_decoder(y,C) (205) Decodes the vector y according to the BMS decoder for the alternant code C.

alternant_matrix(h,a,r) (180) Creates the *alternant control matrix* of order r associated to the vectors h and a (which must be of the same length).

append(x,a) (228) Yields the result of appending the object a to the list or vector x. It can also be expressed as x|{a} or x|[a] if x if x is a list or vector, respectively.

assignment expression See assignment.

asymptotically bad / good families (94) See Remark 1.79.

asymptotic bounds (96) Any function of $\delta = d/n$ that bounds above $\limsup k/n$ (when we let $n \to \infty$ keeping d/n fixed) is called an asymptotic upper bound (asymptotic lower bounds are defined similarly). To each of the upper (lower) bounds for $A_q(n, d)$ there is a corresponding asymptotic upper (lower) bound.

McEliece, Rodemich, Rumsey and Welsh (page 95) have obtained the best upper bound. Corresponding to the Gilbert–Varshamov lower bound for $A_q(n, d)$ there is an asymptotic Gilbert–Varshamov bound (see E.1.55).

B

Bassalygo–Elias bound See *Elias bound*.

BCH Acronym for Bose–Chaudhuri–Hocquenghem.

BCH(α,d,l) (187) Creates the code $BCH_\alpha(d, l)$ of designed distance d and offset l based on the element α.

BCH bounds See *BCH codes*.

BCH codes The *BCH* code of length n, designed distance δ and offset ℓ over $\mathbb{F}_q$, which is denoted $C = BCH_\omega(\delta, \ell)$, is the cyclic code with roots $\omega^{\ell+i}$, for $i = 0, \ldots, \delta - 2$, where ω is a primitive n-th root of unity in a suitable extension of $\mathbb{F}_q$. The minimum distance of C is at least δ (*BCH bound of the minimum distance*, Theorem 3.15). The dimension k of C satisfies $n - (\delta - 1) \geqslant k \geqslant n - m(\delta - 1)$, where m is the degree of ω over $\mathbb{F}_q$ (*BCH bounds of the dimension*, Proposition 3.18). In the binary case, the dimension lower bound can be improved to $k \geqslant n - mt$ if $\delta = 2t + 1$ (Proposition 3.19). When $\ell = 1$, we write $BCH_\omega(\delta)$ instead of $BCH_\omega(\delta, 1)$ and we say that these are strict *BCH* codes. Primitive *BCH* codes are *BCH* codes with $n = q^m - 1$ for some m.

begin S end (232) This construct builds an expression out of statement S.

BER Acronym for *bit error rate*.

Berlekamp algorithm [for factoring a polynomial over $\mathbb{F}_q$] In this text we have only developed a factorization algorithm for $X^n - 1$ (page 158). There are also algorithms for the factorization of arbitrary polynomials. For an introduction to this subject, including detailed references, see [14], Chapter 2.

bezout(m,n) This function can be called on integers or univariate polynomials m, n. It returns a vector $[d, a, b]$ such that $d = \gcd(m, n)$ and $d = am + bn$ (Bezout's identity). In fact a and b also satisfy $|a| < |n|$ and $|b| < |m|$ in the integer case, and $deg(a) < deg\,n$ and $deg(b) < deg\,m$ in the polynomial case.

binary alphabet (15) The set $\{0, 1\}$ of binary digits or bits. Also called binary numbers.

binary digit (2,15) Each of the elements of the binary alphabet $\{0, 1\}$. With addition and multiplication modulo 2, $\{0, 1\}$ is the field $\mathbb{Z}_2$ of binary numbers.

binary symmetric channel (2) In this channel, the symbols sent and received are *bits* and it is characterized by a probability $p \in [0, 1]$ that a bit is altered (the same for

0 and 1). Since p denotes the average proportion of erroneous bits at the receiving end, it is called the bit error rate (*BER*) of the channel.

bit (2, 15) Acronym for *binary digit*. It is also the name of the fundamental unit of information.

bit error rate (2) See *binary symmetric channel*.

block (221) In a **WIRIS** session, each of the sets of statements delimited by a variable height '['. Each statement in a block is delimited with a variable height '|'.

block code (16) If T is the *channel alphabet* and $q = |T|$, a q-ary block code of length n is a subset C of T^n.

block encoder (15) An *encoder* $f : S \rightarrow T^*$ such that $f(S) \subseteq T^n$ for some n (called the length of the block encoder).

blow(h,F) (183) For a finite field F=K[x]/(f) and a vector h with entries in F, it creates a vector with entries in K by replacing each entry of h by the sequence of its components with respect to the basis $1, x, \ldots, x^{r-1}$ of F as a K-vector space. If H is a matrix with entries in F, then blow(H,F) is the result of replacing each column h of H by blow(h,F).

BMS (203) Acronym for *Berlekamp–Massey–Sugiyama*.

BMS algorithm (204) A decoding algorithm for alternant codes based on Sugiyama's method for solving the *key equation* and on *Forney's formula* for the determination of the error values.

Boolean expression (230) An expression whose value is true or false.

Boolean function (230) An function whose value is true or false.

BSC Acronym for *binary symmetric channel*.

C

capacity (4) It measures the maximum fraction of source information that is available at the receiving end of a communication channel. In the case of a *binary symmetric channel* with bit error probability p, it is a function $C(p)$ given by Shannon's formula $C(p) = 1 + p \log_2(p) + (1 - p) \log_2(1 - p)$.

cardinal(R) (224) The cardinal of a ring R.

Cauchy matrix (194) A matrix of the form $(1/(\alpha_i - \beta_j))$, where $\alpha_1, \ldots, \alpha_n$ and $\beta_1, \ldots, \beta_n$ belong to some field and $\alpha_i \neq \beta_j$ for $1 \leq i, j \leq n$.

cauchy_matrix(a,b) Given vectors a and b, this function constructs the *Cauchy matrix* $(1/(a_i - b_j))$, provided that $a_i \neq b_j$ for all i, j with i in range(a) and all j in range(b).

channel Short form of *communication channel*.

channel alphabet (14) The finite set T of symbols (called channel symbols) with which the information stream fed to the *communication channel* is composed. It is also called transmission alphabet. Although we do make the distinction in this text, let us point out that in some channels the input alphabet may be different from the output alphabet.

channel coding theorem (4) If C is the *capacity* of a *BSC*, and R and ε are positive real numbers with $R < C$, then there exist block codes for this channel with *rate* at least R and *code error rate* less than ε.

channel symbol See *symbol*.

characteristic [of a ring or field] See *characteristic homomorphism*.

characteristic(R) (224) The characteristic of a finite ring R.

characteristic homomorphism (109) For a ring A, the unique ring homomorphism $\phi_A : \mathbb{Z} \to A$. Note that necessarily $\phi_A(n) = n \cdot 1_A$. The image of $\mathbb{Z}$ is the minimum subring of A (which is called the prime ring of A). If $\ker(\phi_A) = \{0\}$, then A is said to have characteristic 0. Otherwise the characteristic of A is the minimum positive integer such that $n \cdot 1_A = 0$. In this case, there is a unique isomorphism from $\mathbb{Z}_n$ onto the prime ring of A. If A is a domain (has no zero-divisors), then the prime ring is $\mathbb{Z}$ if the characteristic is 0 and the field $\mathbb{Z}_p$ if the characteristic p is positive (p is necessarily prime, because otherwise $\mathbb{Z}_p$ would not be a domain).

check b(X) (232) In the definition of a function, between the argument sequence (X) and the operator := preceeding the body of the function, it is allowed to include a directive of the form check b(X), where b(X) stands for a Boolean expression formed with the arguments of the function. This mechanism extends the filtering mechanism with types (A.16, page 232).

check matrix (40) For a linear code C, a matrix H such that $x \in C$ if and only if $x H^T = 0$. Also called parity-check matrix, or control matrix. For an introductory illustration, see Example 1.9.

classical Goppa codes (191) Alternant codes $C = A_K(h, \alpha, r)$, $h, \alpha \in \bar{K}^n$, such that there exists a polynomial $g \in \bar{K}[X]$ of degree r with $g(\alpha_i) \neq 0$ and $h_i = 1/g(\alpha_i)$, $i = 1, \ldots, n$. As shown in P.4.6, this definition gives the same codes as Goppa's original definition, which looks quite different at a first glance. The alternant bounds give $d_C \geqslant r + 1$ and $n - r \geqslant k_C \geqslant n - rm$, $m = [\bar{K} : K]$. Strict *BCH* codes are classical Goppa codes (Proposition 4.5). In the case of binary classical Goppa codes we have $d_C \geqslant 2r + 1$ if g does not have multiple roots (P.4.7).

clear X Assuming that X is a sequence of identifiers, this command undoes all the assignments made earlier to one or more of these identifiers.

code In this text, short form of *block code*.

code efficiency See *code rate*.

code error (3) It occurs when the received vector either cannot be decoded (decoder error) or else its decoded vector differs from the sent vector (undetectable error). The number of erroneous symbols at the destination due to code errors is usually bounded above using the observation that for a code of dimension k a code error produces at most k symbol errors.

code error rate (3) The probability of a *code error*. For an introductory example, see **Rep(3)**.

code rate (16) For a code C, the quotient k/n, where k and n denote the *dimension* and the *length* of C, respectively. It is also called code efficiency or transmission rate. For an introductory example, see **Rep(3)**.

code vector (16) A *code word*.

code word (16) An element of the image of an *encoder*, or an element of a *block code*. It is also called a code vector.

coefficient(f,x,j) For a polynomial f, the coefficient of x^j. If f is multivariate, it is considered as a polynomial in x with coefficients which are polynomials in the remaining variables.

coefficient_ring(f) This function gives the ring to which the coefficients of a polynomial f belong.

collect(f,X) (143) Assuming X is a list of variables and f a multivariate polynomial, this function returns f seen as a polynomial in the variables X (with coefficients which are polynomials in the variables of f which are not in X. The argument X can be a single variable x, and in this case the call is equivalent to collect(f,{x}).

communication channel (14) The component of a *communication system* that effects the transmission of the source information to the destination or receiver.

communication system According to *Shannon*'s, widely accepted model, it has an information source, which generates the messages to be sent to the information destination, and a communication channel, which is the device or medium that effects the transmission of that information.

complete decoder A *full decoder*.

Complex (232) The type associated to complex numbers. Denoted $\mathbb{C}$ in palette style.

conference matrix (70) A square matrix C with zeroes on its diagonal, ± 1 entries otherwise, and such that $CC^T = (n-1)I_n$, where n is the order of C.

conference_matrix(F) (70) Computes the conference matrix associated to a finite field F of odd cardinal.

conjugates (135) The set of conjugates of $\alpha \in L$, L a finite field, over a subfield K of L is

$$\{\alpha, \alpha^q, \alpha^{q^2}, \dots, \alpha^{q^{r-1}}\},$$

where $q = |K|$ and r is the least positive integer such that $\alpha^{q^r} = \alpha$. If K is not specified, it is assumed that it is the prime field of L.

conjugates(a,K) (135) Yields the list of conjugates of a over K. The element a is assumed to belong to a finite field that contains K as a subfield. The call conjugates(a) is equivalent to conjugates(a,K), where K is the prime field of the field of a.

constant_list(n,a) (228) The list of length n all of whose terms are equal to a.

constant_matrix(m,n,a) (230) The matrix with m rows and n columns all whose entries are equal to a. constant_matrix(n,a) is equivalent to constant_matrix(n,n,a).

constant_vector(n,a) (228) The vector a_n (the vector of length n all of whose components are equal to a).

constrained_maximum(u,C) Computes the maximum of a linear polynomial u with respect to the constraints C (a list of linear inequalities). This function is crucial for the computation of the linear programming bound LP(n,s).

control matrix See *check matrix*.

control polynomial (146) For a cyclic code C over K, the polynomial $(X^n - 1)/g$, where n is the length of C and g its *generator polynomial*. The control polynomial is also called check polynomial.

correcting capacity (20) A *decoder* $g : D \to C$ for a q-ary block code $C \subseteq T^n$ has correcting capacity t (t a positive integer) if for any $x \in C$ and any $y \in T^n$ such that $hd(x, y) \leqslant t$ we have $y \in D$ and $g(y) = x$.

covering radius If $C \subseteq T^n$, where T is finite set, the convering radius of C is the integer

$$r = \max_{y \in T^n} \min_{x \in C} hd(y, x).$$

Thus r is the minimum positive integer such that $\bigcup_{x \in C} B(x, r) = T^n$. The Gilbert bound (Theorem 1.17) says that if C is optimal then $r \leqslant d - 1$, where d is the minimum distance of C.

crossover probability In a *binary symmetric channel*, the bit error probability.

cyclic code (142) Said of a linear code $C \subseteq \mathbb{F}_q^n$ such that $(a_n, a_1, \ldots, a_{n-1}) \in C$ when $(a_1, \ldots, a_{n-1}, a_n) \in C$.

cyclic_control_matrix(h,n) (150) Constructs the control matrix of the cyclic code of length n whose control polynomial is h.

cyclic_matrix(g,n) (148) Constructs a generating matrix of the cyclic code of length n whose generator polynomial is g.

cyclic_normalized_control_matrix(g,n) (149) Constructs the normalized control generating matrix of the cyclic code of length n whose generator polynomial is g corresponding to cyclic_normalized_matrix(g,n).

cyclic_normalized_matrix(g,n) (149) Constructs a normalized generating matrix of the cyclic code of length n whose generator polynomial is g.

cyclic_product(a,b,n,x) (143) Equivalent to cyclic_reduction(a·b,n,x), assuming that a and b are polynomials, n a positive integer and x a variable. We may omit x for univariate polynomials. Finally the call cyclic_product(a,b) works for two vectors a and b of the same length, say n, and it yields the coefficient vector of cyclic_reduction(a′·b′,n), where a′ and b′ are univariate polynomials whose coefficient vectors a and b. The result is padded with zeros at the end to make a vector of length n.

cyclic_reduction(f,n,x) (142) Replaces any monomial x^j in the polynomial f by $x^{[j]_n}$. The polynomial f can be multivariate. The call cyclic_reduction(f,n) works for univariate polynomials and is equivalent to cyclic_reduction(f,n,x), where x is the variable of f.

ciclic_shift(x) (18) If x is the vector $(x_1, \ldots, x_n)$, the vector $(x_n, x_1, \ldots, x_{n-1})$.

ciclic_shifts(x) (18) If x is a vector, the list of cyclic permutations of x.

constant_matrix(m,n,a) (230) The $m \times n$ matrix whose entries are all set equal to a. The call constant_matrix(n,a) is equivalent to constant_matrix(n,n,a).

cyclotomic class (157) For an integer q such that $gcd(q, n) = 1$, the q-cyclotomic class of an integer j modulo n is the set $\{jq^i \bmod n\}_{0 \leqslant i < r}$, where r is the least positive integer such that $jq^r \equiv j \bmod n$.

cyclotomic_class(j,n,q) (157) The value of this function is the q-cyclotomic class of an integer j mod a positive integer n, where q is required to be a positive intger relatively prime with n.

cyclotomic_classes(n,q) (157) For two relatively prime positive integers n and q, it gives the list of the q-cyclotomic classes mod n.

cyclotomic polynomial (160) The n-th cyclotomic polynomial over $\mathbb{F}_q$ (see P.3.7) is $Q_n = \prod_{gcd(j,n)=1}(X - \omega^j)$, where ω is a primitive n-th root of unity in a suitable extension field of $\mathbb{F}_q$ (for example $\mathbb{F}_{q^m}$, $m = ord_n(q)$).

cyclotomic_polynomial(n,T) (162) Provided n is a positive integer and T a variable, it returns the n-th *cyclotomic polynomial* $Q_n(T) \in \mathbb{Z}[T]$.

D

decoder A *decoding function*.

decoder error (20) It occurs when the received vector is *non-decodable*.

decoder_trial(C,s,K) (209) For a code C over a finite field K and a positive integer s, this function generates a random vector x of C, a random error vector e of weight s, and calls alternant_decoder(x+e,C). If the result x' coincides with x, $[x, e, x + e]$ is returned. If $x' \neq x$, but otherwise is a vector, $[x, e, x + e, x']$ is returned.

If x' is not a vector, $[x, e]$ is returned. The call decoder_trial(C,s) is equivalent decoder_trial(C,s,K), where K is the base field of C.

decoding function (19) For a block code $C \subseteq T^n$, a map $g : D \rightarrow C$ such that $C \subseteq D \subseteq T^n$ and $g(x) = x$ for all $x \in C$. The elements of D, the domain of g, are said to be g-decodable.

degree (134) **1** The degree of a finite field L over a subfield K, usually denoted $[L : K]$, is the dimension of L as a K-vector space. If K is not specified, it is assumed to be the prime field of L. **2** The degree of $\alpha \in L$ over a subfield K of L is $[K[\alpha] : K]$. It coincides with degree of the *minimum polynomial* of α over K. If K is not specified, it is assumed to be the prime field of L.

degree(f) Yields the degree of a polynomial f.

designed distance See *BCH codes*.

determinant(A) (230) Assuming A is a square matrix, the determinant of A ($|A|$ in palette style).

diagonal_matrix(a) (180) Creates the diagonal matrix whose diagonal entries are the components of the vector a.

dimension (16) The dimension of a q-ary block code C of length n is $k_C = \log_q(M)$, where $M = |C|$.

dimensions(A) (229) For a matrix A with m rows and n columns, the sequence m,n.

discrete logarithm See *exponential representation*.

Divisor (232) The type associated to divisors.

divisor (225) A **WIRIS** divisor is a sequence of terms of the form $a \rightarrow b$, where a and b are arbitrary objects, enclosed in brakets. Divisors can be thought of as a kind of function. Indeed, if D is a divisor and x any object, then D(x) returns the sum (possibly 0) of the values y such that x→y appears in D. Divisors can be added. The sum D+D' of two divisors D and D' is characterized by the relation (D+D')(a)=D(a)+D'(a),. for all a.

divisors(n) (106) The list of positive divisors of a nonzero integer n.

domain A ring with no zero-divisors. In other words, a ring in which $x \cdot y = 0$ can only occur when $x = 0$ or $y = 0$.

domain(R) (225) For a relation R, the set of values a for which there is a term of the form a→b in R. A similar definition works for the other pairing flavours. In case R is a divisor, it coincides with support(R).

dual code (40) If C is a linear code over $\mathbb{F}_q$, the code

$$C^\perp = \{h \in \mathbb{F}_q^n \mid \langle x|h \rangle = 0 \text{ for all } x \in C\}.$$

dual Hamming code (54) Denoted $Ham_a^\vee(r)$, it is the dual of the Hamming code $Ham_q(r)$. It is equidistant with minimum distance q^{r-1} (Proposition 1.43). These codes satisfy equality in the *Plotkin bound* (Example 1.65).

E

Element(Field) The type of objects that happen to be an element of some field.

element(j,R) (224) For a finite ring R of cardinal n, this function yields the elements of R when j runs over the range 0..(n-1).

elias(x) (95, 96) Computes the *asymptotic Elias upper bound*.

Elias bound (86) For positive integers q, n, d, r with $q \geqslant 2$, $r \leqslant \beta n$ (where $\beta = (q-1)/q$) and $r^2 - 2\beta nr + \beta nd > 0$,

$$A_q(n, d) \leqslant \frac{\beta nd}{r^2 - 2\beta nr + \beta nd} \cdot \frac{q^n}{vol_q(n, r)}.$$

encoder (15) In a *communication system*, an injective map $f : S \to T^*$ from the *source alphabet* S to the set T^* of finite sequences of *channel symbols*.

entropy (94) The q-ary entropy function is

$$H_q(x) = x \log_q(q-1) - x \log_q(x) - (1-x) \log_q(1-x)$$

for $0 \leqslant x \leqslant 1$. For $q = 2$, $H_2(x) = 1 - C(x)$, where $C(x)$ is the *capacity* function. The entropy plays a basic role in the expression of some of the *asymptotic bounds*.

entropy(x,q) (94, 96) Computes the entropy function $H_q(x)$. The call entropy(x) is equivalent to entropy(x,2).

epsilon_vector(n,j) (228) If j is in the range 1..n, the vector of length n whose components are all 0, except the j-th, which is 1. Otherwise, the vector 0_n.

equidistant code (54) A code such that $hd(x, y)$ is the minimum distance for all distinct code words x and y.

equivalent codes See *equivalence criteria*.

equivalence criteria (18, 19) Two q-ary block codes C and C' of length n are said to be equivalent if there exist permutations $\sigma \in S_n$ and $\tau_i \in S_q$ ($i = 1, \ldots, n$) such that C' is the result of applying τ_i to the symbols in position i for all words of C, $i = 1, \ldots, n$, followed by permuting the n components of each vector according to σ. In the case where the q-ary alphabet T is a finite field and each τ_i is the permutation of T given by the multiplication with a nonzero element of T, the codes are said to be scalarly equivalent.

erf(n,t,p) 24 Computes the *error-reduction factor* of a code of length n and error correcting capacity t for a symmetric channel with symbol error probability p.

error detection (18, 32) A block code of *minimum distance* d can detect up to $d-1$ errors. It can also simultaneously correct t errors and detect $s \geqslant t$ errors if $t + s < d$ (P.1.1).

error-locator polynomial (196, 215) [in relation to an alternant code $A_K(h, \alpha, r)$ and an error pattern e] **1** In the *BMS* decoding approach, the polynomial $\sigma(z) = \prod_{i=1}^{s}(1 - \eta_i z)$, where $\eta_1, \ldots, \eta_s$ are the *error locators* corresponding to the error pattern e. **2** In the PGZ approach, the polynomial $\sigma(z) = \prod_{i=1}^{s}(z - \eta_i)$. This definition can also be used, with suitable modifications, in the BMS approach (Remark 4.7 and P.4.9).

error locators (195) [in relation to an alternant code $A_K(h, \alpha, r)$ and an error pattern e] The values $\eta_i = \alpha_{m_i}$ ($1 \leqslant i \leqslant s$) where $\{m_1, \ldots, m_s\} \subseteq \{1, \cdots, n\}$ is the *support* of e (the set of error positions).

error-evaluator polynomial (196) [in relation to an alternant code $A_K(h, \alpha, r)$ and an error pattern e] The polynomial $\epsilon(z) = \sum_{i=1}^{s} h_{m_i} e_{m_i} \prod_{j=1, j \neq i}^{s}(1 - \eta_i z)$, where $\eta_1, \ldots, \eta_s$ are the *error locators* corresponding to the error pattern e.

error pattern (195) See *error vector*.

error-reduction factor (3, 24) The quotient of the symbol error rate obtained with a coding/decoding scheme by the symbol error rate without coding.

error vector (195) Also called error pattern, it is defined as the difference $y - x$ between the sent code vector x and the received vector y.

Euler's function $\varphi(n)$ (104) For an integer $n > 1$, $\varphi(n)$ counts the number of integers k such that $1 \leqslant k \leqslant n - 1$ and $gcd\,(k, n) = 1$. It coincides with the cardinal of the group $\mathbb{Z}_n^*$. The value of $\varphi(1)$ is conventionally defined to be 1.

evaluate(f,a) Yields the value of a polynomial f at a. Here a is a list with as many terms as the number of variables of f, as in evaluate(x^3+y^5,{3,2}). In the case of a single variable, the braces surrounding the unique object can be omitted (thus evaluate(x^3+x^5,{17}) and evaluate(x^3+x^5,17) are equivalent).

evaluation (222) In a **WIRIS** session, the action and result of clicking on the red arrow (or by pressing *Ctrl+Enter*). The value of each statement of the active block is computed and displayed, preceded by a red arrow icon, to the right of the statement.

even?(n) (231) This Boolean function tests whether the integer n is even.

exponent [of a monic irreducible polynomial] See *primitive polynomial*.

exponential representation (128) The expression $x = \alpha^i$ of a nonzero element x of a finite field L as a power of a given primitive element α. The exponent i is called the index (or discrete logarithm) of x with respect to α, and is denoted $ind_\alpha(x)$. If we know $i = ind_\alpha(x)$ and $j = ind_\alpha(y)$, then $xy = \alpha^k$, where $k = i + j \bmod q - 1$ (q the cardinal of L). This scheme for computing the product, which is equivalent to say that $ind_\alpha(xy) = ind_\alpha + ind_\alpha(y) \bmod q - 1$, is called index calculus (with respect to α). In practice, the index calculus is implemented by compiling a table, called an index table, whose entries are the elements x of L (for example in *additive notation*) and whose values are the corresponding indices $ind_\alpha(x)$.

expression (221) A formula or an assignment (see *statement*).

extension(K,f) (113, 224) Given a ring K and a polynomial f∈K[x] whose leading coefficient is invertible (a monic polynomial in particular), then extension(K,f), or the synomym expression K[x]/(f), constructs the quotient $K[x]/(f)$ and assigns the class of x to x itself (thus the result is equivalent to the ring $K[x]$, with $f(x) = 0$. If we want another name α for the class of x, then we can use the call extension(K,α,f).

F

factor(f,K) Given a field K and a polynomial f with coefficients in K, finds the factorization of f into irreducible factors.

factorization algorithm (158) When $gcd\,(q, n) = 1$, the irreducible factors of the polynomial $X^n - 1 \in \mathbb{F}_q[X]$ are $f_C = \prod_{j \in C}(X - \omega^j)$, where C runs over the q-cyclotomic classes mod n and ω is a primitive n-th root of unity in $\mathbb{F}_{q^m}$, $m = ord_n(q)$.

false (230) Together with true forms the set of Boolean values.

Fermat's little theorem It asserts that if p is a prime number and a is any integer not divisible by p, then $a^{p-1} \equiv 1 \pmod{p}$. This is equivalent to say that any nonzero element α of the field $\mathbb{Z}_p$ satisfies $\alpha^{p-1} = 1$. We have established two generalizations of this statement: one for the group $\mathbb{Z}_n^*$ (see E.2.5) and another for the multiplicative group of any finite field (see E.2.6).

Field (232) The type associated to fields.

field (101) A commutative ring for which all nonzero elements have a multiplicative inverse.

field(a) Assuming a is an element of some field F, this function returns F. For example, the value of *field(2/3)* is $\mathbb{Q}$, and the value of field(5:$\mathbb{Z}_{11}$) is $\mathbb{Z}_{11}$.

filter (232) Any Boolean expression F, which may include types, used in the form x:F to check that an argument x of a function has the required properties.

Finite_field(K) (224) This Boolean function, or its synonym GF(K), tests whether K is a Galois field.

finite fields (111) A finite field (also called a Galois field) is a field K whose cardinal is finite. If $q = |K|$, then q must be a prime power (Proposition 2.8): $q = p^r$, p the *characteristic* of K and r a positive integer. Conversely, if p is a prime number and r a positive integer, then there is a finite field of p^r elements (Corollary 2.18). Furthermore, any two fields whose cardinal is p^r are isomorphic (Theorem 2.41). This field is denoted $\mathbb{F}_q$, or $GF(q)$. Note that $\mathbb{F}_p = \mathbb{Z}_p$, the field of integers mod p. For any divisor s of r there is a unique subfield of $\mathbb{F}_{p^r}$ whose cardinal is p^s, and these are the only subfields of $\mathbb{F}_{p^r}$ (E.2.15).

flip(x,l) The components x_i of the vector x indicated by the elements i of the list l are replaced by $1 - x_i$. For a binary vector this amounts to complement the bits at the positions indicated by l. We remark that l can be a range or a vector. On the other hand, *flip(x)* is equivalent to flip(x,1..n), n the length of x.

Float (232) The type of decimal numbers.

formula (222) See *statement*.

formula expression (222) See *statement*.

Forney's formula (196) Expression for the error values in terms of the error-locator and error-evaluator polynomials (Proposition 4.8).

Frobenius automorphism (111) **1** If L is a finite field of characteristic p, the map $L \to L$ such that $x \mapsto x^p$ (absolute Frobenius automorphism of L). **2** If K is a subfield of L with $|K| = q$, the map $L \to L$ such that $x \mapsto x^q$ is an automorphism of L over K (Frobenius automorphism of L over K).

frobenius(x) (111) If x lies in a field of characteristic p, the value x^p. The call frobenius(K,x), where K is a finite field of q elements, yields x^q.

full decoder (19) A *decoder* $g : D \to C$ such that $D = T^n$ (hence all length n vectors are decodable).

Function (232) The type associated to functions, in the sense of A.15, page 232.

G

gcd(m,n) The greatest common divisor of two integers (or of two univariate polynomials) m and n.

geometric_progression See (geometric_series).

geometric_series(x,n) (103) Given expression x and a non-negative integer f, returns the vector $[1, x, \ldots, x^{n-1}]$ (note that n is its length). The call geometric_series(a,x,n), where a is an expression, is equivalent to a·geometric_series(x,n). Synonym: geometric_progression.

gilbert(x) (94, 96) Computes the *asymptotic Gilbert lower bound*.

g-decodable See *decoding function*.

Galois field (34) See *finite fields*.

Gauss algorithm (131) A fast procedure for finding a *primitive element* of a finite field (P.2.10).

generalized Reed–Solomon codes (185) Alternant codes $A_K(h, \alpha, r)$ for which $\bar{K} = K$ (and hence such that $h, \alpha \in K^n$).

generating matrix (37) For a linear code C, a matrix G whose rows form a linear basis of C. In particular, $C = \langle G \rangle$.

generator polynomial (146) For a cyclic code C over K, the unique monic divisor g of $X^n - 1$ such that $C = C_g$ (Proposition 3.3).

GF(K) (224) Synonym of Finite_field(K), which tests whether K is a Galois field.

Gilbert lower bound (30) $A_q(n, d) \geqslant q^n / vol_q(n, d - 1)$.

Gilbert–Varshamov bound (51) $A_q(n, d) \geqslant q^k$ if (n, k, d) satisfy the *Gilbert–Varshamov condition*.

Gilbert–Varshamov condition (50) For positive integers n, k, d such that $k \leqslant n$ and $2 \leqslant d \leqslant n + 1$, the relation $vol_q(n - 1, d - 2) < q^{n-k}$. It implies the existence of a linear $[n, k, d]_q$ code (Theorem 1.38).

Golay codes (151, 169) The binary Golay code is a cyclic $[23, 12, 7]$ code (see Example 3.21, page 169). The ternary Golay code is a cyclic $[11, 6, 5]_3$ code (see Example 3.6, page 151). For the weight enumerators of these codes, and their parity completions, see P.3.3 (ternary) P.3.10 (binary).

Goppa codes See *classical Goppa codes*.

Griesmer bound (80) If $[n, k, d]$ are the parameters of a linear code, then we have (Proposition 1.58) $G_q(k, d) \leqslant n$, where $G_q(k, d) = \sum_{i=0}^{k-1} \lceil d/q^i \rceil$ (this expression is called Griesmer function).

griesmer(k,d,q) (80, 81) Computes the Griesmer function $G_q(k, d) = \sum_{i=0}^{k-1} \left\lceil \dfrac{d}{q^i} \right\rceil$, which is a lower bound on the length n of a linear code $[n, k, d]_q$.

Griesmer function See *Griesmer bound*.

GRS Acronym for *Generalized Reed–Solomon*.

GRS(h,a,r) (186) Creates the *GRS* code of codimension r associated to the vectors h and a, which must have the same length.

H

Hadamard codes (72) If H is a *Hadamard matrix* of order n, the binary code $C_H \sim (n, 2n, n/2)$ formed with the rows of H and $-H$ after replacing -1 by 0 is the Hadamard code associated to H. A Hadamard $(32, 64, 16)$ code, which is equivalent to the *Reed–Muller code* $RM_2(5)$, was used in the Mariner missions (1969-1972). For the decoding of Hadamard codes, see Proposition 1.55 (page 74).

hadamard_code(H) (72) Returns the *Hadamard code* associated to the *Hadamard matrix* H.

hadamard_code(n) (72) For positive integer n, this function returns the *Hadamard code* associated to the *Hadamard matrix* $H^{(n)}$.

hadamard_code(F) (72) For a finite field F of odd cardinal, we get the *Hadamard code* associated to the *Hadamard matrix* of F.

Hadamard matrix (63, 71) A square matrix H whose entries are ± 1 and such that $HH^T = nI_n$, where n is the order of H.

hadamard_matrix(n) (64) The $2^n \times 2^n$ *Hadamard matrix* $H^{(n)}$.

hadamard_matrix(F) (71) The *Hadamard matrix* of a finite field F of odd cardinal.

Hadamard decoder (74) The decoder for *Hadamard codes* based on Proposition 1.55.

hadamard_decoder(y,H) (74) Decodes vector y by the *Hadamard decoder* of the *Hadamard code* associated to the *Hadamard matrix* H.

hamming(x) (95, 96) Computes the *asymptotic Hamming upper bound*.

Hamming code (21, 51) Any linear code over $\mathbb{F} = \mathbb{F}_q$ that has a check matrix $H \in M_n^r(\mathbb{F})$ whose columns form a maximal set with the condition that no two of them are proportional. They are perfect codes with length $n = (q^r - 1)/(q - 1)$ and minimum distance 3, and are denoted $Ham_q(r)$. The check matrix H is said to be a Hamming matrix. The binary Hamming code $Ham_2(3)$ is the $[7, 4, 3]$ code studied in Example 1.9. For the weight enumerator of the Hamming codes, see Example 1.46. The Hamming codes are cyclic (Example 3.14, page 165).

Hamming distance (17) Given a finite set T and $x, y \in T^n$, it is the number of indices i in the range $1..n$ such that $x_i \neq y_i$ and it is denoted $hd(x, y)$.

hamming_distance(x,y) (229) The *Hamming distance* between the vectors x and y. For convenience can be called with the synonym hd(x,y).

Hamming matrix (52) See *Hamming code*.

Hamming's upper bound (28) $A_q(n, d) \leqslant q^n/vol_q(n, t), t = \lfloor (d - 1)/2 \rfloor$.

hamming_weight_enumerator(q,r,T) (59) The weight enumerator, expressed as a polynomial in the variable T, of the codimension r Hamming code over a field of cardinal q. The call hamming_weight_enumerator(r,T) is equivalent to hamming_weight_-enumerator(2,r,T).

Hankel matrix (215) The matrix (S_{i+j}), $0 \leqslant i, j \leqslant \ell - 1$, associated to the vector $(S_0, \ldots, S_{\ell-1})$.

hankel(S) (215) Produces the Hankel matrix associated to the vector S.

hd(x,y) (229) Synonym of hamming_distance(x,y), which yields the *Hamming distance* between the vectors x and y.

I

icon (221) Each of the choices in a palette menu of the **WIRIS** user interface, often represented with an ideogram.

ideal (102) An additive subgroup I of a commutative ring A such that $a \cdot x \in I$ for all $a \in A$ and $x \in I$.

identifier (223) A letter, or a letter followed by characters that can be a letter, a digit, an underscore or a question mark. Capital and lower case letters are distinct. Identifiers are also called variables.

identity_matrix(r) (230) The identity matrix of order r, I_r.

if b then S end (233) Evaluates statement S when the value of the Boolean expression b is true, otherwise it does nothing.

if b then S else T end (233) Evaluates statement S when the value of the Boolean expression b is true, otherwise evaluates statement T.

incomplete decoding (50) A correction-detection scheme for linear codes used when asking for retransmission is possible. It corrects error-patterns of weight up to the error-correcting capacity and dectects all error-patterns of higher weight which are not code vectors.

index(a,x) (227) For a vector or list x, 0 if the object a does not belong to x, otherwise the index of first occurrence of a in x.

index calculus See *exponential representation*.

index table See *exponential representation*.

ind_table(a) (129) Given an element a of a finite field, it returns the relation $\{a^j \to j\}_{0 \leqslant j \leqslant n-1}$, n the order of a. In order to incluce 0, we also add the pair {0→_} (here we use the underscore character, but often the symbol ∞ is used instead).

infimum(f,r) (96) For a function f and a range r, it returns the minimum of the values f(t) when t runs over r. If f is continuous on a closed interval $[a, b]$, we can approximate the minimum of f in $[a, b]$, as closely as wanted, by choosing r=a..b..ε with ε small enough.

information source (14) The generator of messages in a *communication system*. A (source) message is a stream of *source symbols*.

inner distribution (89) For a a code $C \sim (n, M)_q$, the set of numbers $a_i = A_i/M$, $i = 1, \ldots, n$, where A_i is the set of pairs $(x, y) \in C \times C$ such that $hd(x, y) = i$. In the case of a linear code it coincides with the *weight distribution*.

input alphabet See *channel alphabet*.

Integer (232) The type associated to integers. Denoted $\mathbb{Z}$ in palette style.

inverse(x) 1 The inverse, x^{-1} or 1/x, of an invertible ring element x. 2 If R is a relation, the relation formed with all pairs b→a such that a→b is a pair of R.

invert_entries(x) (181) Gives the vector whose components are the inverses of the components of the vector x.

invertible?(x) (231) For a ring element x, it yields true when x is invertible in its ring, and false otherwise.

is?(x,T) (232) Tests whether x has type T or not. Takes the form x∈T in palette style.

ISBN (35) Acronym for International Standard Book Number (see Example 1.19).

is_prime_power?(n) (35) For an integer n, decides whether it is a prime power or not.

irreducible?(f,K) (231) For a field K and a polynomial f with coefficients in K, tests whether f is irreducible over K or not.

irreducible_polynomial(K,r,T) (121) Given a finite field K, a positive integer r and a variable T, returns a monic irreducible polynomial of degree r in the variable T and with coefficients in K. Since it makes random choices, the result may vary from call to call.

item(i,s) (225) The i-th term of a sequence s.

J

Jacobi logarithm See *Zech logarithm*.

Johnson's upper bound (87, 88) An upper bound for binary codes which improves Hamming's upper bound (Theorem 1.72, and also Proposition 1.71).

join(x,y) (227) The concatenation of x and y (both vectors or both lists). It coincides with x|y.

K

keyboard style (221) Conventions for representing the action of palette icons by means of a keyboard character string. For example, the symbol $\mathbb{Z}$ for the ring of integers that can be input from the Symbol palette has the same value as Integer. Similarly, the fraction $\frac{a}{b}$ composed by clicking the fraction icon in the **Operations** menu has the same value as a/b.

key equation (197) For alternant codes defined by an alternant matrix of order r, the error-evaluator polynomial $\epsilon(z)$ is, modulo z^r, the product of the error-locator polynomial $\sigma(z)$ times the syndrome polynomial $S(z)$ (Theorem 4.9).

Krawtchouk polynomials (90) The q-ary degree j Krawtchouk polynomial for the length n is

$$K_j(x, n, q) = \sum_{s=0}^{j} (-1)^s \binom{x}{s} \binom{n-x}{j-s} (q-1)^{j-s}.$$

As shown by Delsarte, these polynomials play a fundamental role in the *linear programming bound* (cf. Propostion 1.74 and Theorem 1.75).

L

lb_gilbert(n,d,q) (31) The value of $\lceil q^n / vol(n, d-1, q) \rceil$, which is the Gilbert lower bound for $A_q(n, d)$. The expression lb_gilbert(n,d) is defined to be lb_gilbert(n,d,2).

lb_gilbert_varshamov(n,d,q) (31) The value q^k, where k is the highest positive integer such that $vol_q(n-1, d-2) < q^{n-k}$ (*Gilbert–Varshamov condition*). The expression lb_gilbert_varshamov(n,d) is defined to be lb_gilbert_varshamov(n,d,2).

lcm(m,n) The least common multiple of two integers (or of two univariate polynomials) m and n.

leaders' table (45) A table $E = \{s \to e_s\}_{s \in \mathbb{F}_q^{n-k}}$ formed with a vector $e_s \in \mathbb{F}_q^n$ of syndrome s with respect to a check matrix H of a code $[n, k]_q$ and which has minimum weight with that condition. The vector e_s is called a leader of the set of vectors whose H-syndrome is s.

Legendre character (66, 67) Given a finite field $\mathbb{F}$ of odd cardinal q, the map

$$\chi : \mathbb{F}^* \to \{\pm 1\}$$

such that $\chi(x) = 1$ if x is a square (also called a quadratic residue) and $\chi(x) = -1$ if x is not a square (also called a quadratic non-residue). Conventionally $\chi(0)$ is defined to be 0.

left_parity_extension(G) See parity_extension(G).

legendre(a,F) (67) Yields the *Legendre charater* of the element a of a finite field F.

length (15) **1** The length of a vector, word, or list is the number of its elements. **2** The length of a *block encoder* or of a *block code* is the length of any of its words.

length(x) (226) The *length* of x (an aggregate).

let (223) The sentence let x=e binds the value of e to x and assigns the name x to this value.

linear codes (35) A code whose alphabet T is a Galois field and which is a linear subspace of T^n.

linear encoding (38) If G is a generating matrix of a linear code C over $\mathbb{F}_q$, the encoding $\mathbb{F}_q^k \to \mathbb{F}_q^n$ given by $u \mapsto uG$.

linear programming bound (91) $A_q(n, d)$ is bounded above by the maximum of the linear function $1 + a_d + \cdots + a_n$ subject to the constraints $a_i \geqslant 0$ ($d \leqslant i \leqslant n$) and

$$\binom{n}{j} + \sum_{i=d}^{n} a_i K_j(i) \geqslant 0 \quad \text{for} \quad j \in \{0, 1, \ldots, n\},$$

where $K_j = K_j(x, n, q)$, $j = 0, \ldots, n$, are the *Krawtchouk polynomials*.

List (232) The type associated to lists.

list(x) (228) If x is a range or vector, the list whose components are the elements of x. The expression {x} is, in both cases, a list of length 1 with x as its only element.

local X (232) A directive to insert at the beginning of a statement a sequence X of local variables for that statement. These variables do not interfere with homonym variables defined outside the statement. The variables introduced with the local directive can be assigned values, as in local x, y=0 (x is left free, while y is assigned the value 0).

LP(n,d) (91, 92) Computes the *linear programming bound* of $A_q(n, d)$. The three-argument call LP(n,d,X), with X a list of inequalities, finds a constrained maximum like LP(n,d), but with the Delsarte constraints (see Theorem 1.75) enlarged with the inequalities X.

M

macwilliams(n,k,q,A,T) (59) The polynomial $q^{-k}(1+(q-1)T)^n \tilde{A}(T)$, where $\tilde{A}(T)$ is the expression obtained after substituting the variable T in A by $(1-T)/(1+(q-1)T)$. The call macwilliams(n,k,A,T) is equivalent to macwilliams(n,k,2,A,T).

MacWilliams identities (59) These determine the weight enumerator of the dual of a linear code C in terms of the weight enumerator of C (see Theorem 7).

main problem (26) See *optimal code*.

Mariner code (72) See also *Hadamard codes*.

Matrix (232) The type associated to matrices. If R is a ring, Matrix(R) is the type of the matrices whose entries are in R.

Mattson–Solomon matrix (190) A matrix of the form $M = (\alpha^{ij})$, $0 \leqslant i, j \leqslant n-1$, where α is an element of a finite field with $ord(\alpha) = n$. The map $x \mapsto xM$ is called a Mattson–Solomon transform.

mattson_solomon_matrix(α) (190) Creates the *Mattson–Solomon matrix* on the element α.

Mattson–Solomon transform (190) See *Mattson–Solomon matrix*.

maximum distance separable (25) Codes $[n, k, d]_q$ that satisfy equality ($k + d = n + 1$) in the Singleton bound ($k + d \leqslant n + 1$).

mceliece(x) (95, 96) Computes the McEliece, Rodemich, Rumsey and Welsh *asymptotic linear programming upper bound*.

MDS Acronym for *maximum distance separable*.

Meggitt algorithm (174) A decoding algorithm for cyclic codes based on a *Meggitt table*.

Meggitt decoder (175) A decoder for cyclic codes that implements the *Meggitt algorithm*.

meggitt(y,g,n) (175) For a positive integer n and a polynomial y=y(x) of degree $< n$ representing the received vector, this function returns the result of decoding y with the *Meggitt decoder* of the cyclic code defined by the monic divisor g of $x^n - 1$. It presupposes that a *Meggitt table* E is available.

Meggitt table (174) Given a cyclic code of length n over $\mathbb{F}_q$, the table $E = \{s \to ax^{n-1} + e\}$, where a is any nonzero element of $\mathbb{F}_q$, e is any polynomial of degree $< n - 1$ and weight $< t$, where t is the error correcting capacity of the code, and s is the syndrome of $ax^{n-1} + e$.

min(x) (228) For a range, list or vector x, the minimum among its terms.

min_weight(X) (47) Given a list of vectors X, the minimum among the weights of the vectors in X.

min_weights(X) (47) The sublist of a list of vectors X whose weight is min_weight(X).

minimum distance (17) Given $C \subseteq T^n$, where T is a finite set, the minimum distance of C is

$$d_C = \min\{hd(x, x') \mid x, x' \in C, x \neq x'\},$$

where $hd(x, x')$ is the *Hamming distance* between x and x'.

minimum distance decoder (20) For a q-ary block code $C \subseteq T^n$, the *decoder* $g : D_C \rightarrow C$ ($D_C = \bigsqcup_{x \in C} B(x, t)$, $t = \lfloor (d-1)/2 \rfloor$) such that $g(y) = x$ for all $y \in B(x, t)$. Its *correcting capacity* is t.

minimum polynomial (134) Given a finite field L and a subfield K, the minimum polynomial of $\alpha \in L$ over K is the unique monic polynomial $p_\alpha \in K[X]$ of least degree that has α as a root. Note that p_α is irreducible.

minimum_polynomial(a,K,X) (134) Calculates the minimum polynomial of a over K, with variable X. The element a is assumed to belong to a finite field that contains K as a subfield. The call minimum_polynomial(a,X) is equivalent to minimum_polynomial(a,K,X) with K set to the prime field of the field of a.

minimum weight (36) For a linear code, the minimum of the weights of its nonzero vectors. It coincides with the minimum distance.

Möbius function $\mu(n)$ (119) If $n > 1$ is an integer, $\mu(n)$ is 0 if n has repeated prime factors. Otherwise is 1 or -1 according to whether the number of prime factors is even or odd. The value of $\mu(1)$ is conventionally defined as 1. If n and n' are positive integers such that $gcd(n, n') = 1$, then $\mu(nn') = \mu(n)\mu(n')$. One useful property of the function μ is that it satisfies the Möbius inversion formula (E.2.18).

Möbius inversion formula See *Möbius function* $\mu(n)$.

mod See remainder.

monic?(f) (231) Tests whether the polynomial f is monic or not.

mu_moebius(n) Computes, for a positive integer n, *Möbius function* $\mu(n)$.

N

n_items(s) (224) The number of terms of the sequence s.

n_columns(A) (229) Abreviation of number_of_columns(A). Yields the number of columns of the matrix A.

n_rows(A) (229) Abreviation of number_of_rows(A). Yields the number of rows of the matrix A.

Nm(α,L,K) (138) For an element a and a subfield K of a finite field L, this function returns $Nm_{L/K}(\alpha)$. The call Nm(α,L) is equivalent to Nm(α,L,K) is with K the prime field of L. The call Tr(α) is equivalent to Tr(α,L) with L=field(α).

noiseless channel (15) A channel that does not alter the symbols sent through it.

noisy channel (15) A channel that is not *noiseless* (symbols sent through it may thus be altered).

non-decodable (20) Given a decoder, said of a vector that is not in the domain of this decoder.

norm (36, 137) **1** The weight is a norm on $\mathbb{F}_q^n$ (E.1.20). **2** The norm of a finite field L over a subfield K is the homomorphism $Nm_{L/K} : L^* \rightarrow K^*$ such that

$Nm_{L/K}(\alpha) = \det(m_\alpha)$, where $m_\alpha : L \to L$ is the multiplication by α (the K-linear map $x \mapsto \alpha x$). For an expression of $Nm_{L/K}(\alpha)$ in terms of the *conjugates* of α, see Theorem 2.43.

normalized Hamming matrix (52) See E.1.33.

normalized_hamming_matrix(K,r) (52) The *normalized Hamming matrix* of r rows with entries in the finite field K.

not b (231) Negation of the Boolean expression b. It is equivalent to not(b).

null (224) The empty sequence (or the only sequence having 0 terms).

number_of_columns(A) (229) The number of columns of the matrix A.

number_of_items(s) (224) The number of terms of the sequence s. Can be shortened to n_items(s).

number_of_rows(A) (229) The number of rows of the matrix A.

O

odd?(n) (231) This Boolean function tests whether the integer n is odd.

offset See *BCH codes*.

optimal code (26) For a given length n and minimum distance d, the highest M such that there exists a code $(n, M, d)_q$. This M is denoted $A_q(n, d)$. The determination of the function $A_q(n, d)$ is often called the main problem of coding theory.

order (124) **1** The order of an element α of a finite group G is the least positive integer r such that $\alpha^r = e$ (e the identity element of G). If $n = |G|$, then $r|n$. **2** In the case of the multiplicative group K^* of a finite field K, the order of α, denoted $\operatorname{ord}(\alpha)$, divides $q - 1$. Conversely, if r divides $q - 1$, then there are exactly $\varphi(r)$ elements of order r in K (Proposition 2.34). **3** If $n > 1$ is an integer, and a is another integer such that $\gcd(a, n) = 1$, the order of $\alpha = [a]_n$ in the group $\mathbb{Z}_n^*$, denoted $e_n(q)$ (or also $\operatorname{ord}_n(a)$), divides $\varphi(n)$. **4** The order of an *alternant control matrix* is the number of rows of this matrix.

order(a) (125) If a is a nonzero element of a finite field, it computes the *order* of a.

order(q,n) (156) The order of the integer q in the group of invertible elements modulo the positive integer n, provided q and n are relatively prime.

output alphabet See *channel alphabet*.

P

pad(a,r,t) (143) From a vector a, say of length n, a positive integer r and an expression t, it constructs a vector of length r as follows. If $n \geqslant r$, the result is equivalent to take(a,r). Otherwise it is equivalent to a|constant_vector(n-r,t). The call pad(a,r) is equivalent to pad(a,r,0).

pairing (225) A relation, a table, a substitution or a divisor.

palette (221) Each of the icon menus of the **WIRIS** user interface. In this release (November 2002) the existing palettes are, in alphabetical order, the following: Analysis, Combinatorics, Edit, Format, Geometry, Greek, Matrix, Operations, Programming, Symbols and Units.

palette style (221) The characteristic typographical features and conventions of input expressions composed with palette icon menus. Designed to be as close as possible to accepted mathematical practices and conventions, it is also the style in which output expressions are presented.

Paley codes (74) See E.1.46.

paley_code(F) (74) Constructs the Paley code (E.1.46) associated to a Finite field.

Paley matrix (68) The Paley matrix of a finite field $\mathbb{F} = \{x_1, \ldots, x_q\}$, q odd, is $(\chi(x_i - x_j)) \in M_q(\mathbb{F})$, $1 \leqslant i, j \leqslant q$, where χ is the *Legendre character* of $\mathbb{F}$.

paley_matrix(F) (68) This function computes the *Paley matrix* of a finite field F of odd cardinal.

parity-check matrix See *check matrix*.

parity completion (151, 38) **1** The result of appending to each vector of a binary code the binary sum of its component bits. Thus the parity completion of an (n, M) code is an $(n + 1, M)$ code with the property that the binary sum of the component bits of each of its elements is zero. **2** More generally, the result of adding to each vector of a code defined over a finite field the negative of the sum of its components. Thus the parity extension of an $(n, M)_q$ code is an $(n + 1, M)_q$ code with the property that the sum of the components of each of its vectors is zero.

parity_completion(G) (151) The argument G is assumed to be a matrix $r \times n$ and the function returns a matrix $r \times (n + 1)$ whose last column is the negative of the sums of the rows of G. The function left_parity_extension(G) is defined in the same way, but the extra column is inserted at the beginning of G.

parity extension *Parity completion*.

perfect codes (29) Codes that satisfy equality in the *Hamming upper bound*. For a review of what is known about these codes, see Remark 1.16, page 30.

period [of a monic irreducible polynomial] See *primitive polynomial*.

period(f,q) (128) Assuming f is a monic irreducible polynomial with coefficients in field of q elements, this function computes the *period* (or exponent) of f.

PGZ Acronym for *Peterson–Gorenstein–Zierler*.

PGZ algorithm (217) A decoding scheme for alternant codes based on a solving a linear system of equations (cf. Proposition 4.17, page 216) for the determination of the error-locator polynomial.

phi_euler(n) (104) Computes *Euler's function* $\varphi(n)$.

plotkin(x) (95) Computes the *asymptotic Plotkin upper bound*.

Plotkin bound (83) If $(n, M, d)_q$ are the parameters of a code, then we have the inequality $d \leqslant \beta nM/(M - 1)$, or $M(d - \beta n) \leqslant d$, where $\beta = (q - 1)/q$. The dual Hamming codes meet the Plotkin bound (Example 1.65). For improvements of the Plotkin bound, see E.1.50 and, for binary codes, E.1.49.

Plotkin construction (33) See P.1.5.

polar representation *Exponential representation*.

Polynomial (232) The type of (possibly multivariate) polynomials.

polynomial(a,X) (142) Given a vector a and an indeterminate X, returns $a_1 + a_2 X + \cdots + a_n X^{n-1}$ (the polynomial in X with coefficients a).

prime?(n) (231) Test whether the integer n is prime or not.

prime ring / field See *characteristic homomorphism*.

primitive BCH codes see *BCH codes*.

primitive element (126) In a finite field $\mathbb{F}_q$, any element of *order* $q - 1$. A primitive element is also called a primitive root. In $\mathbb{F}_q$ there are exactly $\varphi(q - 1)$ primitive elements (Proposition 2.34).

primitive_element(K) (126) Returns a primitive element of the finite field K.

primitive polynomial (127) Any monic irreducible polynomial $f \in K[X]$, where K is a finite field, such that $\alpha = [X]_f$ is a *primitive element* of the field $K[X]/(f)$.

Since the order of α coincides with the *period* (or exponent) of f (E.2.21), which by definition is the least divisor d of $q^r - 1$ such that f divides $X^d - 1$ ($r = deg(f)$), f is primitive if and only if f does not divide $X^d - 1$ for any proper divisor d of $q^r - 1$.

primitive root See *primitive element*.

primitive RS codes See *Reed–Solomon codes*.

principal ideal (102) Given $a \in A$, A a ring, the set $\{xa \mid x \in A\}$ is an ideal of A. It is denoted $(a)_A$, or just (a) if A can be understood, and is called the principal ideal generated by a.

prune(H) (183) The result of dropping from the matrix H the rows that are linear combination of the privious rows.

Q

q-ary See *block code*.

q-ary symmetric channel See *symmetric channel*.

QNR(F) (67) When F is a finite field of odd cardinal, it yields the list of quadratic non-residues of F.

QR(F) (67) When F is a finite field of odd cardinal, it yields the list of quadratic residues of F.

quadratic residue See *Legendre character*

quadratic non-residue See *Legendre character*

quotient(m,n) Given integers m and n (or univariate polynomials with coefficients in a field), computes the quotient of the Euclidean division of m by n.

quotient_and_remainder(m,n) Given integers m and n (or univariate polynomials with coefficients in a field), yields a list with the quotient and the remainder of the Euclidean division of m by n.

R

Range (227, 232) The type associated to ranges.

range (225) A construct of the form a..b..d. It represents the sequence of real numbers x in the interval $[a, b]$ that have the form $x = a + jd$ for some nonnegative integer j. The construct a..b usually is interpreted as a..b..1, but it can be interpreted as different sequence of values in the interval $[a, b]$ when it is passed as an argument to some functions.

range(x) (227) For a vector, list or pairing of length n, the range 1..n.

rank(A) Returns the rank of the matrix A.

rate (16) Short form of *code rate*.

Rational (232) The type associated to rational numbers. Denoted $\mathbb{Q}$ in palette style.

rd(K) (207) Picks an element at random from the finite field K. The call rd(s,K) picks s elements at random from K.

rd_choice(X,m) (207) Given a set X of cardinal n and a positive integer m less than n, makes a (pseudo)random choice of m elements of X. The call rd_choice(n,m), where now n is an integer, chooses m elements at random of the set $\{1, \ldots, n\}$.

rd_error_vector(n,s,K) (208) Produces a random vector of weight s in K^n.

rd_linear_combination(G,K) (208) Generates a random linear combination of the rows of the matrix G with coefficients from the finite field K.

rd_nonzero(K) (207) Picks a nonzero element at random from the finite field K. The call rd_nonzero(s,K) picks s nonzero elements at random from K.

Real (232) The type associated to real numbers. Denoted $\mathbb{R}$ in palette style.

Reed–Muller codes (62) The first order Reed–Muller codes are introduced in P.1.16. They are denoted $RM_q^x(m)$. This code is the result of evaluating the polynomials of degree $\leqslant 1$ in m variables $X_1, \ldots, X_m$ at n points $x = x^1, \ldots, x^n \in \mathbb{F}_q^m$ ($q^{m-1} < n \leqslant q^m$). The Reed–Muller codes of order s, which are not studied in this book, are defined in a similar way, but using polynomials of degree $\leqslant s$ in $X_1, \ldots, X_m$ (see, for example, [25]).

Reed–Solomon codes (39) Introduced in Example 1.24, they are denoted $RS_\alpha(k)$. This code is the result of evaluating the polynomials of degree $< k$ in one variable X over $\mathbb{F}_q$ on n distinct elements $\alpha = \alpha_1, \ldots, \alpha_n$ of $\mathbb{F}_q$. This code has type $[n, k, n-k+1]_q$, and hence is **MDS**. If $n = q-1$ and $\alpha_1, \ldots, \alpha_n$ are the nonzero elements of $\mathbb{F}_q$, $RS_\alpha(k)$ is said to be a primitive *RS* code. Primitive *RS* codes are *BCH* codes (Proposition 3.22).

Relation (232) The type associated to relations.

relation (225) A **WIRIS** relation is a sequence of objects of the form a→b enclosed in braces, where a and b are arbitrary objects. Relations can be thought as a kind of functions. Indeed, if R is a relation and x any object, R(x) returns the sequence of values y (which may be null) such that x→y appears in the definition of R.

relative distance (17) For a block code C of *length* n and *minimum distance* d, the quotient $\delta_C = d/n$.

remainder(m,n) Given integers m and n (or univariate polynomials with coefficients in a field), computes the remainder of the Euclidean division of m by n.

Rep(3) (2) The binary *repetition code* of length 3, or $Rep_2(3)$.

repetition code (25) The code $Rep_q(n)$, of type $[n, 1, n]_q$, that consists of the q constant vectors $t_n, t \in T$.

residue code (79) The residue of a linear code $C \subseteq \mathbb{F}_q^n$ with respect to a $z \in C$ is the image of C under the linear map $\rho_z : \mathbb{F}_q^n \to \mathbb{F}_q^{n-s}$ which extracts, for each $y \in \mathbb{F}_q^n$, the $n-s$ components whose index lies outside the *support* of z (hence s denotes the length of this support).

reverse(x) (227) This yields, for a list or vector x, the list or vector with the objects of x written from the end to the beginning. The call reverse(a..b..d) gives the range b..a..-d.

reverse_print(b) (142) By default, polynomials are displayed by decreasing degree of its terms. This order can be reversed with reverse_print(true). The command reverse_print(false) restitutes the default order.

Ring (232) The type associated to rings.

ring (101) An abelian group $(A, +)$ endowed with a product $A \times A \to A$, $(x, y) \mapsto x \cdot y$, that is associative, distributive with respect to the sum on both sides, and with a unit element 1_A which we always assume to be different from the zero element 0_A of the group $(A, +)$. If $x \cdot y = y \cdot x$ for all $x, y \in A$ we say that the ring is commutative.

ring(x) Returns the ring where the element x belongs.

ring homomorphism (102) A map $f : A \to A'$ from a ring A to a ring A' such that $f(x + y) = f(x) + f(y)$ and $f(x \cdot y) = f(x) \cdot f(y)$, for all $x, y \in A$, and $f(1_A) = 1_{A'}$.

RM Acronym for *Reed–Muller codes*.

roots (164) For a cyclic code C of length n over $\mathbb{F}_q$, the roots of the generator polynomial of C in the splitting field of $X^n - 1$ over $\mathbb{F}_q$.

roots(f,K) (136) For a univariate polynomial f and field K, it returns the list of roots of f in K.

RS Acronym for *Reed–Solomon codes*.

RS(a,k) (185) Creates the Reed–Solomon code of dimension k associated to the vector a.

Ruffini's rule (108) A polynomial $f \in K[X]$, K a field, is divisible by $X - \alpha$ if and only if $f(\alpha) = 0$. It follows that f has a factor $h \in K[X]$ of degree 1 if and only if f has a root in K.

Rule The type of rules. See Substitution.

S

say(t) This command displays a text t at the message window at the bottom of the WIRIS screen. The text string corresponding to an expression e can be obtained with string(e). It may be useful to note that if t is a string and e an expression, then t|e is equivalent to t|string(e).

scalarly equivalent codes (19) See *equivalent codes*.

sequence(x) (224) If x is a range, list or vector, the sequence formed with the components of x.

set(L) (229) From a list L, it forms an ordered list with the distinct values of the terms of L (repetitions are discarded).

Shannon (4) Claude Elwood Shannon (1916-2001) is considered the father of the Information Age. One of his fundamental results is the celebrated *channel coding theorem*. "Look at a compact disc under a microscope and you will see music represented as a sequence of bits, or in mathematical terms, as a sequence of 0's and 1's, commonly referred to as bits. The foundation of our Information Age is this transformation of speech, audio, images and video into digital content, and the man who started the digital revolution was Claude Shannon" (from the obituary by Robert Calderbank and Neil J. A. Sloane).

shortening (84) If T is a q-ary alphabet and $\lambda \in T$, the code $\{x \in T^{n-1} \mid (x|\lambda) \in C\}$ is said to be the λ-shortening of the code $C \subseteq T^n$ with respect to the last position. Shortenings with respect to other positions are defined likewise.

shorthand(X,E) (211) Represents the entries of the vector or matrix X by the corresponding indices on table E (Zech logarithms). The element 0 is represented by the underscore character.

Singleton bound (25) A code $[n, k, d]_q$ satisfies $k + d \leqslant n + 1$ (Proposition 1.11, page 25).

source alphabet (14) The finite set S, whose elements are called source symbols, used to compose source messages. A source message is a stream $s_1, s_2, \ldots$ with $s_i \in S$ (cf. *information source*).

source symbol See *symbol*.

sphere-packing upper bound The *Hamming's upper bound*.

sphere upper bound The *Hamming's upper bound*.

splitting field (116) If K is a field and $f \in K[X]$ a polynomial of degree r, there exists a field L that contains K and elements $\alpha_1, \ldots, \alpha_r \in L$ such that $f =$

$\prod_{i=1}^{r}(X - \alpha_i)$ and $L = K(\alpha_1, \ldots, \alpha_r)$. This field L is the splitting field of f over K. It can be seen that it is unique up to isomorphisms. For the splitting field of $X^n - 1$, which plays a key role in the theory of cyclic codes, see Proposition 3.9.

square_roots(a) (125) If a is a field element, it returns a list with its square roots.

statement (221) Either an expression or a succession of two or more expressions separated by semicolons (the semicolon is not requiered if next expression begins on a new line). An expression is either a formula or an assignment. A formula is a syntactic construct intended to compute a new value by combining operations and values in some explicit way. An assignment is a construct of one of the forms x=e, x:=e or let x=e, where x is an identifier and e is an expression.

strict BCH codes BCH codes with offset 1 (see *BCH codes*).

string(e) See say(t).

subextension?(K,F) The value of this Boolean function is true if and only if K coincides with one of the subextensions of F. The subextensions are defined recursively as follows: if E is a field and F=E[x]/(f), then the subextensions of F are F and the subextensions of E.

subfield (101) A subring that is a field.

subring (101) An addtive subgroup of a ring which is closed with respect to the product and contains the unit of the ring.

Substitution (232) The type of substitutions.

substitution (225) **1** A **WIRIS** substitution is a sequence of objects of the form x⇒a enclosed in braces, where x is an identifier and a is any object. If an identifier x appears more than once, only its last occurrence is retained. Subtitutions can be thought as a kind of functions. Indeed, if S is a substitution and b is any object, S(b) returns the result of replacing any occurrence of x in b by the value a corresponding to x in S. **2** If some x of the pairs $x \Rightarrow a$ is not an identifier, then **WIRIS** classifies S as a Rule. Rules work like substitutions, but they look for patterns to be substituted, rather than variables. The type of rules is Rule.

such_that (234) A directive to force that a Boolean condition is satisfied.

Sugiyama algorithm (198) Solves the *key equation*. It is variation of the extended Euclidean algorithm (see Remark 4.11) for the determination of the *gcd* of two polynomials.

sum(x) (228) For a range, list or vector x, the sum of its terms.

support (79) The support of a vector x is the set of indices i such that $x_i \neq 0$.

symbol (14) An element of the *source alphabet* (source symbol) or of the *channel alphabet* (channel symbol).

symmetric channel (21) A *channel* with q-ary alphabet T is symmetric if any symbol in T has the same probability of being altered and if any of the remaining $q - 1$ symbols are equally likely.

syndrome (22, 44, 173, 195) **1** If H is a *check matrix* for a linear code C of length n, the syndrome with respect to H of a vector y of length n is the vector $s = yH^T$. **2** If g is the *generator polynomial* of a cyclic code, the syndrome of a received vector y is the remainder of the Euclidean division of y by g (interpreting y as a polynomial).

syndrome-leader decoder (45) The decoder $y \mapsto y - E(yH^T)$, where H is a check matrix of a linear code C and E is a leader's table for C.

syndrome polynomial (195) If $S = (S_0, \ldots, S_{r-1})$ is the syndrome of a received vector, the polynomial $S(z) = S_0 + S_1 z + \cdots + S_{r-1} z^{r-1}$.

systematic encoding (39) An encoding of the form $f : T^k \to T^n$ with the property that $u \in T^k$ is the subword of $x = f(u)$ formed with the components of x at some given set of k indices. For example, if the generating matrix G of a linear code of the form $G = (I_k|P)$, then the linear encoding $u \mapsto uG$ is systematic (with respect to the first k positions) because $uG = u|uP$.

T

Table (232) The type of tables.

table (225) A **WIRIS** table is a sequence of objects of the form x=a enclosed in brakets (braces are also allowed), where x is an identifier and a is any object. Tables can be thought as a kind of functions. Indeed, if x is any identifier, T(x) returns the sequence of values a such that x=a is on T.

tail(x) (227) Equivalent to x,n-1 if x is a list, vector or range of length n.

take(x,k) (227) If x is a list or vector of length n, and k is in the range 1..n, the list or vector formed with the first k terms of x. If instead of k we write -k, we get the list or vector of the last k terms of x. This function also makes sense when x is a range, and in this case it yields the list of the first k terms of x (or the last k for the call take(x,-k)).

tensor(A,B) (64) Returns the tensor product of the matrices A and B.

Tr(α,L,K) (138) For an element a and a subfield K of a finite field L, this function returns $Tr_{L/K}(\alpha)$. The call Tr(α,L) is equivalent to Tr(α,L,K) with K the prime field of L. The call Tr(α) is equivalent to Tr(α,L) with L=field(α).

trace (137) The trace of a finite field L over a subfield K is the K-linear map $Tr_{L/K} : L \to K$ such that $Tr_{L/K}(\alpha) = Tr(m_\alpha)$, where $m_\alpha : L \to L$ is the multiplication by α (the K-linear map $x \mapsto \alpha x$). For an expression of $Tr_{L/K}(\alpha)$ in terms of the *conjugates* of α, see Theorem 2.43.

trace(A) (230) Assuming A is a square matrix, the trace of A (the sum of its diagonal entries).

transmission alphabet See *channel alphabet*.

transmission rate See *code rate*.

transpose(A) (229) The matrix obtained by writing the rows of A as columns. Denoted A^T.

trivial perfect codes (29) The total q-ary codes T^n, the one-word codes and the odd-length binary repetion codes.

true (230) Together with false, they form the set of Boolean values.

ub_elias(n,d,q) (86) Gives the *Elias upper bound* for $A_q(n,d)$.

ub_griesmer(n,d,q) (80, 81) The Griesmer upper bound for the function $B_q(n,d)$. This bound has the form q^k, where k is the highest non-negative integer such that

$$G_q(k,d) \leqslant n \text{ and } G_q(k,d) = \sum_{i=0}^{k-1} \left\lceil \frac{d}{q^i} \right\rceil \text{ (the Griesmer function).}$$

ub_johnson(n,d) (88) Gives the *Johnson upper bound* for $A_2(n,d)$. There is also the call ub_johnson(n,d,w), which computes the upper bound introduced in Proposition 1.71 for the function $A(n,d,w)$. This function is, by definition, the maximum M for binary codes (n,M,d) all whose words have weight w.

ub_sphere(n,d,q) (28) The value of $\lfloor q^n/vol(n,t,q) \rfloor$, $t = \lfloor (d-1)/2 \rfloor$, which is the sphere-packing upper bound for $A_q(n,d)$. The expression ub_sphere(n,d) is defined to be ub_sphere(n,d,2).

undetectable code error See *code error*.

Univariate_polynomial(f) (148) The value of this Boolean funtion is true if f is a univariate polynomial polynomial and false otherwise.

V

Vandermonde determinant (41) The determinant, denoted $D(\alpha_1, \ldots, \alpha_n)$, of a square Vandermonde matrix $V_n(\alpha_1, \ldots, \alpha_n)$.

Vandermonde matrix (39) The Vandermonde matrix of r rows on the elements $\alpha = \alpha_1, \ldots, \alpha_n$ is (α_j^i), with $0 \leqslant i \leqslant r - 1$ and $1 \leqslant j \leqslant n$. It is denoted $V_r(\alpha)$.

vandermonde_matrix(a,r) (180) Creates the *Vandermonde matrix* of order r associated to the vector a.

vanlint(x) (96, 96) Computes the *asymptotic van Lint upper bound*.

variable *Identifier*.

variable(f) Retrieves the variable of a univariate polynomial.

variables(f) Retrieves the list of variables of a polynomial.

Vector (232) The type of vectors. If R is a ring, the type of the vectors with components in R is Vector(R).

vector See *code word*.

vector(x) (228) If x is a range or list, the vector whose components are the elements of x. In the case in which x is a range, vector(x) is the same as [x]. When x is a list, however, [x] is a vector of length 1 with x as its only component.

vector(p) (142) If p is a univariate polynomial $a_1 + a_2 X + \cdots + a_n X^{n-1}$, the vector $[a_1, \ldots, a_n]$.

volume(n,r,q) (27) The value of $\sum_{i=0}^{r} \binom{n}{i}(q - 1)^i$ (for a q-ary alphabet, this is the number of length n vectors that lie at a Hamming distance at most r of any given length n vector. The value of volume(n,r) coincides with volume(n,r,2).

W

weight (17, 36) Given a vector x, the number of nonzero components of x. It is denoted $|x|$ or $wt(x)$.

weight(x) (229) Yields the *weight* of the vector x. Synomym expression: wt(x).

weight distribution See *weight enumerator*.

weight enumerator (54) Given a linear code C of length n, the polynomial $a(t) = \sum_{i=0}^{n} a_i t^i$, where a_i is the number of vectors in C which have weight i. The sequence $a_0, a_1, \ldots, a_n$ is called the weight distribution of the code.

where Synonym of such_that.

WIRIS/cc (i, 7) The general mathematical system used to carry out the computations sollicited by the examples in this book. It includes the interface that takes care of the editing (mathematics and text). It is introduced gradually along the text. For a summary, see the Appendix. The examples in this book can be accessed and run at http://www.wiris.com/cc/.

with (234) A directive to make a variable or variables run over an aggregate or list of aggregates. For more details see A.17.

word See *code word*.

wt(x) Synonym of **weight**, yields the weight of the vector x.

Z

zero?(x) (229) If x is a vector, returns true if x is zero and false otherwise.

zero_positions(f,a) (205) Given a univariate polynomial f and a vector a of length n, finds the list of indices i in the range $1..n$ such that $f(a_i) = 0$.

zero_vector(n) (228) The vector 0_n.

Zn(p) (223) Ring $\mathbb{Z}_p$ of of integers mod p (a field if p is prime, also denoted $\mathbb{F}_p$.

Zech logarithms (130) In the *exponential representation* with respect to a primitive element α of a finite field, the Zech (or Jacobi) logarithm $Z(j)$ of an exponent j is defined to be $ind_\alpha(1 + \alpha^j)$, and so $1 + \alpha^j = \alpha^{Z(j)}$. Thus a table $\{j \mapsto Z(j)\}$ of Zech logarithms allows to find the exponential representation of a sum of terms also expressed in the exponential representation.

zero-parity code (38) For a given length n, the subcode of $\mathbb{F}_q^n$ formed with the vectors the sum of whose components is zero. For its weight enumerator see Example 1.45.

Universitext

Edwards, R. E.: A Formal Background to Higher Mathematics IIa, and IIb

Emery, M.: Stochastic Calculus in Manifolds

Endler, O.: Valuation Theory

Erez, B.: Galois Modules in Arithmetic

Everest, G.; Ward, T.: Heights of Polynomials and Entropy in Algebraic Dynamics

Farenick, D. R.: Algebras of Linear Transformations

Foulds, L. R.: Graph Theory Applications

Frauenthal, J. C.: Mathematical Modeling in Epidemiology

Friedman, R.: Algebraic Surfaces and Holomorphic Vector Bundles

Fuks, D. B.; Rokhlin, V. A.: Beginner's Course in Topology

Fuhrmann, P. A.: A Polynomial Approach to Linear Algebra

Gallot, S.; Hulin, D.; Lafontaine, J.: Riemannian Geometry

Gardiner, C. F.: A First Course in Group Theory

Gårding, L.; Tambour, T.: Algebra for Computer Science

Godbillon, C.: Dynamical Systems on Surfaces

Goldblatt, R.: Orthogonality and Spacetime Geometry

Gouvêa, F. Q.: p-Adic Numbers

Gustafson, K. E.; Rao, D. K. M.: Numerical Range. The Field of Values of Linear Operators and Matrices

Hahn, A. J.: Quadratic Algebras, Clifford Algebras, and Arithmetic Witt Groups

Hájek, P.; Havránek, T.: Mechanizing Hypothesis Formation

Heinonen, J.: Lectures on Analysis on Metric Spaces

Hlawka, E.; Schoißengeier, J.; Taschner, R.: Geometric and Analytic Number Theory

Holmgren, R. A.: A First Course in Discrete Dynamical Systems

Howe, R., Tan, E. Ch.: Non-Abelian Harmonic Analysis

Howes, N. R.: Modern Analysis and Topology

Hsieh, P.-F.; Sibuya, Y. (Eds.): Basic Theory of Ordinary Differential Equations

Humi, M., Miller, W.: Second Course in Ordinary Differential Equations for Scientists and Engineers

Hurwitz, A.; Kritikos, N.: Lectures on Number Theory

Iversen, B.: Cohomology of Sheaves

Jacod, J.; Protter, P.: Probability Essentials

Jennings, G. A.: Modern Geometry with Applications

Jones, A.; Morris, S. A.; Pearson, K. R.: Abstract Algebra and Famous Inpossibilities

Jost, J.: Compact Riemann Surfaces

Jost, J.: Postmodern Analysis

Jost, J.: Riemannian Geometry and Geometric Analysis

Kac, V.; Cheung, P.: Quantum Calculus

Kannan, R.; Krueger, C. K.: Advanced Analysis on the Real Line

Kelly, P.; Matthews, G.: The Non-Euclidean Hyperbolic Plane

Kempf, G.: Complex Abelian Varieties and Theta Functions

Kitchens, B. P.: Symbolic Dynamics

Kloeden, P.; Ombach, J.; Cyganowski, S.: From Elementary Probability to Stochastic Differential Equations with MAPLE

Kloeden, P. E.; Platen; E.; Schurz, H.: Numerical Solution of SDE Through Computer Experiments

Kostrikin, A. I.: Introduction to Algebra

Krasnoselskii, M. A.; Pokrovskii, A. V.: Systems with Hysteresis

Luecking, D. H., Rubel, L. A.: Complex Analysis. A Functional Analysis Approach

Ma, Zhi-Ming; Roeckner, M.: Introduction to the Theory of (non-symmetric) Dirichlet Forms

Mac Lane, S.; Moerdijk, I.: Sheaves in Geometry and Logic

Druck und Bindung: Strauss GmbH, Mörlenbach